Library of Congress Cataloging-in-Publication Data

Kasap, S. O. (Safa O.)
Optoelectronic devices and photonics : principles and practices / Safa O. Kasap.
p. cm.
Includes bibliographical references and index.
ISBN 0-201-61087-6
1. Optoelectronic devices. 2. Photonics I. Title.
TK0834 .K37 2000
621.381'045--dc21

00-050188

Vice President and Editorial Director, ECS: *Marcia J. Horton*
Publisher: *Tom Robbins*
Associate Editor: *Alice Dworkin*
Vice President and Director of Production and Manufacturing, ESM: *David W. Riccardi*
Executive Managing Editor: *Vince O'Brien*
Managing Editor: *David A. George*
Production Editor: *Scott Disanno*
Director of Creative Services: *Paul Belfanti*
Creative Director: *Carole Anson*
Art Director: *Jayne Conte*
Cover Designer: *Bruce Kenselaar*
Cover Art: *S. O. Kasap*
Art Editor: *Adam Velthaus*
Manufacturing Manager: *Trudy Pisciotti*
Manufacturing Buyer: *Pat Brown*
Senior Marketing Manager: *Holly Stark*

Published by Prentice-Hall, Inc.
Upper Saddle River, New Jersey 07458

Printed in the United States of America
10 9 8 7 6 5 4 3 2 1

ISBN 0-201-61087-6

Prentice-Hall International (UK) Limited, *London*
Prentice-Hall of Australia Pty. Limited, *Sydney*
Prentice-Hall Canada Inc., *Toronto*
Prentice-Hall Hispanoamericana, S.A., *Mexico*
Prentice-Hall of India Private Limited, *New Delhi*
Prentice-Hall of Japan, Inc., *Tokyo*
Pearson Education Asia Pte. Ltd., *Singapore*
Editora Prentice-Hall do Brasil, Ltda., *Rio de Janeiro*

Selected Semiconductors

Crystal structure, lattice parameter a, bandgap energy E_g at 300 K, type of bandgap (D = Direct and I = Indirect), change in E_g per unit temperature change (dE_g/dT), bandgap wavelength λ_g and refractive index n close to λ_g. (A = Amorphous, D = Diamond, W = Wurtzite, ZB = Zinc blende.)

Semiconductors	Crystal	a nm	E_g eV	Type	dE_g/dT meV K^{-1}	λ_g (μm)	n around λ_g
Group IV							
Ge	D	0.5658	0.66	I	−0.37	1.87	4
Si	D	0.5431	1.12	I	−0.23	1.11	3.45
a-Si:H	A		1.7–1.8			0.73	
SiC			2.9	I		0.42	3.1
III–V Compounds							
AlAs	ZB	0.5661	2.16	I	−0.5	0.57	3.2
AlP	ZB	0.5451	2.45	I	−0.35	0.52	3
AlSb	ZB	0.6135	1.58	I	−0.3	0.75	3.7
GaAs	ZB	0.5653	1.42	D	−0.45	0.87	3.6
$GaAs_{0.88}Sb_{0.12}$	ZB		1.15	D		1.08	
GaN	W	0.3190 a 0.5190 c	3.44	D	−0.45	0.36	
GaP	ZB	0.5451	2.24	I	−0.54	0.55	3.4
GaSb	ZB	0.6096	0.73	D	−0.35	1.7	4
$In_{0.53}Ga_{0.47}As$ on InP	ZB	0.5869	0.75	D		1.65	
$In_{0.58}Ga_{0.42}As_{0.9}P_{0.1}$ on InP	ZB	0.5870	0.80	D		1.55	
$In_{0.72}Ga_{0.28}As_{0.62}P_{0.38}$ on InP	ZB	0.5870	0.95	D		1.3	
InP	ZB	0.5869	1.35	D	−0.46	0.91	3.4–3.5
InAs	ZB	0.6058	0.35	D	−0.28	3.5	3.8
InSb	ZB	0.6479	0.18	D	−0.3	7	4.2
II–VI Compounds							
ZnSe	ZB	0.5668	2.7	D	−0.72	0.46	2.3
ZnTe	ZB	0.6101	2.25	D		0.55	2.7

Physical Constants

c	Speed of light in vacuum	2.9979×10^{8} m s^{-1}
h	Planck's constant	6.6261×10^{-34} J s
$\hbar$	$\hbar = h/2$	1.0546×10^{-34} J s
e	Electronic charge	1.60218×10^{-19} C
ε_o	Absolute permittivity	8.8542×10^{-12} F m^{-1}
k_B	Boltzmann constant ($k_B = R/N_A$)	1.3807×10^{-23} J K^{-1}
k_BT/e	Thermal voltage at 300 K	0.02585 V
m_e	Electron mass in free space	9.10939×10^{-31} kg
μ_o	Absolute permeability	$4\pi \times 10^{-7}$ H m^{-1}
N_A	Avogadro's number	6.0221×10^{23} mol^{-1}
R	Gas constant (N_Ak)	8.31451 J mol^{-1} K^{-1}

Useful Information

Wavelength ranges and colors as usually specified for LEDs

Color	Blue	Emerald Green	Green	Yellow	Amber	Orange	Red-Orange	Red	Deep red	Infrared
λ (nm)	$\lambda < 500$	530–564	565–579	580–587	588–594	595–606	607–615	616–632	632–700	$\lambda > 700$

$1\ \text{eV} = 1.60218 \times 10^{-19}$ J
$1\ \text{Å} = 10^{-10}\ \text{m} = 0.1$ nm

Materials Data

Extensive materials data available at

Web-*Materials*

http://Materials.Usask.Ca

Optoelectronics and Photonics: Principles and Practices

S.O. Kasap
Professor of Electrical Engineering
University of Saskatchewan
Canada

Prentice Hall
Upper Saddle River, NJ 07458

"We have a habit in writing articles published in scientific journals to make the work as finished as possible, to cover up all the tracks, to not worry about the blind alleys or describe how you had the wrong idea first, and so on. So there isn't any place to publish, in a dignified manner, what you actually did in order to get to do the work."

Richard P. Feynman
Nobel Lecture, 1966

Preface

This textbook represents a first course in optoelectronic materials and devices suitable for a half- or one-semester semester course at the undergraduate level in electrical engineering, engineering physics, and materials science and engineering departments. It can also be used at the graduate level as an introductory course by including some of the selected topics in the CD-ROM. Normally, the students would not have covered Maxwell's equations. Although Maxwell's equations are mentioned in the text to alert the student, they are not used in developing the principles. It is assumed that the students would have taken a basic first- or second-year physics course, with modern physics, and would have seen rudimentary concepts in geometrical optics, interference, and diffraction, but not Fresnel's equations and concepts, such as group velocity and group index. Typically, an optoelectronics course would either be given after a semiconductor devices course or concurrently. Students would have been exposed to elementary quantum mechanical concepts, perhaps in conjunction with a basic semiconductor science course.

I tried to keep the general treatment and various proofs at a semiquantitative level without going into detailed physics. Most topics are initially introduced through intuitive explanations to allow the concept to be grasped first before any mathematical development. The mathematical level is assumed to include vectors, complex numbers, and partial differentiation, but excludes Fourier transforms. On the one hand, we are required to cover as much as possible and, on the other hand, professional engineering accreditation requires students to solve numerical problems and carry out "design calculations." In preparing the text, I tried to satisfy engineering degree accreditation requirements in as much breadth as possible. Obviously one cannot solve numerical problems, carry out design calculations, and derive each equation at the same time without expanding the size of the text to an unacceptable level. I have missed many topics but I have also covered many; though, undoubtedly, my own biased selection.

The book has a CD-ROM that contains the figures as large color diagrams in a common portable document format (PDF). They can be printed on nearly any color printer to make overhead projector transparencies for the instructor and class-ready notes for the students so they do not have to draw the diagrams during the lectures. The diagrams have been also put into PowerPoint for directly delivering the lecture material from a computer. In addition, there are numerous selected topics and other educational features on the CD-ROM that follows a web-format. Both instructors and students will find the selected topics very useful. These selected topics have been prepared by various authors and specialists in optoelectronics as stand-alone chapters, and they cover a wide range of topics. Although some of these topics are treated at the graduate level and review a particular area, there are also numerous selected topics at the elementary level for undergraduate students. In addition, some of these topics appear as color reprints of interesting articles taken, with permission, from various educational journals such as *Physics Today*, *Physics World, IEEE Spectrum*, *American Journal of Physics, Laser Focus, Photonics,* and various other magazines and journals.

A number of colleagues took time to read portions of the manuscripts and provided many useful suggestions that made this a better book. My special thanks go to Professor Charbel Tannous (Brest University, France) and Dr. Yann Boucher (RESO Laboratory, École Nationale d'Ingénieurs de Brest, France), both of whom kept challenging me with their incisive criticisms and dedication to accuracy. It's a pleasure to thank Professors Dave Dodds (University of Saskatchewan), Jai Singh (Northern Territory University, Australia), Harry Ruda (University of Toronto), Fary Ghassemlooy (Sheffield-Hallam University), John McClure (University of Texas, El Paso), Rajendra Singh (Clemson University), Drs. Costas Saravanos (Siecor, Texas), Ray DeCorby, Chris Haugen (both at TRLabs, Edmonton), Don Scansen (Semiconductor Insights, Ottawa), Brad Polischuk (Anrad, Montreal), and Daniel DeForest for their valuable comments. I also would like to thank the reviewers who were commissioned by Addison-Wesley and Prentice-Hall for their helpful suggestions. And, not least, my wife Nicolette, who was always cheerfully ready whenever I needed her help.

No textbook is perfect and I can only improve the text with your input. Please feel free to write to me with your comments. Although I may not be able to reply to each individual comment and suggestion, I do read all my email messages and take note of suggestions and comments.

S.O. Kasap
Kasap@Engr.Usask.Ca
http:// Optoelectronics.Usask.Ca
http://ElectronicMaterials.Usask.Ca

Contents

CHAPTER 1

Wave Nature of Light

"Physicists use the wave theory on Mondays, Wednesdays and Fridays and the particle theory on Tuesdays, Thursdays and Saturdays."
—Sir William Henry Bragg

Augustin Jean Fresnel (1788–1827) was a French physicist and a civil engineer for the French government who was one of the principal proponents of the wave theory of light. He made a number of distinct contributions to optics including the well-known Fresnel lens that was used in lighthouses in the 19th century. He fell out with Napoleon in 1815 and was subsequently put under house-arrest until the end of Napoleon's reign. During his enforced leisure time he formulated his wave ideas of light into a mathematical theory. *(Photo: Smithsonian Institution, courtesy of AIP Emilio Segrè Visual Archives.)*

"If you cannot saw with a file or file with a saw, then you will be no good as an experimentalist." —Augustin Fresnel

1.1 LIGHT WAVES IN A HOMOGENEOUS MEDIUM

A. Plane Electromagnetic Wave

The wave nature of light, quite aside from its photonic behavior, is well recognized by such phenomena as interference and diffraction. We can treat light as an electromagnetic wave with time varying electric and magnetic fields, that is E_x and B_y respectively, which

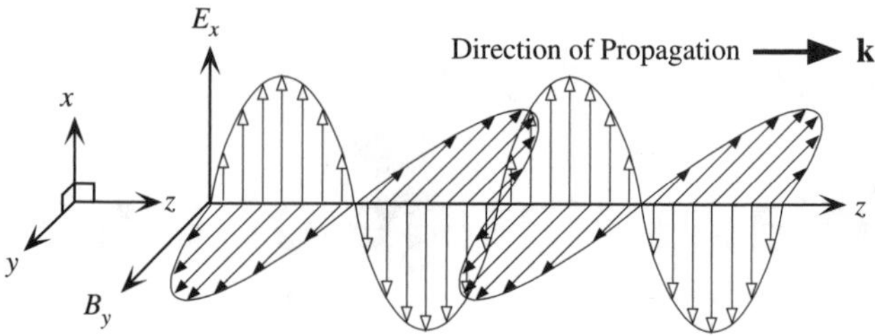

FIGURE 1.1 An electromagnetic wave is a traveling wave which has time varying electric and magnetic fields which are perpendicular to each other and the direction of propagation, z.

are propagating through space in such a way that they are always perpendicular to each other and the direction of propagation z as depicted in Figure 1.1. The simplest traveling wave is a sinusoidal wave that, for propagation along z, has the general mathematical form,

Traveling wave along z

$$E_x = E_o \cos(\omega t - kz + \phi_o) \tag{1}$$

in which E_x is the electric field at position z at time t, k is the **propagation constant**, or **wave number**, given by $2\pi/\lambda$, where λ is the wavelength, ω is the angular frequency, E_o is the amplitude of the wave and ϕ_o is a phase constant, which accounts for the fact that at $t = 0$ and $z = 0$, E_x may or may not necessarily be zero depending on the choice of origin. The argument $(\omega t - kz + \phi_o)$ is called the **phase** of the wave and denoted by ϕ. Equation (1) describes a **monochromatic plane wave** of infinite extent traveling in the positive z direction as depicted in Figure 1.2. In any plane perpendicular to the direction of propagation (along z), the phase of the wave, according to Eq. (1) is constant, which means that the field in this plane is also constant. A surface over which the phase of a wave is constant is referred to as a **wavefront**. A wavefront of a plane wave is obviously a plane perpendicular to the direction of propagation as shown in Figure 1.2.

We know from electromagnetism that time varying magnetic fields result in time varying electric fields (Faraday's law) and vice versa. A time varying electric field would set up a time varying magnetic field with the same frequency. According to electromagnetic principles,[1] a traveling electric field E_x as represented by Eq. (1) would always be accompanied by a traveling magnetic field B_y with the same wave frequency and propagation constant (ω and k) but the directions of the two fields would be orthogonal as in Figure 1.1. Thus, there is a similar traveling wave equation for the magnetic field component B_y. We generally describe the interaction of a light wave with a non-conducting matter (conductivity, $\sigma = 0$) through the electric field component E_x rather than B_y because it is the electric field that displaces the electrons in molecules or ions in the crystal and thereby gives rise to the polarization of matter. However, the two fields are linked, as in Figure 1.1, and there is an intimate relationship between them. The **optical field** refers to the electric field E_x.

We can also represent a traveling wave using the exponential notation since $\cos\phi = \text{Re}[\exp(j\phi)]$ in which **Re** refers to the real part. We then need to take the real part of any complex result at the end of calculations. Thus, we can write Eq. (1) as

[1] Maxwell's equations formulate electromagnetic phenomena and provide relationships between the electric and magnetic fields and their space and time derivatives. We need only to use a few selected results from Maxwell's equation without delving into their derivations. The magnetic field B is also called the magnetic induction or magnetic flux density. The magnetic field *intensity* H and magnetic field B in a non-magnetic material are related by $B = \mu_o H$ in which μ_o is the absolute permeability.

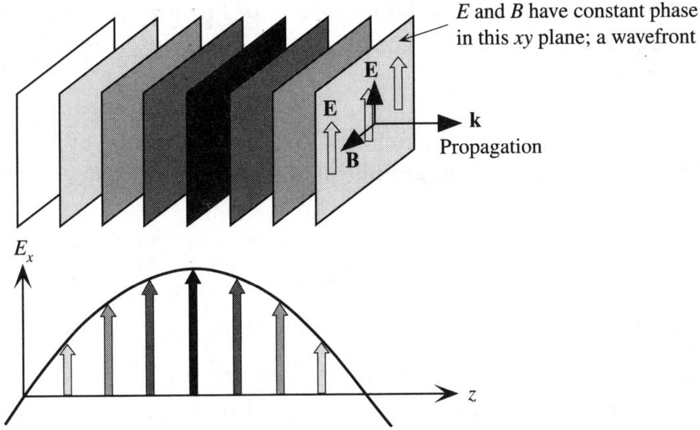

FIGURE 1.2 A plane EM wave traveling along z, has the same E_x (or B_y) at any point in a given xy plane. All electric field vectors in a given xy plane are therefore in phase. The xy planes are of infinite extent in the x and y directions.

$$E_x(z, t) = \text{Re}\left[E_o \exp(j\phi_o) \exp j(\omega t - kz)\right]$$

or

Traveling wave along z

$$E_x(z, t) = \text{Re}\left[E_c \exp j(\omega t - kz)\right] \quad (2)$$

in which $E_c = E_o \exp(j\phi_o)$ is a complex number that represents the amplitude of the wave and includes the constant phase information ϕ_o.

We indicate the direction of propagation with a vector $\mathbf{k}$, called the **wave vector**, whose magnitude is the propagation constant, $k = 2\pi/\lambda$. It is clear that $\mathbf{k}$ is perpendicular to constant phase planes as indicated in Figure 1.2. When the electromagnetic (EM) wave is propagating along some arbitrary direction $\mathbf{k}$, as indicated in Figure 1.3, then the electric field $E(\mathbf{r}, t)$ at a point $\mathbf{r}$ on a plane perpendicular to $\mathbf{k}$ is

$$E(\mathbf{r}, t) = E_o \cos(\omega t - \mathbf{k} \cdot \mathbf{r} + \phi_o) \quad (3)$$

because the dot product $\mathbf{k} \cdot \mathbf{r}$ is along the direction of propagation similar to kz. The dot product is the product of $\mathbf{k}$ and the projection of $\mathbf{r}$ onto $\mathbf{k}$, which is $\mathbf{r}'$ in Figure 1.3, so that $\mathbf{k} \cdot \mathbf{r} = kr'$. Indeed, if propagation is along z, $\mathbf{k} \cdot \mathbf{r}$ becomes kz. In general, if $\mathbf{k}$ has components k_x, k_y and k_z along x, y and z, then from the definition of the dot product, $\mathbf{k} \cdot \mathbf{r} = k_x x + k_y y + k_z z$.

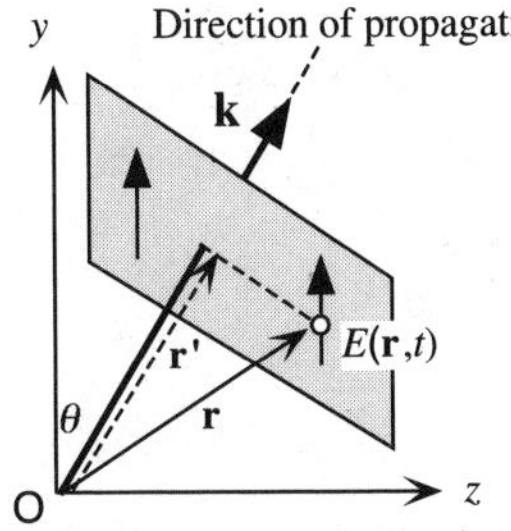

FIGURE 1.3 A traveling plane EM wave along a direction **k**.

The relationship between time and space for a given phase, ϕ for example, that corresponding to a maximum field, according to Eq. (1) is described by

$$\phi = \omega t - kz + \phi_o = \text{constant}$$

During a time interval δt, this constant phase (and hence the maximum field) moves a distance δz. The phase velocity of this wave is therefore $\delta z/\delta t$. Thus the **phase velocity** v is

Phase velocity

$$v = \frac{dz}{dt} = \frac{\omega}{k} = \upsilon\lambda \tag{4}$$

in which υ is the frequency ($\omega = 2\pi\upsilon$).

We are frequently interested in the phase difference $\Delta\phi$ at a given time between two points on a wave (Figure 1.1) that are separated by a certain distance. If the wave is traveling along z with a wavevector k, as in Eq. (1), then the phase difference between two points separated by Δz is simply $k\Delta z$ since ωt is the same for each point. If this phase difference is 0 or multiples of 2π then the two points are **in phase**. Thus the phase difference $\Delta\phi$ can be expressed as $k\Delta z$ or $2\pi\Delta z/\lambda$.

B. Maxwell's Wave Equation and Diverging Waves

Consider the plane EM wave in Figure 1.2. All constant phase surfaces are xy planes that are perpendicular to the z-direction. A cut of a plane wave parallel to the z-axis is shown in Figure 1.4 (a) in which the parallel dashed lines at right angles to the z-direction are wavefronts. We normally show wavefronts that are separated by a phase of 2π or a whole wavelength λ as in the figure. The vector that is normal to a wavefront surface at a point such as P represents the direction of wave propagation (**k**) at that point P. Clearly, the propagation vectors everywhere are all parallel and the plane wave propagates without the wave diverging; *the plane wave has no divergence.* The amplitude of the planar wave E_o does not depend on the distance from a reference point, and it is the same at all points on a given plane perpendicular to **k**, *i.e.* independent of x and y. Moreover, as

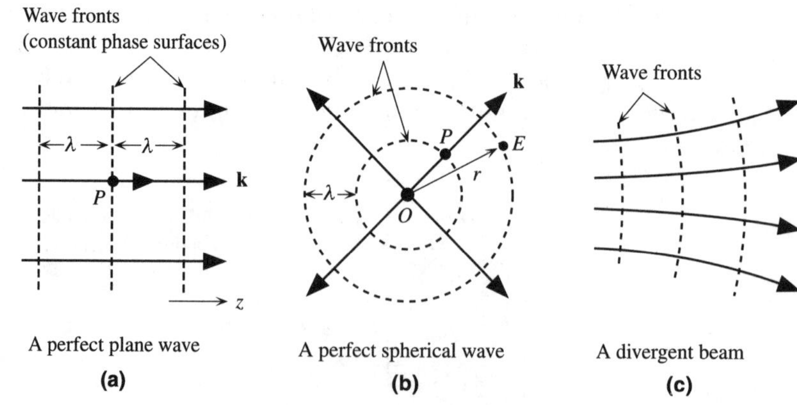

FIGURE 1.4 Examples of possible EM waves

these planes extend to infinity there is infinite energy in the plane-wave. A plane wave such as that in Figure 1.4 (a) is an idealization that is useful in analyzing many wave phenomena. In reality, however, the electric field in a plane at right angles to **k** does not extend to infinity since the light beam would have a finite cross sectional area and finite power. We would need an infinitely large EM source with infinite power to generate a perfect plane wave!

In practice there are many types of possible EM waves. All these possible EM waves must obey a special wave equation that describes the time and space dependence of the electric field. In an isotropic and linear dielectric medium, *i.e.* relative permittivity (ε_r) is the same in all directions and that it is independent of the electric field, the field E must obey **Maxwell's EM wave equation,**

$$\frac{\partial^2 E}{\partial x^2} + \frac{\partial^2 E}{\partial y^2} + \frac{\partial^2 E}{\partial z^2} = \varepsilon_o \varepsilon_r \mu_o \frac{\partial^2 E}{\partial t^2} \quad \textbf{(5)}$$

Wave equation

in which μ_o is the absolute permeability, ε_o is the absolute permittivity and ε_r is the relative permittivity of the medium. Equation (5) assumes that the conductivity of the medium is zero. To find the time and space dependence of the field, we must solve Eq. (5) in conjunction with the initial and boundary conditions. We can easily show that the plane wave in Eq. (1) satisfies Eq. (5). There are many possible waves that satisfy Eq. (5) that can therefore exist in nature.

A **spherical wave** is described by a traveling field that emerges from a point EM source and whose amplitude decays with distance r from the source. At any point r from the source, the field is given by

$$E = \frac{A}{r}\cos(\omega t - kr) \quad \textbf{(6)}$$

Spherical wave

in which A is a constant. We can substitute Eq. (6) into Eq. (5) to show that Eq. (6) is indeed a solution of Maxwell's equation (transformation from Cartesian to spherical coordinates would help). A cut of a spherical wave is illustrated in Figure 1.4 (b) where it can be seen that *wavefronts are spheres* centered at the point source O. The direction of propagation **k** at any point such as P is determined by the normal to the wavefront at that point. Clearly **k** vectors diverge out and, as the wave propagates, the constant phase surfaces become larger. **Optical divergence** refers to the angular separation of wavevectors on a given wavefront. The spherical wave has 360° of divergence. It is apparent that plane and spherical waves represent two extremes of wave propagation behavior from perfectly parallel to fully diverging wavevectors. They are produced by two extreme sizes of EM wave source; an infinitely large source for the plane wave and a point source for the spherical wave. In reality, an EM source is neither of infinite extent nor in point form, but would have a finite size and finite power. Figure 1.4 (c) shows a more practical example in which a light beam exhibits some inevitable divergence while propagating; the wavefronts are slowly bent away thereby spreading the wave. Light rays of geometric optics are drawn to be normal to constant phase surfaces (wavefronts). Light rays therefore follow the wavevector directions. Rays in Figure 1.4 (c) slowly diverge away from each other. The reason for favoring plane waves in many optical explanations is that, at a distance far away from a source, over a *small spatial region*, the

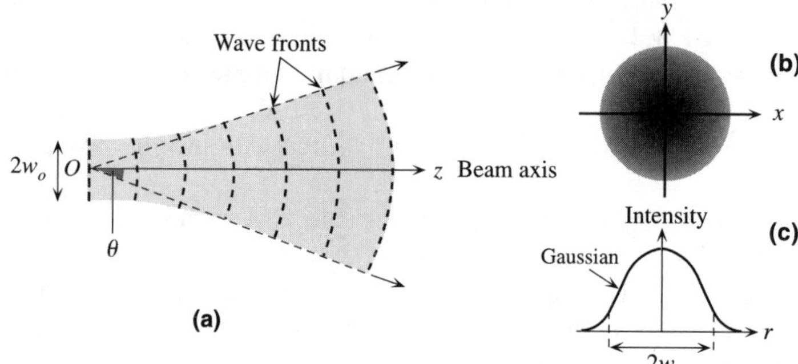

FIGURE 1.5 (a) Wavefronts of a Gaussian light beam. (b) Light intensity across beam cross section. (c) Light irradiance (intensity) vs. radial distance r from beam axis (z).

wavefronts will appear to be plane even if they are actually spherical. Figure 1.4 (a) may be a small part of a "huge" spherical wave.

Many light beams, such as the output from a laser, can be described by assuming that they are Gaussian beams. Figure 1.5 illustrates a Gaussian beam traveling along the z-axis. The beam still has an $\exp j(\omega t - kz)$ dependence to describe propagation characteristics but the amplitude varies spatially away from the beam axis and also *along* the beam axis. Such a beam has similarities to that in Figure 1.4 (c); it slowly diverges and is the result of radiation from a source of finite extent. The light intensity distribution across the beam cross-section anywhere along z is Gaussian. The **beam diameter** $2w$ at any point z is defined in such a way that the cross sectional area πw^2 at that point contains 85% of the beam power. Thus the beam diameter $2w$ increases as the beam travels along z.

The Gaussian beam that is shown in Figure 1.5 starts from O with a finite width $2w_o$ where the wavefronts are parallel and then the beam slowly diverges as the wavefronts curve out during propagation along z. The finite width $2w_o$ where the wavefronts are parallel is called the **waist** of the beam; w_o is the **waist radius** and $2w_o$ is the **spot size.** Far away from the source, the beam diameter $2w$ increases linearly with distance z. The increase in beam diameter $2w$ with z makes an angle 2θ at O, as shown in Figure 1.5, which is called the **beam divergence.** The greater the waist, the narrower the divergence. The two are related by

Beam divergence

$$2\theta = \frac{4\lambda}{\pi(2w_o)} \tag{7}$$

Suppose that we reflect the Gaussian beam back on itself so that the beam is traveling in the $-z$ direction and converging towards O; simply reverse the direction of travel in Figure 1.5 (a). The wavefronts would be "straightening out," and at O they would be parallel again. The beam would still have the same finite diameter $2w_o$ (waist) at O. From then on, the beam again diverges out just as it did traveling in $+z$ direction. The relationship in Eq. (7) therefore defines a minimum spot size to which a Gaussian beam can be focused.

EXAMPLE 1.1.1 A diverging laser beam

Consider a HeNe laser beam at 633 nm with a spot size of 10 mm. Assuming a Gaussian beam, what is the divergence of the beam?

Solution Using Eq. (7), we find,

$$2\theta = \frac{4\lambda}{\pi(2w_o)} = \frac{4(633 \times 10^{-9}\text{ m})}{\pi(10 \times 10^{-3}\text{ m})} = 8.06 \times 10^{-5}\text{ rad} = 0.0046°$$

1.2 REFRACTIVE INDEX

When an EM wave is traveling in a dielectric medium, the oscillating electric field polarizes the molecules of the medium at the frequency of the wave. Indeed, the EM wave propagation can be considered to be the propagation of this polarization in the medium. The field and the induced molecular dipoles become coupled. The net effect is that the polarization mechanism delays the propagation of the EM wave. Put differently, it slows down the EM wave with respect to its speed in a vacuum where there are no dipoles with which the field can interact. The stronger the interaction between the field and the dipoles, the slower the propagation of the wave. The relative permittivity ε_r measures the ease with which the medium becomes polarized and hence it indicates the extent of interaction between the field and the induced dipoles. For an EM wave traveling in a nonmagnetic dielectric medium of relative permittivity ε_r, the phase velocity v is given by

$$v = \frac{1}{\sqrt{\varepsilon_r \varepsilon_o \mu_o}} \quad (1)$$

Phase velocity in a medium with ε_r

If the frequency υ is in the optical frequency range, then ε_r will be due to electronic polarization as ionic polarization will be too sluggish to respond to the field. However, at the infrared frequencies or below, the relative permittivity also includes a significant contribution from ionic polarization and the phase velocity is slower. For an EM wave traveling in free space, $\varepsilon_r = 1$ and $v_{\text{vacuum}} = 1/\sqrt{[\varepsilon_o \mu_o]} = c = 3 \times 10^8\text{ m s}^{-1}$, the velocity of light in vacuum. The ratio of the speed of light in free space to its speed in a medium is called the **refractive index** n of the medium,

$$n = \frac{c}{v} = \sqrt{\varepsilon_r} \quad (2)$$

Definition of refractive index

If k is the wave vector ($k = 2\pi/\lambda$) and λ is the wavelength, both in free space, then in the medium $k_{\text{medium}} = nk$ and $\lambda_{\text{medium}} = \lambda/n$. Equation (2) is in agreement with our intuition that light propagates more slowly in a denser medium that has a higher refractive index. We should note that the frequency υ remains the same. The refractive index of a medium is not necessarily the same in all directions. In noncrystalline materials such as glasses and liquids, the material structure is the same in all directions and n does not depend on the direction. The refractive index is then **isotropic**. In crystals, however, the atomic arrangements and interatomic bonding are different along different directions. Crystals, in general, have nonisotropic, or *anisotropic*, properties. Depending on the crystal structure, the relative permittivity ε_r is different along different crystal directions. This means that, in general, the refractive index n seen by a propagating electromagnetic wave in a crystal will depend on the value of ε_r along the direction of the oscillating electric field (that is along the direction of polarization). For example, suppose that the wave in Figure 1.1 is traveling along the z-direction in a particular crystal with its electric field oscillating along the x-direction. If the relative permittivity along this x-direction is ε_{rx}, then $n_x = \sqrt{(\varepsilon_{rx})}$. The wave therefore propagates with a phase velocity that is c/n_x. The variation of n with direction of propagation and the direction of the electric field depends on the particular crystal structure. With the exception of cubic crystals (such as diamond), all crystals exhibit a degree of optical anisotropy that leads to a number of important applications as discussed in Chapter 7. Typically, noncrystalline solids such as glasses and liquids, and cubic crystals are **optically isotropic**; they possess only one refractive index for all directions.

EXAMPLE 1.2.1 Relative permittivity and refractive index

Relative permittivity ε_r or the dielectric constant of materials, in general, depends on the frequency of the electromagnetic wave. The relationship $n = \sqrt{(\varepsilon_r)}$ between the refractive index n and ε_r must be applied at the same frequency for both n and ε_r. The relative permittivity for many materials can be vastly different at high and low frequencies because different polarization mechanisms operate at these frequencies.[2] At low frequencies all polarization mechanisms present can contribute to ε_r, whereas at optical frequencies only the electronic polarization can respond to the oscillating field. Table 1.1 lists the relative permittivity ε_r(LF) at low frequencies (*e.g.* 60 Hz or 1 kHz as would be measured for example using a capacitance bridge in the laboratory) for various materials. It then compares $\sqrt{[\varepsilon_r(\text{LF})]}$ with n.

For silicon and diamond there is an excellent agreement between ε_r(LF) and n. Both are covalent solids in which electronic polarization (electronic bond polarization) is the only polarization mechanism at low and high frequencies. Electronic polarization involves the displacement of light electrons with respect to heavy positive ions of the crystal. This process can readily respond to the field oscillations up to optical or even ultraviolet frequencies.

TABLE 1.1 Low frequency (LF) relative permittivity ε_r(LF) and refractive index n.

Material	ε_r(LF)	$\sqrt{[\varepsilon_r(\text{LF})]}$	n (optical)	Comment
Si	11.9	3.44	3.45 (at 2.15 μm)	Electronic bond polarization up to optical frequencies
Diamond	5.7	2.39	2.41 (at 590 nm)	Electronic bond polarization up to uv light
GaAs	13.1	3.62	3.30 (at 5 μm)	Ionic polarization contributes to ε_r(LF)
SiO_2	3.84	2.00	1.46 (at 600 nm)	Ionic polarization contributes to ε_r(LF)
Water	80	8.9	1.33 (at 600 nm)	Dipolar polarization contributes to ε_r(LF), which is large

[2] Ch. 7 in *Principles of Electronic Materials and Devices, Second Edition*, S.O. Kasap (McGraw-Hill, 2001)

For GaAs and SiO_2 $\sqrt{[\varepsilon_r(\mathrm{LF})]}$ is larger than n because at low frequencies both of these solids possess a degree of ionic polarization. The bonding is not totally covalent and there is a degree of ionic bonding that contributes to polarization at frequencies below far-infrared wavelengths.

In the case of water, the $\varepsilon_r(\mathrm{LF})$ is dominated by orientational or dipolar polarization, which is far too sluggish to respond to high frequency oscillations of the field at optical frequencies.

It is instructive to consider what factors affect n. The simplest (and approximate) expression for relative permittivity is

$$\varepsilon_r \approx 1 + N\alpha/\varepsilon_o$$

in which N is the number of molecules per unit volume and α is the polarizability per molecule. Both atomic concentration, or density, and polarizability therefore increase n. For example, glasses of given type but with greater density tend to have higher n.

1.3 GROUP VELOCITY AND GROUP INDEX

Since there are no perfect monochromatic waves in practice, we have to consider the way in which a group of waves differing slightly in wavelength will travel along the z-direction as depicted in Figure 1.6. When two perfectly harmonic waves of frequencies $\omega - \delta\omega$ and $\omega + \delta\omega$ and wavevectors $k - \delta k$ and $k + \delta k$ interfere, as shown in Figure 1.6, they generate a **wave packet** that contains an oscillating field at the mean frequency ω that is amplitude modulated by a slowly varying field of frequency $\delta\omega$. The maximum amplitude moves with a wavevector δk and thus with a velocity that is called the **group velocity** that is given by

$$v_g = \frac{d\omega}{dk} \tag{1}$$

Group velocity

The **group velocity** therefore defines the speed with which energy or information is propagated since it defines the speed of the envelope of the amplitude variation. The maximum electric field in Figure 1.6 advances with a velocity v_g whereas the phase variations in the electric field are propagating at the phase velocity v.

Inasmuch as $\omega = vk$ and the phase velocity $v = c/n$, the group velocity in a medium can be readily evaluated from Eq. (1). In vacuum, obviously v is simply c and independent of the wavelength or k. Thus for waves traveling in vacuum, $\omega = ck$ and the group velocity is,

$$v_g(\text{vacuum}) = \frac{d\omega}{dk} = c = \text{phase velocity} \tag{2}$$

Group velocity in vacuum

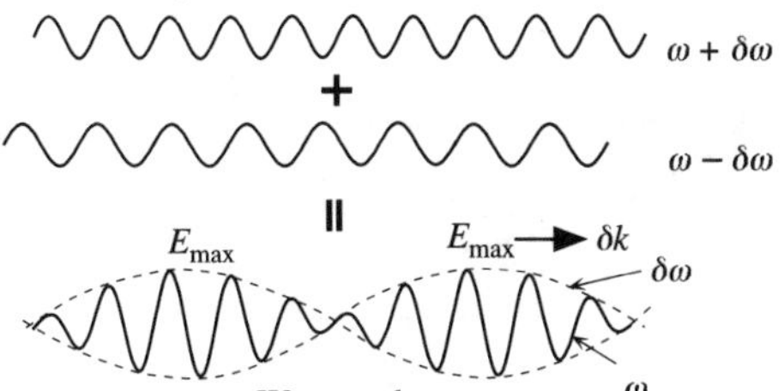

FIGURE 1.6 Two slightly different wavelength waves traveling in the same direction result in a wave packet that has an amplitude variation which travels at the group velocity.

On the other hand, suppose that v depends on the wavelength or k by virtue of n being a function of the wavelength as in the case for glasses. Then,

$$\omega = vk = \left[\frac{c}{n(\lambda)}\right]\left[\frac{2\pi}{\lambda}\right] \quad \textbf{(3)}$$

in which $n = n(\lambda)$ is a function of the wavelength. The group velocity v_g in a medium, from differentiating Eq. (3) in Eq. (1), is approximately given by,

$$v_g(\text{medium}) = \frac{d\omega}{dk} = \frac{c}{n - \lambda\frac{dn}{d\lambda}}$$

This can be written as

Group velocity in a medium

$$v_g(\text{medium}) = \frac{c}{N_g} \quad \textbf{(4)}$$

in which

Group index

$$N_g = n - \lambda\frac{dn}{d\lambda} \quad \textbf{(5)}$$

is defined as the **group index of the medium**. Equation (5) defines the group refractive index N_g of a medium and determines the effect of the medium on the group velocity via Eq. (4).

In general, for many materials the refractive index n and hence the group index N_g depend on the wavelength of light by virtue of ε_r being frequency dependent. Then both the phase velocity v and the group velocity v_g depend on the wavelength and the medium is called a **dispersive medium**. Refractive index n and the group index N_g of pure SiO_2 (silica) glass are important parameters in optical fiber design in optical communications. Both of these parameters depend on the wavelength of light as shown in Figure 1.7. Around 1300 nm, N_g is minimum, which means that for wavelengths close to 1300 nm, N_g is wavelength independent. Thus, light waves with wavelengths around 1300 nm travel with the same group velocity and do not experience dispersion. This phenomenon is significant in the propagation of light in optical fibers as discussed in Chapter 2.

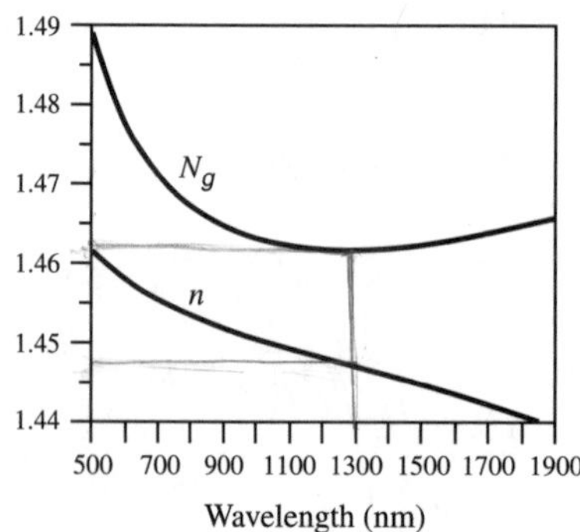

FIGURE 1.7 Refractive index n and the group index N_g of pure SiO_2 (silica) glass as a function of wavelength.

EXAMPLE 1.3.1 Group velocity

Consider two sinusoidal waves that are close in frequency, that is, waves of frequencies $\omega - \delta\omega$ and $\omega + \delta\omega$ as in Figure 1.6. Their wave vectors will be $k - \delta k$ and $k + \delta k$. The resultant wave will be

$$E_x(z,t) = E_o \cos[(\omega - \delta\omega)t - (k - \delta k)z] + E_o \cos[(\omega + \delta\omega)t - (k + \delta k)z]$$

By using the trigonometric identity $\cos A + \cos B = 2\cos[\frac{1}{2}(A - B)]\cos[\frac{1}{2}(A + B)]$ we arrive at

$$E_x(z,t) = 2E_o \cos[(\delta\omega)t - (\delta k)z]\cos[\omega t - kz]$$

As depicted in Figure 1.6, this represents a sinusoidal wave of frequency ω, which is amplitude modulated by a very slowly varying sinusoidal of frequency $\delta\omega$. The system of waves, that is, the modulation, travels along z at a speed determined by the modulating term, $\cos[(\delta\omega)t - (\delta k)z]$. The maximum in the field occurs when $[(\delta\omega)t - (\delta k)z] = 2m\pi$ = constant (m is an integer), which travels with a velocity

$$\frac{dz}{dt} = \frac{\delta\omega}{\delta k} \quad \text{or} \quad v_g = \frac{d\omega}{dk}$$

This is the group velocity of the waves, as stated in Eq. (1), since it determines the speed of propagation of the maximum electric field along z.

EXAMPLE 1.3.2 Group and phase velocities

Consider a light wave traveling in a pure SiO_2 (silica) glass medium. If the wavelength of light is 1 μm and the refractive index at this wavelength is 1.450, what is the phase velocity, group index (N_g) and group velocity (v_g)?

Solution The phase velocity is given by

$$v = c/n = (3 \times 10^8 \text{ m s}^{-1})/(1.450)$$

$$= 2.069 \times 10^8 \text{ m s}^{-1}.$$

From Figure 1.7, at $\lambda = 1$ μm, $N_g = 1.460$, so that

$$v_g = c/N_g = (3 \times 10^8 \text{ ms}^{-1})/(1.460)$$

$$= 2.055 \times 10^8 \text{ m s}^{-1}.$$

The group velocity is about ~0.7% slower than the phase velocity.

1.4 MAGNETIC FIELD, IRRADIANCE AND POYNTING VECTOR

Although we have considered the electric field component E_x of the electromagnetic (EM) wave, we should recall that the magnetic field (magnetic induction) component B_y always accompanies E_x in an EM wave propagation. In fact, if v is the phase velocity of an EM wave in an isotropic dielectric medium and n is the refractive index, then according to electromagnetism, at all times and anywhere in an EM wave,[3]

[3] This is actually a statement of Faraday's law for EM waves. In vector notation it is often expressed as $\omega\mathbf{B} = \mathbf{k} \times \mathbf{E}$.

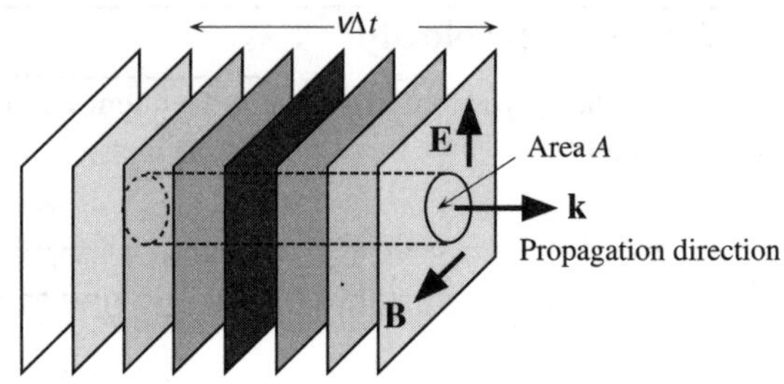

FIGURE 1.8 A plane EM wave travelling along **k** crosses an area A at right angles to the direction of propagation. In time Δt, the energy in the cylindrical volume $Av\Delta t$ (shown dashed) flows through A.

Fields in an EM wave

$$E_x = vB_y = \frac{c}{n}B_y \tag{1}$$

in which $v = (\varepsilon_o\varepsilon_r\mu_o)^{-1/2}$ and $n = \sqrt{\varepsilon_r}$. Thus, the two fields are simply and intimately related for an EM wave propagating in an isotropic medium. Any process that alters E_x also intimately changes B_y in accordance with Eq. (1).

As the EM wave propagates in the direction of the wave vector **k** as shown in Figure 1.8, there is an energy flow in this direction. The wave brings with it electromagnetic energy. A small region of space in which the electric field is E_x has an energy density, that is, energy per unit volume, given by $(1/2)\varepsilon_o\varepsilon_r E_x^2$. Similarly, a region of space where the magnetic field is B_y has an energy density $(1/2\mu_o)B_y^2$. Since the two fields are related by Eq. (1), the energy densities in the E_x and B_y fields are the same,

Energy densities in an EM wave

$$\frac{1}{2}\varepsilon_o\varepsilon_r E_x^2 = \frac{1}{2\mu_o}B_y^2 \tag{2}$$

The total energy density in the wave is therefore $\varepsilon_o\varepsilon_r E_x^2$. Suppose that an ideal "energy meter" is placed in the path of the EM wave so that the receiving area A of this meter is perpendicular to the direction of propagation. In a time interval Δt, a portion of the wave of spatial length $v\Delta t$ crosses A. Thus, a volume $Av\Delta t$ of the EM wave crosses A in time Δt. The energy in this volume consequently becomes received. If S is the EM power flow per unit area, then

$$S = \text{Energy flow per unit time per unit area}$$

giving

$$S = \frac{(Av\Delta t)(\varepsilon_o\varepsilon_r E_x^2)}{A\Delta t} = v\varepsilon_o\varepsilon_r E_x^2 = v^2\varepsilon_o\varepsilon_r E_x B_y \tag{3}$$

In an isotropic medium, the energy flow is in the direction of wave propagation. If we use the vectors **E** and **B** to represent the electric and magnetic fields in the EM wave, then the wave propagates in a direction **E** × **B** because this direction is perpendicular to both **E** and **B**. The EM power flow per unit area in Eq. (3) can be written as,

Poynting vector

$$\mathbf{S} = v^2\varepsilon_o\varepsilon_r \mathbf{E} \times \mathbf{B} \tag{4}$$

in which **S**, called the **Poynting vector**, represents the energy flow per unit time per unit area in a direction determined by $\mathbf{E} \times \mathbf{B}$ (direction of propagation). Its magnitude, power flow per unit area, is called the **irradiance**.[4]

The field E_x at the receiver location (say, $z = z_1$) varies sinusoidally, which means that the energy flow also varies sinusoidally. The irradiance in Eq. (3) is the **instantaneous irradiance**. If we write the field as $E_x = E_o \sin(\omega t)$ and then calculate the average irradiance by averaging S over one period we would find the **average irradiance**,

$$I = S_{\text{average}} = \tfrac{1}{2} v \varepsilon_o \varepsilon_r E_o^2 \qquad \textbf{(5)}$$

Average irradiance (intensity)

Since $v = c/n$ and $\varepsilon_r = n^2$ we can write Eq. (5) as,

$$I = S_{\text{average}} = \tfrac{1}{2} c \varepsilon_o n E_o^2 = (1.33 \times 10^{-3}) n E_o^2 \qquad \textbf{(6)}$$

Average irradiance (intensity)

The instantaneous irradiance can be measured only if the power meter can respond more quickly than the oscillations of the electric field, and since this is in the optical frequencies range, all practical measurements invariably yield the average irradiance. This is because all detectors have a response rate much slower than the frequency of the wave.

EXAMPLE 1.4.1 Electric and magnetic fields in light

The intensity (irradiance) of the red laser beam from a He-Ne laser at a certain location has been measured to be about 1 mW cm^{-2}. What are the magnitudes of the electric and magnetic fields? What are the magnitudes if this beam is in a glass medium with a refractive index $n = 1.45$?

Solution Using Eq. (6) for the average irradiance, the field in air is

$$E_o = \sqrt{\frac{2I}{c\varepsilon_o n}} = \sqrt{\frac{2(1 \times 10^{-3} \times 10^4 \text{ W m}^{-2})}{(3 \times 10^8 \text{ m s}^{-1})(8.85 \times 10^{-12} \text{ F m}^{-1})(1)}}$$

so that

$$E_o = 87 \text{ V m}^{-1} \quad \text{or} \quad 0.87 \text{ V cm}^{-1}.$$

The corresponding magnetic field is

$$B_o = E_o/c = (87 \text{ V m}^{-1})/(3 \times 10^8 \text{ m s}^{-1}) = 0.29\ \mu\text{T}.$$

If this 1 mW cm^{-2} beam were in a glass medium of $n = 1.45$ and still had the same intensity, then

$$E_o(\text{medium}) = \sqrt{\frac{2I}{c\varepsilon_o n}} = \sqrt{\frac{2(1 \times 10^{-3} \times 10^4 \text{ W m}^{-2})}{(3 \times 10^8 \text{ m s}^{-1})(8.85 \times 10^{-12} \text{ F m}^{-1})(1.45)}}$$

or

$$E_o(\text{medium}) = 72 \text{ V m}^{-1}$$

and

$$B_o(\text{medium}) = nE_o(\text{medium})/c = (1.45)(72 \text{ V m}^{-1})/(3 \times 10^8 \text{ m s}^{-1}) = 0.35\ \mu\text{T}.$$

---

[4] The term *intensity* is widely used and interpreted by many engineers as power flow per unit area even though the strictly correct term is *irradiance*. Many optoelectronic data books simply use intensity to mean irradiance.

1.5 SNELL'S LAW AND TOTAL INTERNAL REFLECTION (TIR)

We consider a traveling plane EM wave in a medium (1) of refractive index n_1 propagating towards a medium (2) with a refractive index n_2. Constant phase fronts are joined with broken lines and the wave vector $\mathbf{k}_i$ is perpendicular to the wave fronts as shown in Figure 1.9. When the wave reaches the plane boundary between the two media, a transmitted wave in medium 2 and a reflected wave in medium 1 appear. The transmitted wave is called the **refracted light**. The angles, $\theta_i, \theta_t, \theta_r$ define the directions of the incident, transmitted and reflected waves respectively with respect to the normal to the boundary plane as shown in Figure 1.9. The wave vectors of the reflected and transmitted waves are denoted as $\mathbf{k}_r$ and $\mathbf{k}_t$. Since both the incident and reflected waves are in the same medium, the magnitudes of $\mathbf{k}_r$ and $\mathbf{k}_i$ are the same, $k_r = k_i$.

Simple arguments based on constructive interference can be used to show that there can be only one reflected wave that occurs at an angle equal to the incidence angle. The two waves along A_i and B_i are in phase. When these waves are reflected to become waves A_r and B_r then they must still be in phase, otherwise they will interfere destructively and destroy each other. The only way the two waves can stay in phase is if $\theta_r = \theta_i$. All other angles lead to the waves A_r and B_r being out of phase and interfering destructively.

The refracted waves A_t and B_t are propagating in a medium of refracted index $n_2(<n_1)$ that is different than n_1. Hence the waves A_t and B_t have different velocities than A_i and B_i. We consider what happens to a wavefront such as AB, corresponding perhaps to the maximum field, as it propagates from medium 1 to 2. We recall that the points A and B on this front are always in phase. During the time it takes for the phase B on

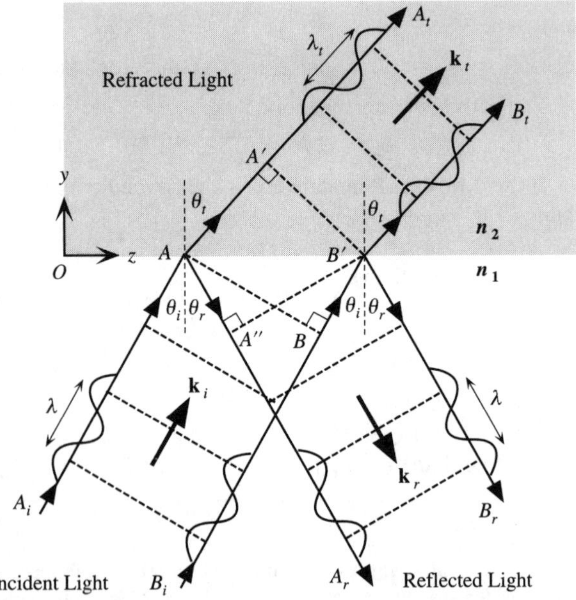

FIGURE 1.9 A light wave traveling in a medium with a greater refractive index $(n_1 > n_2)$ suffers reflection and refraction at the boundary.

wave B_i to reach B', phase A on wave A_t has progressed to A'. The wavefront AB thus becomes the front $A'B'$ in medium 2. Unless the two waves at A' and B' still have the same phase, there will be no transmitted wave. A' and B' points on the front are in phase only for one particular transmitted angle, θ_t.

If it takes time t for the phase at B on wave B_i to reach B', then $BB' = v_1 t = ct/n_1$. During this time t, the phase A has progressed to A' where $AA' = v_2 t = ct/n_2$. A' and B' belong to the same front just like A and B so that AB is perpendicular to $\mathbf{k}_i$ in medium 1 and $A'B'$ is perpendicular to $\mathbf{k}_t$ in medium 2. From geometrical considerations, $AB' = BB'/\sin\theta_i$ and $AB' = AA'/\sin\theta_t$ so that

$$AB' = \frac{v_1 t}{\sin\theta_i} = \frac{v_2 t}{\sin\theta_t}$$

or

$$\frac{\sin\theta_i}{\sin\theta_t} = \frac{v_1}{v_2} = \frac{n_2}{n_1} \qquad (1)$$

Snell's Law

This is **Snell's law**,[5] which relates the angles of incidence and refraction to the refractive indices of the media.

If we consider the reflected wave, the wave front AB becomes $A''B'$ in the reflected wave. In time t, phase B moves to B' and A moves to A''. Since they must still be in phase to constitute the reflected wave, BB' must be equal to AA''. Suppose it takes time t for the wavefront B to move to B' (or A to A''). Then, since $BB' = AA'' = v_1 t$, from geometrical considerations,

$$AB' = \frac{v_1 t}{\sin\theta_i} = \frac{v_1 t}{\sin\theta_r}$$

so that $\theta_i = \theta_r$. Angles of incidence and reflection are the same.

When $n_1 > n_2$ then obviously the transmitted angle is greater than the incidence angle as apparent in Figure 1.9. When the refraction angle θ_t reaches 90°, the incidence angle is called the **critical angle** θ_c, which is given by

$$\sin\theta_c = \frac{n_2}{n_1} \qquad (2)$$

Total internal reflection, TIR

When the incidence angle θ_i exceeds θ_c then there is no transmitted wave but only a reflected wave. The latter phenomenon is called **total internal reflection** (TIR). The effect of increasing the incidence angle is shown in Figure 1.10. It is the TIR phenomenon that leads to the propagation of waves in a dielectric medium surrounded by a medium of smaller refractive index as shown in Ch. 2. Although Snell's law for $\theta_i > \theta_c$ shows

[5] Willebrord van Roijen Snell (1580–1626), a Dutch physicist and mathematician, was born in Leiden and eventually became a professor at Leiden University. He obtained his refraction law in 1621 and it was published by Réne Descartes in France in 1637; it is not known whether Descartes knew of Snell's law or formulated it independently.

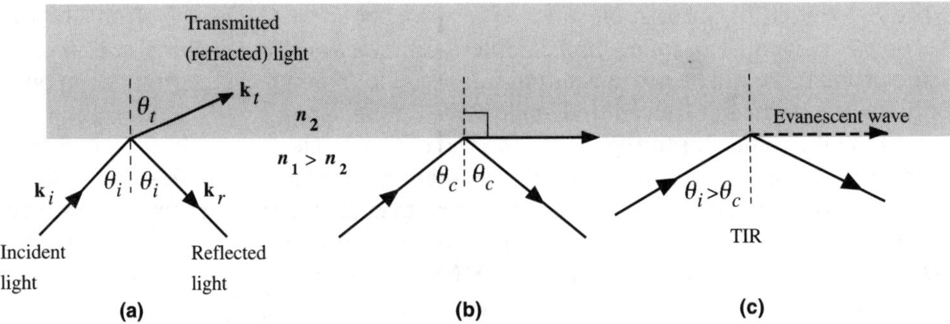

FIGURE 1.10 Light wave traveling in a more dense medium strikes a less dense medium. Depending on the incidence angle with respect to θ_c, which is determined by the ratio of the refractive indices, the wave may be transmitted (refracted) or reflected. (a) $\theta_i > \theta_c$ (b) $\theta_i = \theta_c$ (c) $\theta_i > \theta_c$ and total internal reflection (TIR).

that $\sin\theta_t > 1$ and hence θ_t is an "imaginary" angle of refraction, there is however a wave that propagates along the boundary called the **evanescent wave** as discussed below.

1.6 FRESNEL'S EQUATIONS

A. Amplitude Reflection and Transmission Coefficients (*r* and *t*)

Although the ray picture with constant phase wave fronts is useful in understanding refraction and reflection, to obtain the magnitude of the reflected and refracted waves and their relative phases, we need to consider the electric field in the light wave. The electric field in the wave must be perpendicular to the direction of propagation as shown in Figure 1.1. We can resolve the field E_i of the incident wave into two components, one

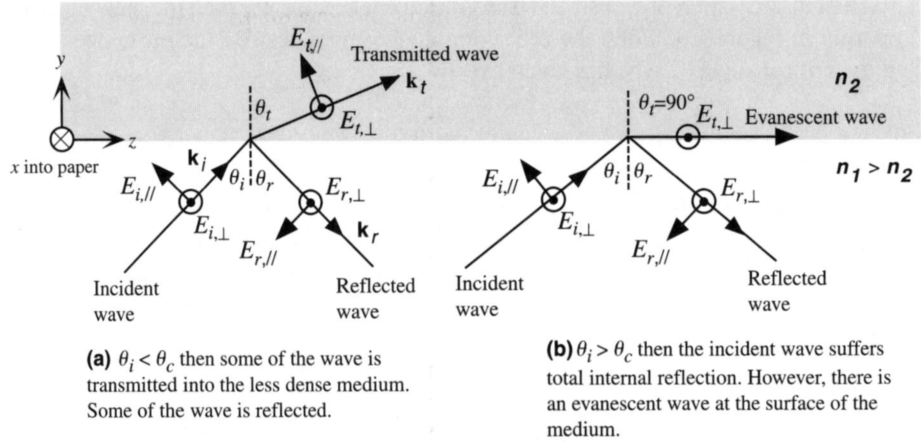

FIGURE 1.11 Light wave travelling in a more dense medium strikes a less dense medium. The plane of incidence is the plane of the paper and is perpendicular to the flat interface between the two media. The electric field is normal to the direction of propagation. It can be resolved into perpendicular (⊥) and parallel (//) components.

in the plane of incidence, $E_{i,//}$ and the other perpendicular to the plane of incidence, $E_{i,\perp}$ as in Figure 1.11. The plane of incidence is defined as the plane containing the incident and the reflected rays, which in Figure 1.11 corresponds to the plane of the paper.[6] Similarly for both the reflected and transmitted waves, we will have field components parallel and perpendicular to the plane of incidence, i.e, $E_{r,//}$, $E_{r,\perp}$, and $E_{t,//}$, $E_{t,\perp}$.

As apparent from Figure 1.11, the incident, transmitted, and reflected wave all have a wave vector component along the z-direction, that is, they have an effective velocity along z. The fields $E_{i,\perp}$, $E_{r,\perp}$ and $E_{t,\perp}$ are all perpendicular to the z-direction. These waves are called **transverse electric field** (TE) waves. On the other hand, waves with $E_{i,//}$, $E_{r,//}$, and $E_{t,//}$ have only their magnetic field components perpendicular to the z-direction, and these are called **transverse magnetic field** (TM) waves.

We will describe the incident, reflected, and refracted waves by the exponential representation of a traveling wave, *i.e.*

$$E_i = E_{io} \exp j(\omega t - \mathbf{k}_i \cdot \mathbf{r})$$ *Incident wave*

$$E_r = E_{ro} \exp j(\omega t - \mathbf{k}_r \cdot \mathbf{r})$$ *Reflected wave*

$$E_t = E_{to} \exp j(\omega t - \mathbf{k}_t \cdot \mathbf{r})$$ *Transmitted wave*

in which $\mathbf{r}$ is the position vector, the wave vectors $\mathbf{k}_i$, $\mathbf{k}_r$, and $\mathbf{k}_t$ describe the directions of the incident, reflected, and transmitted waves, and E_{io}, E_{ro}, and E_{to} are the respective amplitudes. Any phase changes such as ϕ_r and ϕ_t in the reflected and transmitted waves with respect to the phase of the incident wave are incorporated into the complex amplitudes, E_{ro} and E_{to}. Our objective is to find E_{ro} and E_{to} with respect to E_{io}.

We should note that similar equations can be stated for the magnetic field components in the incident, reflected, and transmitted waves but these will be perpendicular to the corresponding electric fields. The electric and magnetic fields anywhere on the wave must be perpendicular to each other as a requirement of electromagnetic wave theory. This means that with $E_{//}$ in the EM wave we have a magnetic field $B_\perp$ associated with it such that $B_\perp = (n/c)E_{//}$. Similarly, $E_\perp$ will have a magnetic field $B_{//}$ associated with it such that $B_{//} = (n/c)E_\perp$.

There are two useful fundamental rules in electromagnetism that govern the behavior of the electric and magnetic fields at a boundary between two dielectric media, which we can arbitrarily label as 1 and 2. These rules are called **boundary conditions**. The first states that the electric field that is tangential to the boundary surface, $E_{\text{tangential}}$, must be continuous across the boundary from medium 1 to 2, *i.e.* at the boundary, $y = 0$ in Figure 1.11,

$$E_{\text{tangential}}(1) = E_{\text{tangential}}(2)$$ *Boundary condition*

---

[6] The definitions of the field components follow those of S.G. Lipson *et al.* (*Optical Physics*, Third Edition, Cambridge University Press, Cambridge, 1995) and Grant Fowles (*Introduction to Modern Optics*, Second Ed., Dover Publications, Inc. New York, 1975), whose clear treatments of this subject are highly recommended. The majority of authors use a different convention, which leads to different signs later in the equations ("they [Fresnel's equations] must be related to the specific electric field directions from which they are derived," Eugene Hecht, *Optics*; *Second Edition* (Addison Wesley Publishing Company, Reading, MA, 1987) Addison Wesley).

The second rule is that the tangential component of the magnetic field, $B_{tangential}$, to the boundary must likewise be continuous from medium 1 to 2, provided that the two media are non-magnetic (relative permeability, $\mu_r = 1$),

Boundary condition

$$B_{tangential}(1) = B_{tangential}(2)$$

Using the boundary conditions above for the fields at $y = 0$ and the relationship between the electric and magnetic fields, we can find the reflected and transmitted waves in terms of the incident wave. The boundary conditions can be satisfied only if the reflection and incidence angles are equal, $\theta_r = \theta_i$, and the angles for the transmitted and incident wave obey Snell's law, $n_1 \sin\theta_1 = n_2 \sin\theta_2$.

Applying the boundary conditions above to the EM wave going from medium 1 to 2, the amplitudes of the reflected and transmitted waves can be readily obtained in terms of n_1, n_2 and the incidence angle θ_i alone.[7] These relationships are called **Fresnel's equations**. If we define $n = n_2/n_1$ as the relative refractive index of medium 2 to that of 1, then the **reflection and transmission coefficients** for $E_\perp$ are

Reflection coefficient

$$r_\perp = \frac{E_{r0,\perp}}{E_{i0,\perp}} = \frac{\cos\theta_i - [n^2 - \sin^2\theta_i]^{1/2}}{\cos\theta_i + [n^2 - \sin^2\theta_i]^{1/2}} \tag{1a}$$

and

Transmission coefficient

$$t_\perp = \frac{E_{t0,\perp}}{E_{i0,\perp}} = \frac{2\cos\theta_i}{\cos\theta_i + [n^2 - \sin^2\theta_i]^{1/2}} \tag{1b}$$

There are corresponding coefficients for the $E_{//}$ fields with corresponding **reflection and transmission coefficients**, $r_{//}$ and $t_{//}$,

Reflection coefficient

$$r_{//} = \frac{E_{r0,//}}{E_{i0,//}} = \frac{[n^2 - \sin^2\theta_i]^{1/2} - n^2\cos\theta_i}{[n^2 - \sin^2\theta_i]^{1/2} + n^2\cos\theta_i} \tag{2a}$$

Transmission coefficient

$$t_{//} = \frac{E_{t0,//}}{E_{i0,//}} = \frac{2n\cos\theta_i}{n^2\cos\theta_i + [n^2 - \sin^2\theta_i]^{1/2}} \tag{2b}$$

Further, the coefficients above are related by

Transmission and reflection

$$r_{//} + nt_{//} = 1 \qquad \text{and} \qquad r_\perp + 1 = t_\perp \tag{3}$$

The significance of these equations is that they allow the amplitudes and phases of the reflected and transmitted waves to be determined from the coefficients $r_\perp, r_{//}, t_\perp$ and $t_{//}$. For convenience we take E_{io} to be a real number so that phase angles of $r_\perp$ and $t_\perp$ correspond to the **phase changes** measured with respect to the incident wave. For example, if $r_\perp$ is a complex quantity then we can write this as $r_\perp = |r_\perp|\exp(j\phi_\perp)$ in which $|r_\perp|$ and $\phi_\perp$ represent the relative amplitude and phase of the reflected wave with respect to the incident wave for the field perpendicular to the plane of incidence. Of course,

---

[7] These equations are readily available in any electromagnetism textbook. Their derivation from the two boundary conditions involves extensive algebraic manipulation, which we will not carry out here. The electric and magnetic field components on both sides of the boundary are resolved tangentially to the boundary surface and the boundary conditions are then applied. We then use such relations as $\cos\theta_t = [1 - \sin\theta_t]^{1/2}$ and $\sin\theta_t$ is determined by Snell's law, *etc.*

when $r_\perp$ is a real quantity, then a positive number represents no phase shift and a negative number is a phase shift of 180° (or π). As with all waves, a negative sign corresponds to a 180° phase shift. Complex coefficients can be obtained only from Fresnel's equations if the terms under the square roots become negative and this can happen only when $n < 1$ (or $n_1 > n_2$), and also when $\theta_i > \theta_c$, the critical angle. Thus phase changes other than 0 or 180° occur only when there is total internal reflection.

Figure 1.12 (a) shows how the magnitudes of the reflection coefficients, $|r_\perp|$ and $|r_{//}|$ vary with the incidence angle θ_i for a light wave traveling from a more dense medium, $n_1 = 1.44$, to a less dense medium, $n_2 = 1.00$, as predicted by Fresnel's equations. Figure 1.12 (b) shows the changes in the phase of the reflected wave, $\phi_\perp$ and $\phi_{//}$, with θ_i. The critical angle θ_c as determined from $\sin\theta_c = n_2/n_1$ in this case is 44°. It is clear that for incidence close to normal (small θ_i), there is no phase change in the reflected wave. For example, putting normal incidence ($\theta_i = 0$) into Fresnel's equations we find,

$$r_{//} = r_\perp = \frac{n_1 - n_2}{n_1 + n_2} \tag{4}$$

Normal incidence

This is a positive quantity for $n_1 > n_2$, which means that the reflected wave suffers no phase change. This is confirmed by $\phi_\perp$ and $\phi_{//}$ in Figure 1.12 (b). As the incidence angle increases, eventually $r_{//}$ becomes zero at an angle of about 35°. We can find this special incidence angle, labeled as θ_p, by solving the Fresnel equation (2a) for $r_{//} = 0$. The field in the reflected wave is then always perpendicular to the plane of incidence and hence well defined. This special angle is called the **polarization angle** or **Brewster's angle** and from Eq. (2a) is given by

$$\tan\theta_p = \frac{n_2}{n_1} \tag{5}$$

Brewster's polarization angle

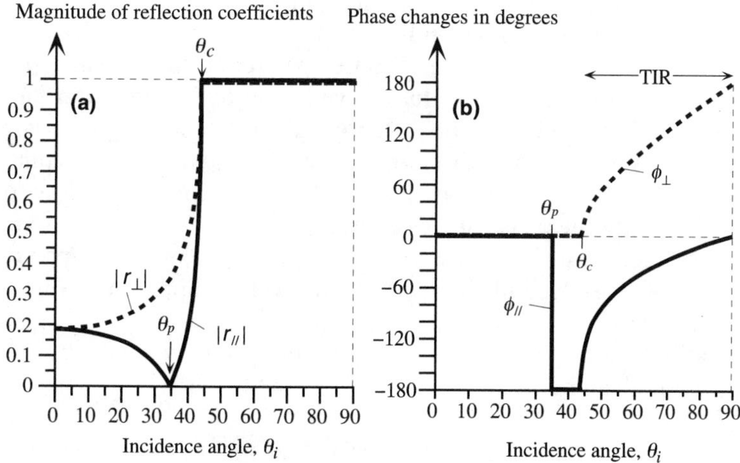

FIGURE 1.12 Internal reflection: (a) Magnitude of the reflection coefficients $r_{//}$ and $r_\perp$ vs. angle of incidence θ_i for $n_1 = 1.44$ and $n_2 = 1.00$. The critical angle is 44°. (b) The corresponding phase changes $\phi_{//}$ and $\phi_\perp$ vs. incidence angle.

The reflected wave is then said to be **linearly polarized** because it contains *electric field oscillations that are contained within a well defined plane*, which is perpendicular to the plane of incidence and also to the direction of propagation. Electric field oscillations in **unpolarized light**, on the other hand, can be in any one of infinite number of directions that are perpendicular to the direction of propagation. In linearly polarized light, however, the field oscillations are contained within a well-defined plane. Light emitted from many light sources, such as a tungsten light bulb or an LED diode, is unpolarized. Unpolarized light can be viewed as a stream or collection of EM waves whose fields are randomly oriented in a direction that is perpendicular to the direction of light propagation.

David Brewster (1781–1868), a British physicist, formulated the polarization law in 1815. *(Courtesy of AIP Emilio Segrè Visual Archives, Zeleny Collection.)*

For incidence angles greater than θ_p but smaller than θ_c, Fresnel's equation (2a) gives a negative number for $r_{//}$, which indicates a phase shift of 180° as shown in $\phi_{//}$ in Figure 1.12 (b). The magnitude of both $r_{//}$ and $r_{\perp}$ increase with θ_i as apparent in Figure 1.12 (a). At the critical angle and beyond (past 44° in Figure 1.12), *i.e.* when $\theta_i \geq \theta_c$, the magnitudes of both $r_{//}$ and $r_{\perp}$ go to unity so that the reflected wave has the same amplitude as the incident wave. The incident wave has suffered **total internal reflection**, TIR. When $\theta_i > \theta_c$, in the presence of TIR, both Eqs. (1a) and (2a) are complex quantities because then $\sin\theta_i > n$ and the terms under the square roots become negative. The reflection coefficients become complex quantities of the type $r_{\perp} = 1 \cdot \exp(j\phi_{\perp})$ and $r_{//} = 1 \cdot \exp(j\phi_{//})$ with the phase angles $\phi_{\perp}$ and $\phi_{//}$ being other than zero or 180°. The reflected wave therefore suffers phase changes, $\phi_{\perp}$ and $\phi_{//}$, in the components $E_{\perp}$ and $E_{//}$. These phase changes depend on the incidence angle, as apparent in Figure 1.12 (b), and on n_1 and n_2.

Examination of Eq. (1a) for $r_{\perp}$ shows that for $\theta_i > \theta_c$, we have $|r_{\perp}| = 1$, but the phase change $\phi_{\perp}$ is given by

Phase change in TIR

$$\tan\left(\tfrac{1}{2}\phi_{\perp}\right) = \frac{\left[\sin^2\theta_i - n^2\right]^{1/2}}{\cos\theta_i} \qquad (6)$$

For the $E_{//}$ component, the phase change $\phi_{//}$ is given by

$$\tan\left(\tfrac{1}{2}\phi_{//} + \tfrac{1}{2}\pi\right) = \frac{\left[\sin^2\theta_i - n^2\right]^{1/2}}{n^2\cos\theta_i} \qquad (7)$$

Phase change in TIR

We can summarize that in internal reflection $(n_1 > n_2)$, the amplitude of the reflected wave from TIR is equal to the amplitude of the incident wave but its phase has shifted by an amount determined by Eqs. (6) and (7).[8] The fact that $\phi_{//}$ has an additional π shift that makes $\phi_{//}$ negative for $\theta_i > \theta_c$ is due to the choice for the direction of the reflected optical field $E_{r,//}$ in Figure 1.11. This π shift can be ignored if we by simply invert $E_{r,//}$. (In many books Eq. (7) is written without the π-shift.)

What happens to the transmitted wave when $\theta_i > \theta_c$? According to the boundary conditions, there must still be an electric field in medium 2; otherwise, the boundary conditions cannot be satisfied. When $\theta_i > \theta_c$, the field in medium 2 is a wave that travels near the surface of the boundary along the z direction as depicted in Figure 1.11 (b). The wave is called an **evanescent wave** and advances along z with its field decreasing as we move into medium 2, *i.e.*

$$E_{t,\perp}(y, z, t) = e^{-\alpha_2 y}\exp j(\omega t - k_{iz}z) \qquad (8)$$

Evanescent wave

in which $k_{iz} = k_i \sin\theta_i$ is the wave vector of the incident wave along the z-axis, and α_2 is an **attenuation coefficient** for the electric field penetrating into medium 2,

$$\alpha_2 = \frac{2\pi n_2}{\lambda}\left[\left(\frac{n_1}{n_2}\right)^2\sin^2\theta_i - 1\right]^{1/2} \qquad (9)$$

Attenuation of evanescent wave

in which λ is the free-space wavelength. According to Eq. (8), the evanescent wave travels along z and has an amplitude that decays exponentially as we move from the boundary into medium 2 (along y). The field of the evanescent wave is e^{-1} in medium 2 when $y = 1/\alpha_2 = \delta$, which is called the **penetration depth**. It is not difficult to show that the evanescent wave is correctly predicted by Snell's law when $\theta_i > \theta_c$. The evanescent wave propagates along the boundary (along z) with the same speed as the z-component velocity of the incident and reflected waves.

The reflection coefficients in Figure 1.12 considered the case in which $n_1 > n_2$. When light approaches the boundary from the higher index side, that is $n_1 > n_2$, the reflection is said to be **internal reflection** and *at normal incidence there is no phase change.* On the other hand, if light approaches the boundary from the lower index side, that is $n_1 < n_2$, then it is called **external reflection**. Thus in external reflection light becomes reflected by the surface of an optically denser (higher refractive index) medium. There is an important difference between the two. Figure 1.13 shows how the reflection coefficients $r_\perp$ and $r_{//}$ depend on the incidence angle θ_i for external reflection ($n_1 = 1$ and $n_2 = 1.44$). At normal incidence, both coefficients are negative, which means that *in external reflection at normal incidence there is a phase shift of 180°*. Further, $r_{//}$ goes through zero at the **Brewster angle** θ_p given by Eq. (5). At this angle of incidence, the reflected wave is polarized in the $E_\perp$ component only. Transmitted light in both internal reflection (when $\theta_i < \theta_c$) and external reflection does not experience a phase shift.

[8] It should be apparent that the concepts and the resulting equations apply to a well-defined linearly polarized light wave.

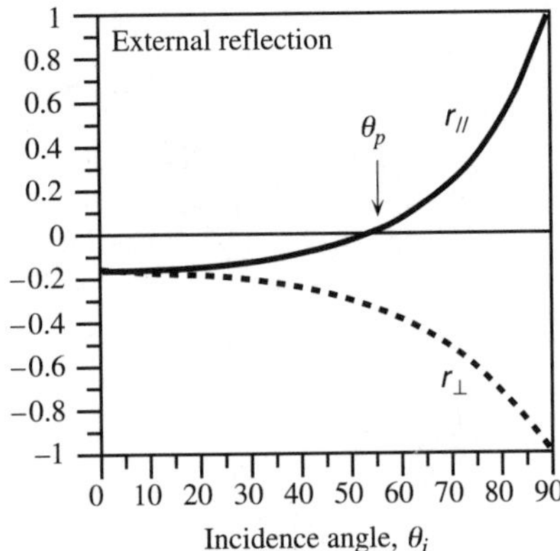

FIGURE 1.13 The reflection coefficients $r_{//}$ and $r_\perp$ vs. angle of incidence θ_i for $n_1 = 1.00$ and $n_2 = 1.44$.

EXAMPLE 1.6.1 Evanescent wave

Total internal reflection (TIR) of light from a boundary between a more dense medium n_1 and a less dense medium n_2 is accompanied by an evanescent wave propagating in medium 2 near the boundary. Find the functional form of this wave and discuss how its magnitude varies with the distance into medium 2.

Solution The transmitted wave has the general form

$$E_{t,\perp} = \mathbf{t}_\perp E_{io,\perp} \exp j(\omega t - \mathbf{k}_t \cdot \mathbf{r})$$

in which $\mathbf{t}_\perp$ is the transmission coefficient. The dot product, examining Figure 1.3, is

$$\mathbf{k}_t \cdot \mathbf{r} = yk_t \cos\theta_t + zk_t \sin\theta_t.$$

However, from Snell's law, when $\theta_i > \theta_c, \sin\theta_t = (n_1/n_2)\sin\theta_i > 1$ and $\cos\theta_t = \sqrt{[1 - \sin^2\theta_t]}$ $= \pm jA_2$ is a purely imaginary number. Thus, taking $\cos\theta_t = -jA_2$

$$\begin{aligned} E_{t,\perp} &= t_\perp E_{io,\perp} \exp j(\omega t - zk_t \sin\theta_t + jyk_tA_2) \\ &= t_\perp E_{io,\perp} \exp(-yk_tA_2) \exp j(\omega t - zk_t \sin\theta_t) \end{aligned}$$

which has an amplitude that decays along y as $\exp(-\alpha_2 y)$ where $\alpha_2 = k_tA_2$. Note that $+jA_2$ is ignored because it implies a light wave in medium 2 whose amplitude and hence intensity grows.

Consider the traveling wave part $\exp j(\omega t - zk_t \sin\theta_t)$. Here, $k_t \sin\theta_t = k_i \sin\theta_i$ (by virtue of Snell's law). But $k_i \sin\theta_i = k_{iz}$, which is the wavevector along z, that is, along the boundary. Thus the evanescent wave propagates along z at the same speed as the incident and reflected waves along z.

Furthermore, for TIR we need $\sin\theta_i > n_2/n_1$. This means that the transmission coefficient,

$$\mathbf{t}_\perp = \frac{n_i \cos\theta_i}{\cos\theta_i + \left[\left(\dfrac{n_2}{n_1}\right)^2 - \sin^2\theta_i\right]^{1/2}} = t_{\perp 0} \exp(j\psi_\perp)$$

must be a complex number as indicated by $t_{\perp 0}\exp(j\psi_\perp)$ in which $t_{\perp 0}$ is a real number and $\psi_\perp$ is a phase change. Note that $\mathbf{t}_\perp$ does not, however, change the general behavior of propagation along z and the penetration along y.

B. Intensity, Reflectance, and Transmittance

It is frequently necessary to calculate the intensity or irradiance[4] of the reflected and transmitted waves when light traveling in a medium of index n_1 is incident at a boundary where the refractive index changes to n_2. In some cases we are simply interested in normal incidence where $\theta_i = 0°$. For example, in laser diodes light is reflected from the ends of an optical cavity where there is a change in the refractive index.

For a light wave traveling with a velocity v in a medium with relative permittivity ε_r, the light intensity I is defined in terms of the electric field amplitude E_o as

$$I = \tfrac{1}{2} v\varepsilon_r\varepsilon_o E_o^2 \qquad \textbf{(10)}$$

Light intensity or irradiance

Here $(1/2)\varepsilon_r\varepsilon_o E_o^2$ represents the energy in the field per unit volume. When multiplied by the velocity v it gives the rate at which energy is transferred through a unit area. Since $v = c/n$ and $\varepsilon_r = n^2$ the intensity is proportional to nE_o^2.

Reflectance R measures the intensity of the reflected light with respect to that of the incident light and can be defined separately for electric field components parallel and perpendicular to the plane of incidence. The reflectances $R_\perp$ and $R_{//}$ are defined by

$$R_\perp = \frac{|E_{ro,\perp}|^2}{|E_{io,\perp}|^2} = |r_\perp|^2 \quad \text{and} \quad R_{//} = \frac{|E_{ro,//}|^2}{|E_{io,//}|^2} = |r_{//}|^2 \qquad \textbf{(11)}$$

Although the reflection coefficients can be complex numbers that can represent phase changes, reflectances are necessarily real numbers representing intensity changes. Magnitude of a complex number is defined in terms of its product with its complex conjugate. For example, when $E_{ro,//}$ is a complex number then

$$|E_{ro,//}|^2 = (E_{ro,//})(E_{ro,//})^*$$

in which $(E_{ro,//})^*$ is the complex conjugate of $(E_{ro,//})$.

From Eqs. (1) and (2) with normal incidence, these are simply given by,

$$R = R_\perp = R_{//} = \left(\frac{n_1 - n_2}{n_1 + n_2}\right)^2 \qquad \textbf{(12)}$$

Reflectance at normal incidence

Since a glass medium has a refractive index of around 1.5 this means that typically 4% of the incident radiation on an air-glass surface will be reflected back.

Transmittance T relates the intensity of the transmitted wave to that of the incident wave in a similar fashion to the reflectance. We must, however, consider that the transmitted wave is in a different medium and also that its direction with respect to the boundary is different from that of the incident wave by virtue of refraction. For normal incidence, the incident and transmitted beams are normal and the transmittances are defined and given by

$$T_\perp = \frac{n_2|E_{to,\perp}|^2}{n_1|E_{io,\perp}|^2} = \left(\frac{n_2}{n_1}\right)|t_\perp|^2 \quad \text{and} \quad T_{//} = \frac{n_2|E_{to,//}|^2}{n_1|E_{io,//}|^2} = \left(\frac{n_2}{n_1}\right)|t_{//}|^2 \qquad \textbf{(13)}$$

or

Transmittance at normal incidence

$$T = T_\perp = T_{//} = \frac{4n_1n_2}{(n_1 + n_2)^2} \qquad \textbf{(14)}$$

Further, the fraction of light reflected and fraction transmitted must add to unity. Thus $R + T = 1$.

EXAMPLE 1.6.2 Reflection of light from a less dense medium (internal reflection)

A ray of light that is traveling in a glass medium of refractive index $n_1 = 1.450$ becomes incident on a less dense glass medium of refractive index $n_2 = 1.430$. Suppose that the free space wavelength (λ) of the light ray is 1 μm.

a. What should the minimum incidence angle for TIR be?

b. What is the phase change in the reflected wave when $\theta_i = 85°$ and when $\theta_i = 90°$?

c. What is the penetration depth of the evanescent wave into medium 2 when $\theta_i = 85°$ and when $\theta_i = 90°$?

Solution

a. The critical angle θ_c for TIR is given by $\sin\theta_c = n_2/n_1 = 1.430/1.450$ so that $\theta_c = 80.47°$.

b. Since the incidence angle $\theta_i > \theta_c$, there is a phase shift in the reflected wave. The phase change in $E_{r,\perp}$ is given by $\phi_\perp$. With $n_1 = 1.450$, $n_2 = 1.430$, and $\theta_i = 85°$,

$$\tan\left(\tfrac{1}{2}\phi_\perp\right) = \frac{\left[\sin^2\theta_i - n^2\right]^{1/2}}{\cos\theta_i} = \frac{\left[\sin^2(85°) - \left(\dfrac{1.430}{1.450}\right)^2\right]^{1/2}}{\cos(85°)} = 1.61447 = \tan\left[\tfrac{1}{2}(116.45°)\right]$$

so that the phase change is 116.45°. For the $E_{r,//}$ component, the phase change is

$$\tan\left(\tfrac{1}{2}\phi_{//} + \tfrac{1}{2}\pi\right) = \frac{\left[\sin^2\theta_i - n^2\right]^{1/2}}{n^2\cos\theta_i} = \frac{1}{n^2}\tan\left(\tfrac{1}{2}\phi_\perp\right)$$

so that

$$\tan\left(\tfrac{1}{2}\phi_{//} + \tfrac{1}{2}\pi\right) = (n_1/n_2)^2\tan(\phi_\perp/2) = (1.450/1.430)^2\tan\left(\tfrac{1}{2}116.45^8\right)$$

which gives $\phi_{//} = -62.1°$. (Note: If we were to invert the reflected field, this phase change would be +117.86°)

We can repeat the calculation with $\theta_i = 90°$ to find $\phi_\perp = 180°$ and $\phi_{//} = 0°$.

Note that as long as $\theta_i > \theta_c$, the magnitude of the reflection coefficients are unity. Only the phase changes.

c. The amplitude of the evanescent wave as it penetrates into medium 2 is

$$E_{t,\perp}(y, t) \sim E_{to,\perp}\exp(-\alpha_2 y)$$

We ignore the z-dependence, $\exp j(\omega t - k_z z)$, as this only gives a propagating property along z. The field strength drops to e^{-1} when $y = 1/\alpha_2 = \delta$, which is called the **penetration depth**. The attenuation constant α_2 is

$$\alpha_2 = \frac{2\pi n_2}{\lambda}\left[\left(\frac{n_1}{n_2}\right)^2\sin^2\theta_i - 1\right]^{1/2}$$

i.e.

$$\alpha_2 = \frac{2\pi(1.430)}{(1.0 \times 10^{-6}\text{ m})}\left[\left(\frac{1.450}{1.430}\right)^2 \sin^2(85°) - 1\right]^{1/2} = 1.28 \times 10^6 \text{ m}^{-1}.$$

so the penetration depth is $\delta = 1/\alpha_2 = 1/(1.28 \times 10^6) = 7.8 \times 10^{-7}$ m, or 0.78 μm. For 90°, repeating the calculation we find $\alpha_2 = 1.5 \times 10^6\text{ m}^{-1}$, so that $\delta = 1/\alpha_2 = 0.66$ μm. We see that the penetration is greater for smaller incidence angles. This will be an important consideration later in analyzing light propagation in optical fibers.

EXAMPLE 1.6.3 Reflection at normal incidence. Internal and external reflection

Consider the reflection of light at normal incidence on a boundary between a glass medium of refractive index 1.5 and air of refractive index 1.

a. If light is traveling from air to glass, what is the reflection coefficient and the intensity of the reflected light with respect to that of the incident light?

b. If light is traveling from glass to air, what is the reflection coefficient and the intensity of the reflected light with respect to that of the incident light?

c. What is the polarization angle in the external reflection in **a** above? How would you make a polaroid device that polarizes light based on the polarization angle?

Solution

a. The light travels in air and becomes partially reflected at the surface of the glass that corresponds to external reflection. Thus $n_1 = 1$ and $n_2 = 1.5$. Then,

$$r_{//} = r_{\perp} = \frac{n_1 - n_2}{n_1 + n_2} = \frac{1 - 1.5}{1 + 1.5} = -0.2$$

This is negative, which means that there is a 180° phase shift. The reflectance (R), which gives the fractional reflected power, is

$$R = r_{//}^2 = 0.04 \quad \text{or} \quad 4\%.$$

b. The light travels in glass and becomes partially reflected at the glass-air interface that corresponds to internal reflection. Thus $n_1 = 1.5$ and $n_2 = 1$. Then,

$$r_{//} = r_{\perp} = \frac{n_1 - n_2}{n_1 + n_2} = \frac{1.5 - 1}{1.5 + 1} = 0.2$$

There is no phase shift. The reflectance is again 0.04 or 4%. In both cases (a) and (b), the amount of reflected light is the same.

c. Light is traveling in air and is incident on the glass surface at the polarization angle. Here $n_1 = 1$, $n_2 = 1.5$ and $\tan\theta_p = (n_2/n_1) = 1.5$ so that $\theta_p = 56.3°$. (We can use Fresnel's equations to readily find the reflected and transmitted amplitudes.)

If we were to reflect light from a glass plate, keeping the angle of incidence at 56.3°, then the reflected light will be polarized with an electric field component perpendicular to the plane of incidence. The transmitted light will have the field greater in the plane of incidence, that is, it will be partially polarized. By using a stack of glass plates one can increase the polarization of the transmitted light. (This type of *pile-of-plates polarizer* was invented by Dominique F.J. Arago in 1812.)

EXAMPLE 1.6.4 Antireflection coatings on solar cells

When light is incident on the surface of a semiconductor, it becomes partially reflected. Partial reflection is an important consideration in solar cells where transmitted light energy into the semiconductor device is converted to electrical energy. The refractive index of Si is about 3.5 at wavelengths around 700–800 nm. Thus the reflectance with $n_1(\text{air}) = 1$ and $n_2(\text{Si}) \approx 3.5$ is

$$R = \left(\frac{n_1 - n_2}{n_1 + n_2}\right)^2 = \left(\frac{1 - 3.5}{1 + 3.5}\right)^2 = 0.309$$

This means that 31% of light is reflected and is not available for conversion to electrical energy; a considerable reduction in the efficiency of the solar cell.

However, we can coat the surface of the semiconductor device with a thin layer of a dielectric material, such as Si_3N_4 (silicon nitride), that has an intermediate refractive index. Figure 1.14 illustrates how the thin dielectric coating reduces the reflected light intensity. In this case $n_1(\text{air}) = 1, n_2(\text{coating}) \approx 1.9$, and $n_3(\text{Si}) = 3.5$. Light is first incident on the air/coating surface and some of it becomes reflected and this reflected wave is shown as A in Figure 1.14. Wave A has experienced a 180° phase change on reflection as this is an external reflection. The wave that enters and travels in the coating then becomes reflected at the coating/semiconductor surface. This wave, which is shown as B in Figure 1.14, also suffers a 180° phase change since $n_3 > n_2$. When wave B reaches A, it has suffered a total delay of traversing the thickness d of the coating twice. The phase difference is equivalent to $k_c(2d)$ in which $k_c = 2\pi/\lambda_c$ is the wavevector in the coating and is given by $2\pi/\lambda_c$ in which λ_c is the wavelength in the coating. Since $\lambda_c = \lambda/n_2$, where λ is the free-space wavelength, the phase difference $\Delta\phi$ between A and B is $(2\pi n_2/\lambda)(2d)$. To reduce the reflected light, A and B must interfere destructively and this requires the phase difference to be π or odd-multiples of π, $m\pi$ in which $m = 1, 3, 5, \ldots$ is an odd-integer. Thus

$$\left(\frac{2\pi n_2}{\lambda}\right)2d = m\pi \qquad \text{or} \qquad d = m\left(\frac{\lambda}{4n_2}\right)$$

Thus, the thickness of the coating must be multiples of the quarter wavelength in the coating and depends on the wavelength.

To obtain a good degree of destructive interference between waves A and B, the two amplitudes must be comparable. It turns out that we need $n_2 = \sqrt{(n_1 n_3)}$. When $n_2 = \sqrt{(n_1 n_3)}$ then

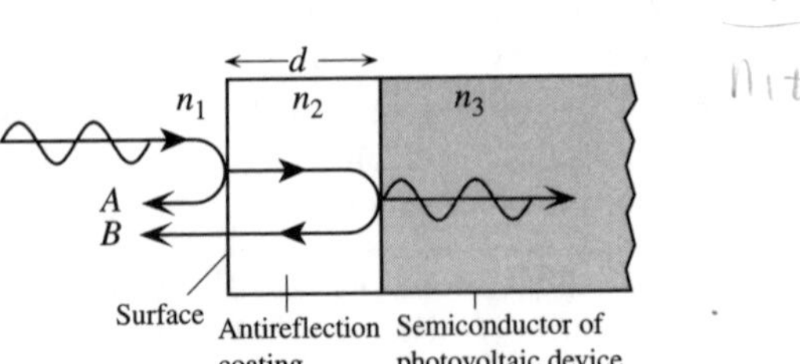

FIGURE 1.14 Illustration of how an antireflection coating reduces the reflected light intensity

the reflection coefficient between the air and coating is equal to that between the coating and the semiconductor. In this case we would need $\sqrt{(3.5)}$ or 1.87. Thus, Si_3N_4 is a good choice as an antireflection coating material on Si solar cells.

Taking the wavelength to be 700 nm, $d = (700\text{ nm})/[4(1.9)] = 92.1$ nm or odd-multiples of d.

EXAMPLE 1.6.5 Dielectric mirrors

A **dielectric mirror** consists of a stack of dielectric layers of alternating refractive indices as schematically illustrated in Figure 1.15, in which n_1 is smaller than n_2. The thickness of each layer is a quarter of wavelength or $\lambda_{\text{layer}}/4$ in which λ_{layer} is the wavelength of light in that layer, or λ_o/n in which λ_o is the free space wavelength at which the mirror is required to reflect the incident light and n is the refractive index of the layer. Reflected waves from the interfaces interfere constructively and give rise to a substantial reflected light. If there are sufficient number of layers, the reflectance can approach unity at the wavelength λ_o. The figure also shows schematically a typical reflectance vs. wavelength behavior of a dielectric mirror with many layers.

The reflection coefficient r_{12} for light in layer 1 being reflected at the 1-2 boundary is $r_{12} = (n_2 - n_1)/(n_1 + n_2)$ and is a positive number indicating no phase change. The reflection coefficient for light in layer 2 being reflected at the 2-1 boundary is $r_{21} = (n_1 - n_2)/(n_2 + n_1)$, which is $-r_{12}$ (negative) indicating a phase change of π. Thus the reflection coefficient alternates in sign through the mirror. Consider two arbitrary waves, A and B, which are reflected at two consecutive interfaces. The two waves are therefore already out of phase by π due to reflections at the different boundaries. Further, wave B travels an additional distance that is twice $(\lambda_2/4)$ before reaching wave A and therefore experiences a phase change equivalent to $2(\lambda_2/4)$ or $\lambda_2/2$, that is π. The phase difference between A and B is then $\pi + \pi$ or 2π. Thus waves A and B are in phase and *interfere constructively*. We can similarly show that waves B and C also interfere constructively and so on, so that all reflected waves from the consecutive boundaries interfere constructively. After several layers (depending on n_1 and n_2) the transmitted intensity will be very small and the reflected light intensity will be close to unity. Dielectric mirrors are widely used in modern vertical cavity surface emitting semiconductor lasers.

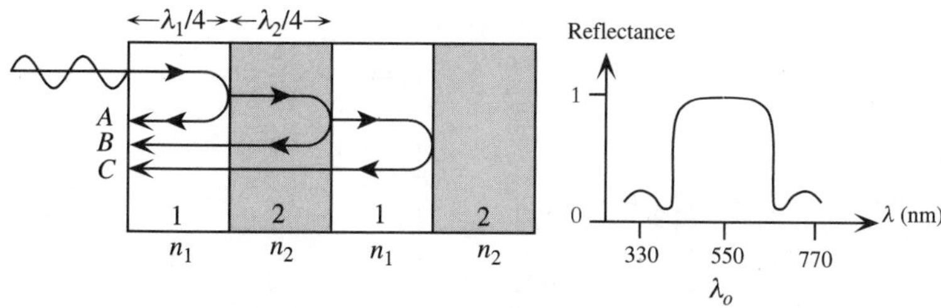

FIGURE 1.15 Schematic illustration of the principle of the dielectric mirror with many low and high refractive index layers and its reflectance.

1.7 MULTIPLE INTERFERENCE AND OPTICAL RESONATORS

Charles Fabry (1867–1945), left, and Alfred Perot (1863–1925), right, were the first French physicists to construct an optical cavity for interferometry (Perot: The Astrophysical Journal, Vol. 64, November 1926, p. 208; Fabry: AIP Emilio Segrè Visual Archives, E. Scott Barr Collection.)

An electrical resonator such as a parallel inductor-capacitor (LC) circuit allows only electrical oscillations at the resonant frequency f_o (determined by L and C) within a narrow bandwidth around f_o. Such an LC circuit thereby stores energy at the same frequency. We know that it also acts as a filter at the resonant frequency f_o, which is how we tune in our favorite radio stations. An **optical resonator** is the optical counterpart of the electrical resonator, storing energy or filtering light only at certain frequencies (wavelengths).

When two flat mirrors are perfectly aligned to be parallel as in Figure 1.16 (a) with free space between them, light wave reflections between the two mirrors M_1 and M_2 lead to constructive and destructive interference of these waves within the cavity. Waves reflected from M_1 traveling towards the right interfere with waves reflected from M_2 traveling towards the left. The result is a series of allowed **stationary or standing EM waves** in the cavity as in Figure 1.16 (b) (just like the stationary waves of a vibrating guitar string stretched between two fixed points). Since the electric field at the mirrors

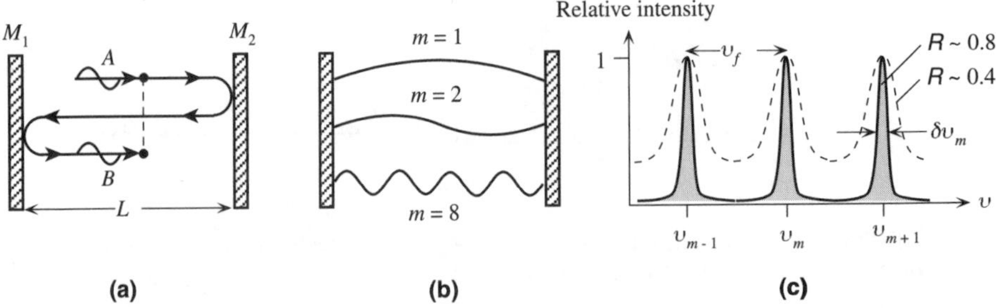

FIGURE 1.16 Schematic illustration of the Fabry-Perot optical cavity and its properties. (a) Reflected waves interfere. (b) Only standing EM waves, *modes*, of certain wavelengths are allowed in the cavity. (c) Intensity vs. frequency for various modes. R is mirror reflectance and lower R means higher loss from the cavity.

(assume metal coated) must be zero, we can only fit in an integer number m of half-wavelengths, $\lambda/2$, into the cavity length L,

$$m\left(\frac{\lambda}{2}\right) = L \qquad m = 1, 2, 3 \ldots \tag{1}$$

Cavity modes

Each particular allowed λ, labeled as λ_m, satisfying Eq. (1) for a given m defines a **cavity mode** as depicted in Figure 1.16 (b). Inasmuch as light frequency υ and wavelength λ are related by $\upsilon = c/\lambda$, the corresponding frequencies υ_m of these modes are the **resonant frequencies** of the cavity,

$$\upsilon_m = m\left(\frac{c}{2L}\right) = m\upsilon_f; \qquad \upsilon_f = c/(2L) \tag{2}$$

Cavity resonant frequencies

in which υ_f is the lowest frequency corresponding to $m = 1$, the fundamental mode, and also the frequency separation of two neighboring modes, $\Delta\upsilon_m = \upsilon_{m+1} - \upsilon_m = \upsilon_f$. It is known as the **free spectral range**. Figure 1.16 (c) illustrates schematically the intensity of the allowed modes as a function of frequency. If there are no losses from the cavity, *i.e.* the mirrors are perfectly reflecting, then the peaks at frequencies υ_m defined by Eq. (2) would be sharp lines. If the mirrors are not perfectly reflecting so that some radiation escapes from the cavity, then the mode peaks are not as sharp and have a finite width, as indicated in Figure 1.16 (c). It is apparent that this simple optical cavity with its mirrors, **etalon**, serves to "store" radiation energy only at certain frequencies and it is called a **Fabry-Perot optical resonator**.

Consider an arbitrary wave such as A traveling towards the right at some instant as shown in Figure 1.16 (a). After one round trip this wave would be again traveling towards the right but now as wave B, which has a phase difference and a different magnitude due to non-perfect reflections. If the mirrors M_1 and M_2 are identical with a reflection coefficient of *magnitude* r, then B has one round-trip phase difference of $k(2L)$ and a magnitude r^2 (two reflections) with respect to A. When A and B interfere, the result is,

$$A + B = A + Ar^2 \exp(-j2kL)$$

Of course, just like A, B will continue on and will be reflected twice, and after one round trip it would be going towards the right again and we will now have three waves interfering and so on. After infinite round-trip reflections, the resultant field E_{cavity} is due to infinite such interferences,

$$\begin{aligned} E_{\text{cavity}} &= A + B + \ldots \\ &= A + Ar^2 \exp(-j2kL) + Ar^4 \exp(-j4kL) + Ar^6 \exp(-j6kL) + \ldots \end{aligned}$$

The sum of this geometric series is easily evaluated as

$$E_{\text{cavity}} = \frac{A}{1 - r^2 \exp(-j2kL)}$$

Once we know the field in the cavity we can calculate the intensity $I_{\text{cavity}} = |E_{\text{cavity}}|^2$. Further, we can use *reflectance* $R = r^2$ to further simplify the expression. The final result after algebraic manipulation is,

Cavity intensity

$$I_{cavity} = \frac{I_o}{(1 - R)^2 + 4R\sin^2(kL)} \quad (3)$$

in which $I_o = A^2$ is the original intensity. The intensity in the cavity is maximum I_{max} whenever $\sin^2(kL)$ in the denominator of Eq. (3) is zero, which corresponds to (kL) being $m\pi$, in which m is an integer. Thus, the intensity vs. k, or equivalently, the intensity vs. frequency spectrum, peaks whenever $kL = m\pi$, as in Figure 1.16 (c). These peaks are located at $k = k_m$ that satisfy $k_m L = m\pi$, which leads directly to Eqs. (1) and (2) that were derived intuitively. For those resonant k_m values, Eq. (3) gives

Maximum cavity intensity

$$I_{max} = \frac{I_o}{(1 - R)^2}; \qquad k_m L = m\pi \quad (4)$$

A smaller mirror reflectance R means more radiation loss from the cavity, which affects the intensity distribution in the cavity. We can show from Eq. (3) that smaller R values result in broader mode peaks and a smaller difference between the minimum and maximum intensity in the cavity as schematically illustrated in Figure 1.16 (c). The **spectral width**[9] $\delta\upsilon_m$ of the Fabry-Perot etalon is the full width at half maximum (FWHM) of an individual mode intensity, as defined in Figure 1.16 (c). It can be calculated in a straightforward fashion when $R > 0.6$ from,

Spectral width

$$\delta\upsilon_m = \frac{\upsilon_f}{F}; \qquad F = \frac{\pi R^{1/2}}{1 - R} \quad (5)$$

in which F is called the **finesse** of the resonator, which increases as losses decrease (R increases). Large finesses lead to sharper mode peaks. Finesse is the ratio of mode separation $(\Delta\upsilon_m)$ to spectral width $(\delta\upsilon_m)$.

The Fabry-Perot optical cavities are widely used in laser, interference filter, and spectroscopic applications. Consider a light beam that is incident on a Fabry-Perot cavity as in Figure 1.17. The optical cavity is formed by partially transmitting and reflecting plates. Part of the incident beam enters the cavity. We know that only special cavity modes are allowed to exist in the cavity since other wavelengths lead to destructive interference. Thus, if the incident beam has a wavelength corresponding to one of the cavity modes, it can sustain oscillations in the cavity and hence lead to a transmitted beam. The output light is a fraction of the light intensity in the cavity and is proportional to Eq. (3). Commercial interference filters are based on this principle except that they typically use two cavities in series formed by dielectric mirrors (a stack of quarter wavelength layers); the structure is more complicated than in Figure 1.17. Further, adjusting the cavity length L provides a "tuning capability" to scan different wavelengths.

Equation (3) describes the intensity of the radiation in the cavity. The intensity of the transmitted radiation in Figure 1.17 can be calculated, as above, by considering that each time a wave is reflected at the right mirror, a portion of it is transmitted, and that these transmitted waves can interfere only constructively to constitute a transmitted beam when $kL = m\pi$. Intuitively, if $I_{incident}$ is the incident light intensity, then a fraction

[9] The spectral width is also called the fringe width and m the fringe order.

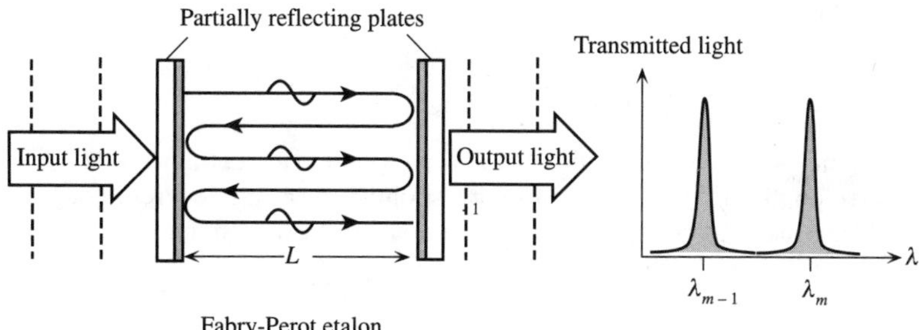

FIGURE 1.17 Transmitted light through a Fabry-Perot optical cavity.

$(1 - R)$ of this would enter the cavity to build up into I_{cavity} in Eq. (3), and a fraction $(1 - R)$ of I_{cavity} would leave the cavity as the transmitted intensity $I_{\text{tansmitted}}$. Thus,

Transmitted mode intensities

$$I_{\text{transmitted}} = I_{\text{incident}} \frac{(1 - R)^2}{(1 - R)^2 + 4R \sin^2(kL)} \tag{6}$$

which is again maximum just as for I_{cavity} whenever $kL = m\pi$ as shown in terms of wavelength in Figure 1.17.

The ideas above can be readily extended to a medium with a refractive index n by using nk for k. Further, if the angle of incidence θ at the etalon face is not normal, then we can resolve k to be along the cavity axis, that is use $k \cos \theta$ instead of k in the discussions above.

EXAMPLE 1.7.1 Resonator modes and spectral width

Consider a Fabry-Perot optical cavity of air of length 100 microns with mirrors that have a reflectance of 0.9. Calculate the cavity mode nearest to 900 nm. Calculate the separation of the modes and the spectral width of each mode.

Solution From Eq. (1),

$$m = \frac{2L}{\lambda} = \frac{2(100 \times 10^{-6})}{(900 \times 10^{-9})} = 222.22$$

so that

$$\lambda_m = \frac{2L}{m} = \frac{2(100 \times 10^{-6})}{(222)} = 900.90 \text{ nm}.$$

Separation of the modes is,

$$\Delta v_m = v_f = \frac{c}{2L} = \frac{(3 \times 10^8)}{2(100 \times 10^{-6})} = 1.5 \times 10^{12} \text{ Hz}.$$

The finesse is

$$F = \frac{\pi R^{1/2}}{1 - R} = \frac{\pi 0.9^{1/2}}{1 - 0.9} = 29.8$$

and each mode width is

$$\delta\upsilon_m = \frac{\upsilon_f}{F} = \frac{1.5 \times 10^{12}}{29.8} = 5.03 \times 10^{10}\ \text{Hz}.$$

Moreover, the mode spectral width $\delta\upsilon_m$ will correspond to a spectral wavelength width $\delta\lambda_m$. The mode wavelength $\lambda_m = 900.90$ nm corresponds to a mode frequency $\upsilon_m = c/\lambda_m = 3.328 \times 10^{14}$ Hz. Since $\lambda_m = c/\upsilon_m$, we can differentiate this expression to relate small changes in λ_m and υ_m,

$$\delta\lambda_m = \left|\delta\left(\frac{c}{\upsilon_m}\right)\right| = \left|-\frac{c}{\upsilon_m^2}\right|\delta\upsilon_m = \frac{(3 \times 10^8)}{(3.33 \times 10^{14})^2}(5.03 \times 10^{10}) = 0.136\ \text{nm}.$$

1.8 GOOS-HÄNCHEN SHIFT AND OPTICAL TUNNELING

A light traveling in an optically more dense medium suffers total internal reflection (TIR) when it is incident on a less dense medium at an angle of incidence greater than the critical angle $(\theta_i > \theta_c)$ as shown in Figure 1.10. Simple ray trajectory analysis gives the impression that the reflected ray emerges from the point of contact of the incident ray with the interface as in Figure 1.10. However, careful optical experiments examining the incident and reflected beams have shown that the reflected wave appears to be laterally shifted from the point of incidence at the interface as illustrated in Figure 1.18. Although the angles of incidence and reflection are the same (as one expects from Fresnel's equation), the reflected beam, nonetheless, is laterally shifted and appears to be reflected from a *virtual plane inside the optically less dense medium*. The lateral shift is known as the **Goos-Haenchen shift**.

The lateral shift of the reflected beam can be understood by considering that the reflected beam experiences a phase change ϕ, as shown in Figure 1.12, and that the electric field extends into the second medium by a penetration depth $\delta = 1/\alpha_2$. We know that phase changes other than 0 or 180° occur only when there is total internal reflection. We can equivalently represent this phase change ϕ and the penetration into the second medium by shifting the reflected wave along the direction of propagation of the evanescent wave, that is, along z, by an amount[10] Δz as shown in Figure 1.18. The lateral shift depends on the angle of incidence and the penetration depth. From simple geometric considerations, $\Delta z = 2\delta \tan\theta_i$. For example, for light with $\lambda = 1$ μm incident at 85° at a glass-glass ($n_1 = 1.450$ and $n_2 = 1.430$) interface and suffering TIR in Example 1.6.2, $\delta = 0.78$ μm, which means $\Delta z \approx 18$ μm.

Total internal reflection occurs whenever a wave propagating in an optically denser medium, such as in A in Figure 1.18, is incident at an angle greater than the critical angle at the interface AB with a medium of lower refractive index B. If we were to shrink the thickness d of medium B, as in Figure 1.19, we would observe that when B is sufficiently thin, an attenuated light beam emerges on the other side of B in C. This phenomenon in which an incident wave is partially transmitted through a medium where it is forbidden in terms of simple geometrical optics is called **optical tunneling** and is a consequence

[10]The actual analysis of the Goos-Haenchen shift is more complicated; see for example J.E. Midwinter, *Optical Fibers for Transmission*(John Wiley and Sons, Chichester, 1979), Ch. 3.

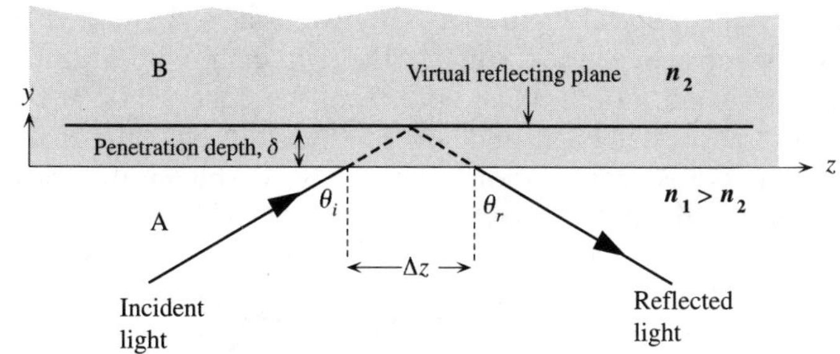

FIGURE 1.18 The reflected light beam in total internal reflection appears to have been laterally shifted by an amount Δz at the interface.

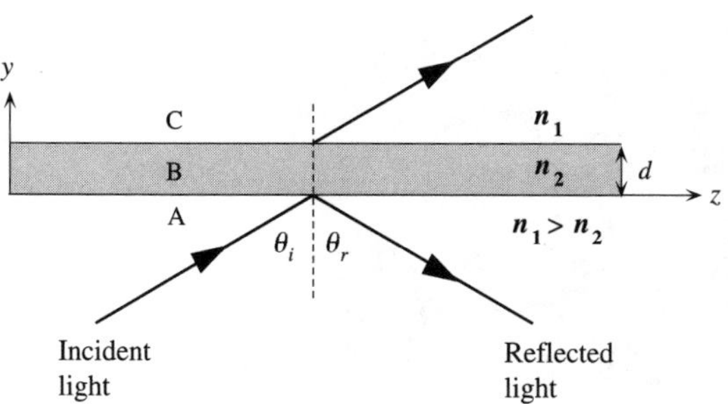

FIGURE 1.19 When medium B is thin (thickness d is small), the field penetrates to the BC interface and gives rise to an attenuated wave in medium C. The effect is the tunneling of the incident beam in A through B to C.

of the electromagnetic wave nature of light. It is due to the fact that the field of the evanescent wave penetrates into medium B and reaches the interface BC. The phenomenon in Figure 1.19 is referred to as **frustrated total internal reflection** (FTIR); the proximity of medium C frustrates TIR. The transmitted beam in C carries some of the light intensity and thus the intensity of the reflected beam is reduced.

Frustrated total internal reflection is fruitfully utilized in beam splitters as shown in Figure 1.20. A light beam entering the glass prism A suffers TIR at the hypotenuse face ($\theta_i > \theta_c$ at the glass-air interface) and becomes reflected; the prism deflects the

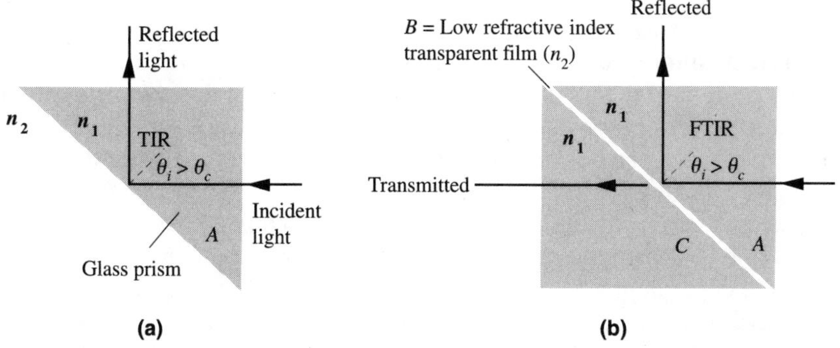

FIGURE 1.20 (a) A light incident at the long face of a glass prism suffers TIR; the prism deflects the light. (b) Two prisms separated by a thin low refractive index film forming a beamsplitter cube. The incident beam is split into two beams by FTIR.

light. In the *beam splitter cube* in Figure 1.20 (b), two prisms, A and C, are separated by a thin film, B, of low refractive index. Some of the light energy is now tunneled through this thin film and transmitted into C and out from the cube. FTIR at the hypotenuse face of A leads to a transmitted beam and hence to the splitting of the incident beam into two beams. The extent of energy division between the two beams depends on the thickness of the thin layer B and its refractive index.

Beam splitter cubes *(Courtesy of Melles Griot)*

1.9 TEMPORAL AND SPATIAL COHERENCE

When we represent a traveling EM wave by a pure sinusoidal wave, for example by

A sinusoidal wave

$$E_x = E_o \sin(\omega_o t - k_o z) \tag{1}$$

with a well defined angular frequency $\omega_o = 2\pi\upsilon_o$ and a wavenumber k_o, we are assuming that the wave extends infinitely over all space and exists at all times as illustrated in Figure 1.21 (a) inasmuch as a sine function extends periodically over all values of its argument. Such a sine wave is *perfectly coherent* because all points on the wave are predictable. **Perfect coherence** is therefore understood to mean that we can predict the phase of any portion of the wave from any other portion of the wave. Temporal coherence measures the extent to which two points, such as P and Q, separated in time at a given location in space can be correlated; that is, one can be reliably predicted from the other. At a given location, for a pure sine wave as in Figure 1.21 (a), any two points such as P and Q separated by any time interval are always correlated because we can predict the phase of one (Q) from the phase of the other (P) for any temporal separation.

Any time dependent arbitrary function $f(t)$ can be represented by a sum of pure sinusoidal waves with varying frequencies, amplitudes, and phases.[11] The **spectrum** of a function $f(t)$ represents the amplitudes of various sinusoidal oscillations that constitute

[11] This is a qualitative statement of what a Fourier transformation does in mathematics.

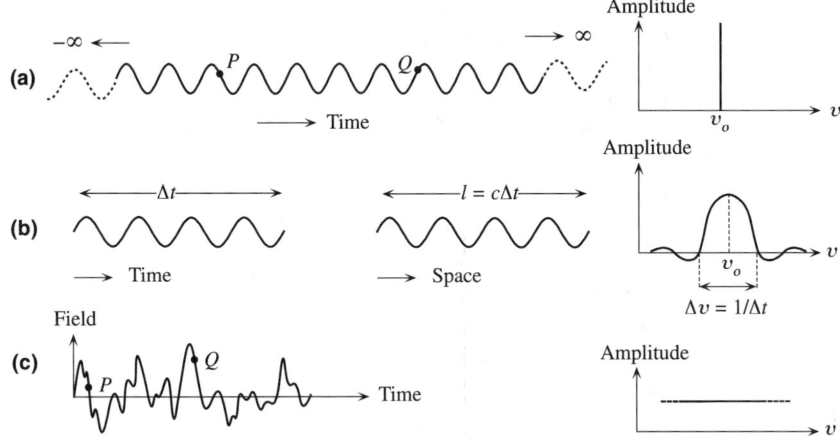

FIGURE 1.21 (a) A sine wave is perfectly coherent and contains a well-defined frequency v_o. (b) A finite wave train lasts for a duration Δt and has a length l. Its frequency spectrum extends over $\Delta v = 1/\Delta t$. It has a coherence time Δt and a coherence length l. (c) White light exhibits practically no coherence.

the function. All these pure sine waves are added, with the right amplitude and phase, to make up $f(t)$. We need only one sine wave at a frequency $v_o = \omega_o/2\pi$ to make up Eq. (1) as illustrated in Figure 1.21 (a).

A pure sine wave is an idealization far from reality and in practice a wave can exist only over a finite time duration Δt, which corresponds to a finite wave-train of length $l = c\Delta t$ as illustrated in Figure 1.21 (b). This duration Δt may be the result of the radiation emission process, modulation of the output from a laser, or some other process (indeed, in practice, the amplitude will not be constant over Δt). It is clear that we can only correlate points in the wave train within the duration Δt or over the spatial extent $l = c\Delta t$. This wave-train has a **coherence time** Δt and a **coherence length** $l = c\Delta t$. Since it is not a perfect sine wave, it contains a number of different frequencies in its spectrum. A proper calculation shows that most of the significant frequencies that constitute this finite wave-train lie, as expected, centered around v_o over a range Δv as shown in Figure 1.21 (b). The spread Δv is the **spectral width** of the wave-train and depends on the temporal coherence length Δt and is given by,

$$\Delta v = \frac{1}{\Delta t} \quad \textbf{(2)}$$

Spectral width of a wave-train

Coherence and spectral width are therefore intimately linked. For example, the orange radiation at 589 nm (both D-lines together) emitted from a sodium lamp has spectral width $\Delta v \approx 5 \times 10^{11}$ Hz. This means that its coherence time $\Delta t \approx 2 \times 10^{-12}$ s or 2 ps, and its coherence length is 6×10^{-4} m or 0.60 mm. On the other hand, the red lasing emission from a He-Ne laser operating in multimode has a spectral width around 1.5×10^9 Hz, which corresponds to a coherence length of 200 mm. Furthermore, a continuous wave laser operating in a single mode will have a very narrow linewidth and the emitted radiation will perhaps have a coherence length of several hundred meters. Typically light waves from laser devices have substantial coherence lengths and are therefore widely used in wave-interference studies and applications.

Suppose that standing at one location in space we measure the field vs. time behavior shown in Figure 1.21 (c) in which the zero crossing of the signal occurs randomly. Given a point P on this "waveform," we cannot predict the "phase" or the signal at any other point Q. Thus P and Q are not in any way correlated for any temporal separation except Q coinciding with P (or being very close to it by an infinitesimally short time interval). There is no coherence in this white light signal and the signal essentially represents white noise; its spectrum typically contains a wide range of frequencies. White light, like an ideal sine wave, is an idealization because it assumes all frequencies are present in the light beam. However, radiation emitted from atoms typically has a central frequency and a certain spectral width, that is, a degree of coherence. The light in the real world lies between (a) and (c) in Figure 1.21.

Coherence between two waves is related to the extent of correlation between two waves. The waves A and B in Figure 1.22 (a) have the same frequency v_o but they coincide only over the time interval Δt and hence they can only give rise to interference phenomena over this time. They therefore have **mutual temporal coherence** over the time interval Δt. This situation can arise, for example, when two identical wave-trains each of coherence length l travel different optical paths. When they arrive at the destination, they can interfere only over a space portion $c\Delta t$. Since interference phenomena can occur only for waves that have mutual coherence, interference experiments, such as Young's double slit experiments, can be used to measure mutual coherence between waves.

Spatial coherence describes the extent of coherence between waves radiated from different locations on a light source as shown in Figure 1.22 (b). If the waves emitted from locations P and Q on the source are in phase, then P and Q are spatially coherent. A spa-

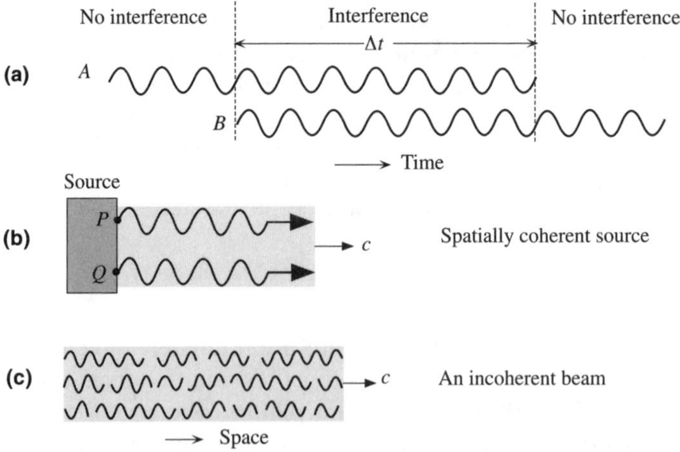

FIGURE 1.22 (a) Two waves can only interfere over the time interval Δt. (b) Spatial coherence involves comparing the coherence of waves emitted from different locations on the source. (c) An incoherent beam.

tially coherent source emits waves that are in phase over its entire emission surface. These waves, however, may have partial temporal coherence. A light beam emerging from a spatially coherent light source will hence exhibit spatial coherence across the beam cross section, that is, the waves in the beam will be in phase over coherence length $c\Delta t$ in which Δt is the temporal coherence. A mostly incoherent beam will contain waves that have very little correlation with each other. The incoherent beam in Figure 1.22 (c) contains waves (across the beam cross section) whose phases change randomly at random times. (Note, however, that there may be a very short time interval over which there is a little bit of temporal coherence.) The quantitative analysis of coherence requires mathematical techniques based on correlation functions and may be found in more advanced textbooks.

1.10 DIFFRACTION PRINCIPLES

A. Fraunhofer Diffraction

An important property of waves is that they exhibit diffraction effects; for example, sound waves are able to bend (deflect around) corners and a light beam can similarly "bend" around an obstruction (though the bending may be very small). Figure 1.23 shows an example of a collimated light beam passing through a circular aperture (a circular opening in an opaque screen). The passing beam is found to be divergent and to exhibit an intensity pattern that has bright and dark rings, called *Airy rings*[12]. The passing beam is said to be diffracted and its light intensity pattern is called a **diffraction pattern**. Clearly, the light pattern of the diffracted beam does *not* correspond to the geometric shadow of the circular aperture. Diffraction phenomena are generally classified into two categories. In **Fraunhofer diffraction**[13], the incident light beam is a plane wave (a collimated light beam) and the observation or detection of the light intensity pattern (by placing a photographic screen, *etc.*) is done far away from the aperture so that the waves received also look like plane waves. Inserting a lens between the aperture and the photographic screen enables the screen to be closer to the aperture. In **Fresnel diffraction**, the incident light beam and the received light waves are not plane waves but have significant wavefront curvatures. Typically, the light source and the photographic screen are both close to the aperture so that the wavefronts are curved. Fraunhofer diffraction is by far the most important.

Diffraction can be understood in terms of the interference of multiple waves emanating from the obstruction[14]. We will consider a plane wave incident on a one-dimensional slit of length a. According to the **Huygens-Fresnel principle**[15], *every*

---

[12] Sir George Airy (1801–1892), Astronomer Royal of Britain from 1835 to 1881.

[13] Joseph von Fraunhofer (1787–1826) was a German physicist who also observed the various dark lines in Sun's spectrum due to hydrogen absorption.

[14] "No one has been able to define the difference between interference and diffraction satisfactorily" [R.P. Feynman, R.B. Leighton, and M. Sands, *The Feynman Lectures on Physics* (Addison-Wesley, Reading MA, 1963)]

[15] Eugene Hecht, *Optics: Second Edition* (Addison-Wesley, Reading, MA, 1987), p. 393.

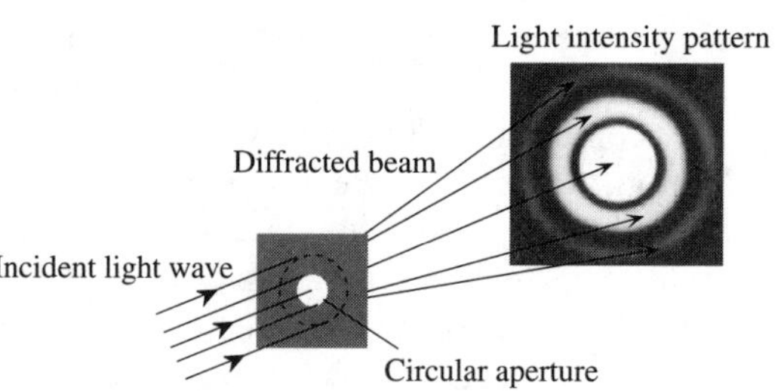

FIGURE 1.23 A light beam incident on a small circular aperture becomes diffracted and its light intensity pattern after passing through the aperture is a diffraction pattern with circular bright rings (called Airy rings). If the screen is far away from the aperture, this would be a Fraunhofer diffraction pattern.

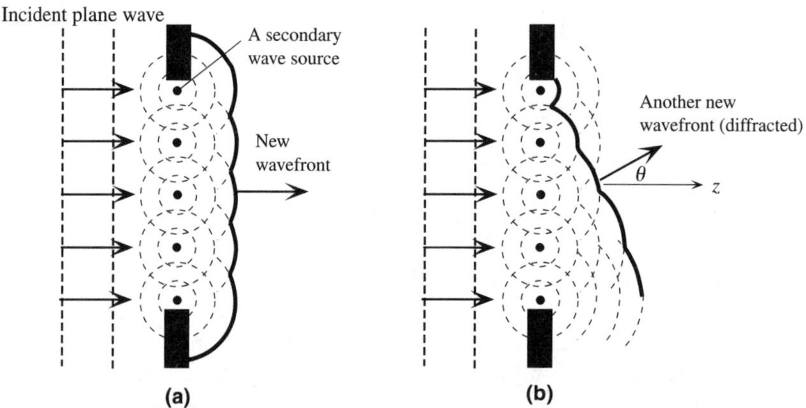

FIGURE 1.24 (a) Huygens-Fresnel principles states that each point in the aperture becomes a source of secondary waves (spherical waves). The spherical wavefronts are separated by λ. The new wavefront is the envelope of the all these spherical wavefronts. (b) Another possible wavefront occurs at an angle θ to the z-direction which is a diffracted wave.

unobstructed point of a wavefront, at a given instant in time, serves as a source of spherical secondary waves (with the same frequency as that of the primary wave). The amplitude of the optical field at any point beyond is the superposition of all these wavelets (considering their amplitudes and relative phases). Figure 1.24 (a) and (b) illustrate this point pictorially showing that, when the plane wave reaches the aperture, points in the aperture become sources of coherent spherical secondary waves. These spherical waves interfere to constitute the new wavefront (the new wavefront is the envelope of the wavefronts of these secondary waves). These spherical waves can interfere constructively not just in the forward direction as in (a) but also in other appropriate directions, as in (b), giving rise to the observed bright patterns (bright rings for a circular aperture) on the observation screen.

We can divide the unobstructed width a of the aperture into a very large number N of coherent "point sources" each of extent $\delta y = a/N$ (obviously δy is sufficiently small to be nearly a point), as in Figure 1.25 (a). Since the aperture a is illuminated uniformly by the plane wave, the strength (amplitude) of each point source would be proportional to $a/N = \delta y$. Each would be a source of spherical waves. In the forward direction ($\theta = 0$), they would all be in phase and constitute a forward wave, along the

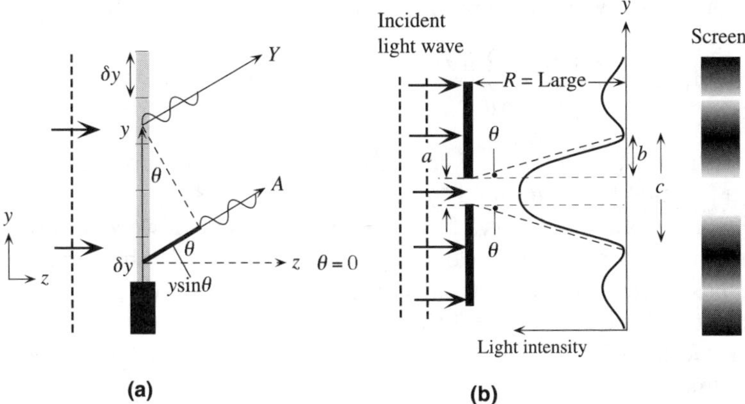

FIGURE 1.25 (a) The aperture is divided into N number of point sources each occupying δy with amplitude $\propto \delta y$. (b) The intensity distribution in the received light at the screen far away from the aperture: the diffraction pattern.

z-direction. But they can also be in phase at some angle θ to the z-direction and hence give rise to a diffracted wave along this direction. We will evaluate the intensity of the received wave at a point on the screen as the *sum of all waves arriving from all point sources in the aperture.* The screen is far away from the aperture so the waves arrive almost parallel at the screen (alternatively a lens can be used to focus the diffracted parallel rays to form the diffraction pattern).

Consider an arbitrary direction θ, and consider the phase of the emitted wave (Y) from an arbitrary point source at y with respect to the wave (A) emitted from source at $y = 0$ as shown in Figure 1.25 (a). If k is the wavevector, $k = 2\pi/\lambda$, the wave Y is out of phase with respect to A by $ky \sin\theta$. Thus the wave emitted from point source at y has a field δE,

$$\delta E \propto (\delta y)\exp(-jky\sin\theta) \tag{1}$$

All of these waves from point sources from $y = 0$ to $y = a$ interfere at the screen and the field at the screen is their sum. Because the screen is far away, a point on the screen is at the same distance from anywhere in the aperture. This means that all the spherical waves from the aperture experience the same phase change and decrease in amplitude in reaching the screen. This simply scales δE at the screen by an amount that is the same for all waves coming from the aperture. Thus, the resultant field $E(\theta)$ at the screen is,

$$E(\theta) = C\int_{y=0}^{y=a} \delta y \exp(-jky\sin\theta) \tag{2}$$

in which C is a constant. Integrating Eq. (2) and using algebraic manipulation we finally obtain,

$$E(\theta) = \frac{Ce^{-j\frac{1}{2}ka\sin\theta}a\sin\left(\frac{1}{2}ka\sin\theta\right)}{\frac{1}{2}ka\sin\theta}$$

The light intensity I at the screen is proportional to $|E(\theta)|^2$. Thus,

Single slit diffraction equation

$$I(\theta) = \left[\frac{C'a\sin\left(\frac{1}{2}ka\sin\theta\right)}{\frac{1}{2}ka\sin\theta}\right]^2 = I(0)\,\mathrm{sinc}^2(\beta); \qquad \beta = \tfrac{1}{2}ka\sin\theta \tag{3}$$

in which C' is a constant and β is a convenient new variable representing θ, and sinc ("sink") is a function that is defined by $\mathrm{sinc}(\beta) = \sin(\beta)/\beta$.

If we were to plot Eq. (3) as a function of θ at the screen we would see the intensity (diffraction) pattern schematically depicted in Figure 1.25 (b). First, observe that the pattern has bright and dark regions, corresponding to constructive and destructive interference of waves from the aperture. Second, the center bright region is wider than the aperture width a, which mean that the transmitted beam must be *diverging*. The zero intensity occurs when, from Eq. (3),

Zero intensity points

$$\sin\theta = \frac{m\lambda}{a}; \qquad m = \pm 1, \pm 2, \ldots \tag{4}$$

A light wave at a wavelength 1300 nm, diffracted by a slit of width $a = 100\ \mu\text{m}$ (about the thickness of this page), has a divergence 2θ of about 1.5°. From Figure 1.25 (b), it is apparent that, using geometry, we can easily calculate the width c of the central bright region of the intensity pattern, given θ from Eq. (4) and the distance R of the screen from the aperture.

The diffraction patterns from two-dimensional apertures such as rectangular and circular apertures are more complicated to calculate but they use the same principle based on the multiple interference of waves emitted from all point sources in the aperture. The diffraction pattern from a circular aperture, known as **Airy rings**, was shown in Figure 1.23, and can be roughly visualized by rotating the intensity pattern in Figure 1.25 (b) about the z-axis (the actual intensity pattern follows a Bessel function not a simply rotated sinc function). The central white spot is called the **Airy disk**; its radius corresponds to the radius of the first dark ring. The angular position θ of the first dark ring, as defined in Figure 1.25 (b), is determined by the diameter $D\,(=a)$ of the aperture and the wavelength λ, and is given by

Angular radius of Airy disk

$$\sin\theta = 1.22\frac{\lambda}{D} \tag{5}$$

The divergence angle from the aperture center to the Airy disk circumference is 2θ. If R is the distance of the screen from the aperture, then the radius of the Airy disk, approximately b, can be calculated from the geometry in Figure 1.25 (b), which gives $b/R = \tan\theta \approx \theta$. If a lens is used to focus the diffracted light waves onto a screen, then $R = f$, focal length of the lens.

The diffraction pattern of a rectangular aperture is shown in Figure 1.26. It involves the multiplication of two individual single slit (sinc) functions, one slit of width a along the horizontal axis, and the other of width b along the vertical axis. (Why is the diffraction pattern wider along the horizontal axis?)

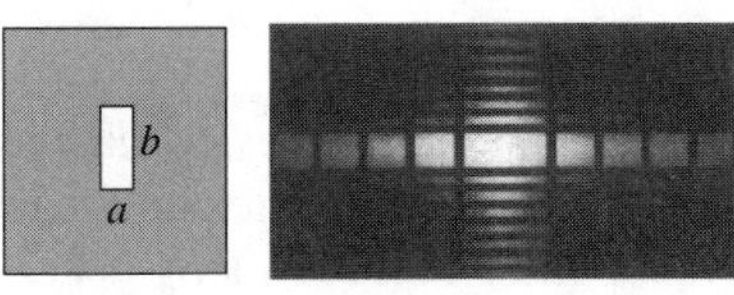

FIGURE 1.26 The rectangular aperture of dimensions $a \times b$ on the left gives the diffraction pattern on the right.

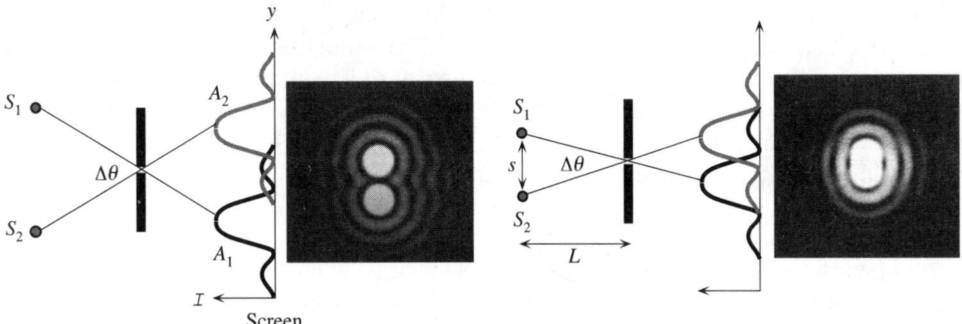

FIGURE 1.27 Resolution of imaging systems is limited by diffraction effects. As points S_1 and S_2 get closer, eventually the Airy disks overlap so much that the resolution is lost.

EXAMPLE 1.10.1 Resolving power of imaging systems

Consider what happens when two neighboring coherent sources are examined through an imaging system with an aperture of diameter D (this may even be a lens). The two sources have an angular separation of $\Delta\theta$ at the aperture. The aperture produces a diffraction pattern of the sources S_1 and S_2, as shown in Figure 1.27. As the points get closer, their angular separation becomes narrower and the diffraction patterns overlap more. According to the **Rayleigh criterion**, the two spots are just resolvable when the principal maximum of one diffraction pattern coincides with the minimum of the other, which is given by the condition,

$$\sin(\Delta\theta_{\min}) = 1.22\frac{\lambda}{D} \tag{6}$$

Angular limit of resolution

The human eye has a pupil diameter of about 2 mm. What would be the minimum angular separation of two points under a green light of 550 nm and their minimum separation if the two objects are 30 cm from the eye? The image will be two diffraction patterns in the eye, and is a result of waves in this medium. If the refractive index $n \approx 1.33$ (water) in the eye, then Eq. (6) is,

$$\sin(\Delta\theta_{\min}) = 1.22\frac{\lambda}{nD} = 1.22\frac{(550 \times 10^{-9})}{(1.33)(2 \times 10^{-3})}$$

giving

$$\Delta\theta_{\min} = 0.0145^\circ$$

Their minimum separation s would be

$$s = 2L\tan(\Delta\theta_{\min}/2) = 2(300\text{ mm})\tan(0.0145^\circ/2) = 0.076\text{ mm} = 76\text{ micron}$$

which is about the thickness of a human hair (or this page).

B. Diffraction grating

A **diffraction grating** in its simplest form is an optical device that has a periodic series of slits in an opaque screen as shown in Figure 1.28 (a). An incident beam of light is diffracted in certain well-defined directions that depend on the wavelength λ and the grating properties. Figure 1.28 (b) shows a typical intensity pattern in the diffracted beam for a finite number of slits. There are "strong beams of diffracted light" along certain directions (θ) and these are labeled according to their occurrence: zero-order (center), first order, either side of the zero order, and so on. If there are infinite number of slits then the diffracted beams have the same intensity. In reality, any periodic variation in the refractive index would serve as a diffraction grating and we will discuss other types later. As in Fraunhofer diffraction we will assume that the observation screen is far away, or that a lens is used to focus the diffracted parallel rays on to the screen (the lens in the observer's eye does it naturally).

We will assume that the incident beam is a plane wave so that the slits become coherent (synchronous) sources. Suppose that the width a of each slit is much smaller than the separation d of the slits as shown in Figure 1.28 (a). Waves emanating at an angle θ from two neighboring slits are out of phase by an amount that corresponds to an optical path difference $d \sin\theta$. Obviously, all such waves from pairs of slits will interfere constructively when this is a multiple of the whole wavelength,

Grating equation

$$d \sin\theta = m\lambda; \qquad m = 0, \pm 1, \pm 2, \ldots \tag{7}$$

which is the well known **grating equation**, also known as the **Bragg**[16] **diffraction condition**. The value of m defines the diffraction order; $m = 0$ being zero-order, $m = 1$ being

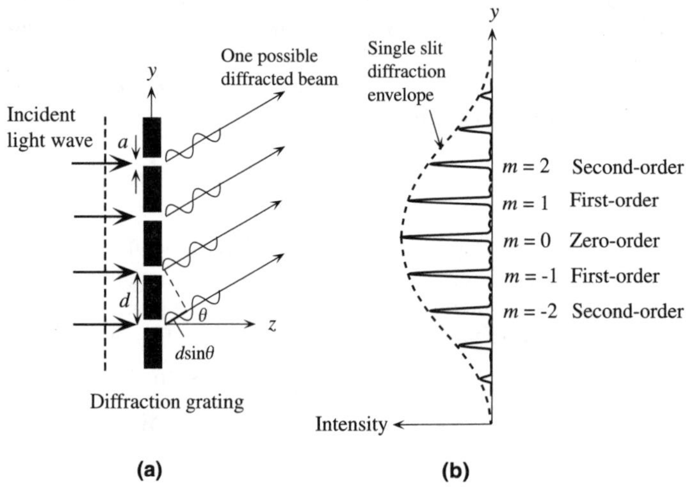

FIGURE 1.28 (a) A diffraction grating with N slits in an opaque screen. (b) The diffracted light pattern. There are distinct beams in certain directions (schematic).

[16] William Lawrence Bragg (1890–1971), Australian—born British physicist, won the Nobel prize with his father William Henry Bragg for his "famous equation" when he was only 25 years old.

first order, *etc.* The problem of determining the actual intensity is more complicated as it involves summing all such waves at the observer and, at the same time, including the diffraction effect of each individual narrow slit. With a smaller than d as in the Figure 1.28 (a), then the amplitude of the diffracted beam is modulated by the diffraction amplitude of a single slit since the latter is spread substantially, as depicted in Figure 1.28 (b). It is apparent that the diffraction grating provides a means of deflecting an incoming light by an amount that depends on its wavelength; the reason for their use in *spectroscopy.*

The diffraction grating in Figure 1.29 (a) is a **transmission grating**. The incident and diffracted beams are on opposite sides of the grating. Typically, parallel thin grooves on a glass plate would serve as a transmission grating as in Figure 1.29 (a). A **reflection grating** has the incident beam and the diffracted beams on the same side of the device as in Figure 1.29 (q). The surface of the device has a periodic reflecting structure, easily formed by etching parallel grooves in a metal film, *etc.* The reflecting unetched surfaces serve as synchronous secondary sources that interfere along certain directions to give diffracted beams of zero-order, first-order, *etc.*

When the incident beam is not normal to the diffraction grating, then Eq. (7) must be modified. If θ_i is the angle of incidence with respect to the grating normal, then the diffraction angle θ_m for the m-th mode is given by,

$$d(\sin\theta_m - \sin\theta_i) = m\lambda; \qquad m = 0, \pm 1, \pm 2, \ldots \tag{8}$$

Grating equation

Diffraction gratings are widely used in spectroscopic applications because of their ability to provide light deflection that depends on the wavelength. In such applications, the undiffracted light that corresponds to the zero-order beam (Figure 1.29) is clearly not desirable because it wastes a portion of the incoming light intensity. Is it possible to shift this energy to a higher order? Robert William Wood (1910) was able to do so by ruling grooves on glass with a controlled shape as in Figure 1.30 where the surface is angled periodically with a spatial period d. The diffraction condition in Eq. (8) applies with respect to the normal to the grating plane, whereas the first order reflection corresponds to reflection from the flat surface, which is at an angle γ. Thus it is possible to "blaze" one of the higher orders (usually $m = 1$) by appropriately choosing γ. Most modern diffraction gratings are of this type.

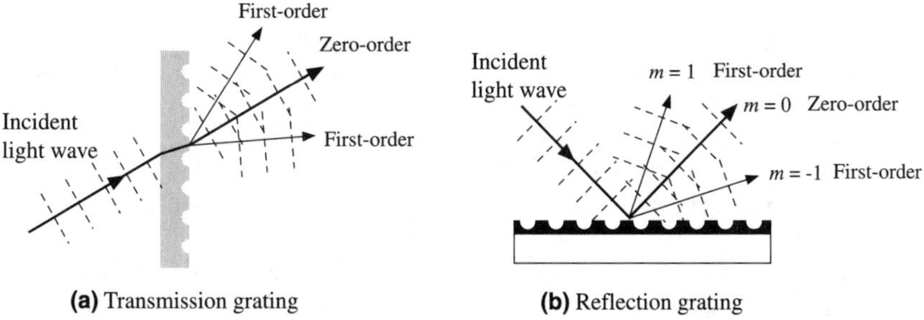

FIGURE 1.29 (a) Ruled periodic parallel scratches on a glass serve as a transmission grating. (b) A reflection grating. An incident light beam results in various "diffracted" beams. The zero-order diffracted beam is the normal reflected beam with an angle of reflection equal to the angle of incidence.

FIGURE 1.30 Blazed (echelette) grating.

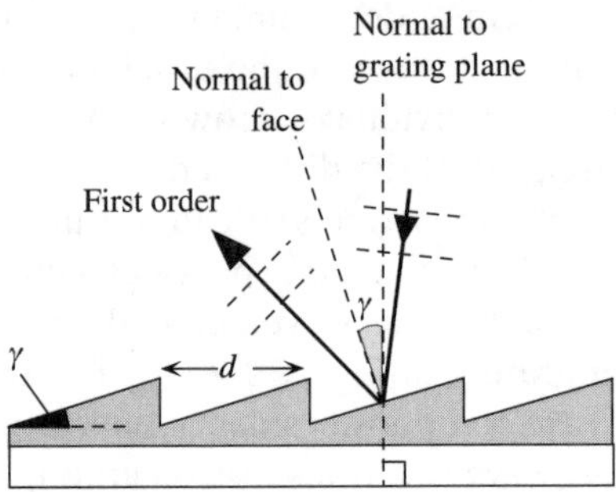

QUESTIONS AND PROBLEMS

1.1 Gaussian beams Two identical spherical mirrors, A and B, have been aligned to be confocal and to directly face each other as in Figure 1.31. The two mirrors and the space in between them (the optical cavity) form an **optical resonator** because only certain light waves with certain frequencies can exit in this optical cavity. The light beam in the cavity is a Gaussian beam. When it starts at A, its wavefront is the same as the curvature of A. Sketch the wavefronts on this beam as it travels towards B, at B, as it is then reflected from B back to A. If $R = 25$ cm, and the mirrors are of diameter 2.5 cm, estimate the divergence of the beam $\Delta\theta$ and its spot size (minimum waist).

FIGURE 1.31 Two confocal spherical mirrors reflect waves to and from each other. F is the focal point and R is the radius. The optical cavity contains a Gaussian beam.

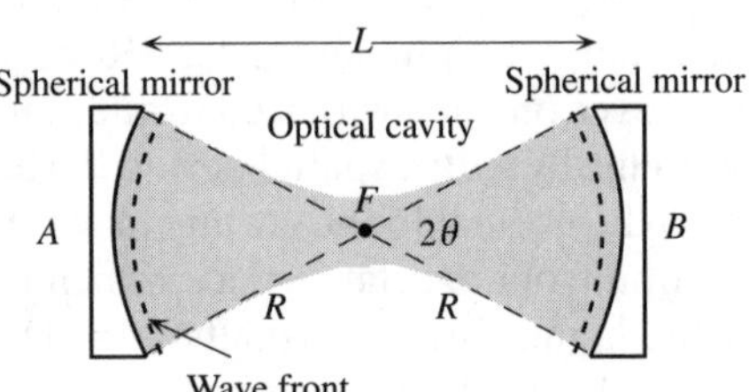

1.2 Refractive index

(a) Consider light of free-space wavelength 1300 nm traveling in pure silica medium. Calculate the phase velocity and group velocity of light in this medium. Is the group velocity ever greater than the phase velocity?

(b) What is the Brewster angle (the polarization angle θ_p) and the critical angle (θ_c) for total internal reflection when the light wave traveling in this silica medium is incident on a silica/air interface. What happens at the polarization angle?

(c) What is the reflection coefficient and reflectance at normal incidence when the light beam traveling in the silica medium is incident on a silica/air interface?

(d) What is the reflection coefficient and reflectance at normal incidence when a light beam traveling in air is incident on an air/silica interface? How do these compare with part (c) and what is your conclusion?

1.3 Refractive index and the Sellmeier dispersion equation The Sellmeier dispersion equation is an empirical expression for the refractive index of glass in terms of the wavelength λ,

$$n^2 - 1 = \frac{G_1\lambda^2}{\lambda^2 - \lambda_1^2} + \frac{G_2\lambda^2}{\lambda^2 - \lambda_2^2} + \frac{G_3\lambda^2}{\lambda^2 - \lambda_3^2} \quad (1)$$

Sellmeier equation

in which G_1, G_2, G_3 and λ_1, λ_2, and λ_3 are constants (called Sellmeier coefficients) that are determined by fitting this expression to the experimental data. The actual Sellmeier formula has more terms in the right-hand summation of the same type *e.g.* $G_i\lambda^2/(\lambda^2 - \lambda_i^2)$ in which $i = 4, 5, \ldots$ but these can generally be neglected in representing n vs. λ behavior over typical wavelengths of interest and ensuring that three terms included in Eq. (1) correspond to the most important or relevant terms in the summation. Table 1.2 gives the Sellmeier coefficients for pure silica (SiO_2) and SiO_2-13.5 mol.% GeO_2. Write a program on your computer or calculator, or use a math software package (*e.g.* Mathcad, Matlab, Mathview *etc.*) or even a spread sheet program (*e.g.* Excel) to obtain the refractive index n as a function of λ from 0.5 μm to 1.8 μ m for both pure silica and SiO_2-13.5% GeO_2. Obtain the group index, N_g, vs. wavelength for both materials and plot it on the same graph. Find the wavelength at which the material dispersion becomes zero in each material.

TABLE 1.2 The Sellmeier coefficients for SiO_2 and SiO_2-13.5% GeO_2. The $\lambda_1, \lambda_2, \lambda_3$ are in μm.

	G_1	G_2	G_3	λ_1	λ_2	λ_3
SiO_2	0.696749	0.408218	0.890815	0.0690660	0.115662	9.900559
SiO_2-13.5% GeO_2	0.711040	0.451885	0.704048	0.0642700	0.129408	9.425478

1.4 Antireflection coating

(a) Consider three dielectric media with flat and parallel boundaries with refractive indices n_1, n_2, and n_3. Show that for normal incidence the reflection coefficient between layers 1 and 2 is the same as that between layers 2 and 3 if $n_2 = \sqrt{[n_1 n_3]}$. What is the significance of this result?

(b) Consider a Si photodiode that is designed for operation at 900 nm. Given a choice of two possible antireflection coatings, SiO_2 with a refractive index of 1.5 and TiO_2 with a refractive index of 2.3, which would you use and what would be the thickness of the antireflection coating? The refractive index of Si is 3.5.

1.5 Reflection at glass-glass and air-glass interfaces A ray of light that is traveling in a glass medium of refractive index $n_1 = 1.460$ becomes incident on a less dense glass medium of refractive index $n_2 = 1.430$. Suppose that the free space wavelength of the light ray is 850 nm.

(a) What should the minimum incidence angle for TIR be?

(b) What is the phase change in the reflected wave when the angle of incidence $\theta_i = 85°$ and when $\theta_i = 90°$?

(c) What is the penetration depth of the evanescent wave into medium 2 when $\theta_i = 85°$ and when $\theta_i = 90°$?

(d) What is the reflection coefficient and reflectance at normal incidence $(\theta_i = 0°)$ when the light beam traveling in the silica medium ($n = 1.455$) is incident on a silica/air interface?

(e) What is the reflection coefficient and reflectance at normal incidence when a light beam traveling in air is incident on an air/silica ($n = 1.455$) interface? How do these compare with part (d) and what is your conclusion?

1.6 TIR and polarization at water-air interface

(a) Given that the refractive index of water is about 1.33, what is the polarization angle for light traveling in air and reflected from the surface of the water?

(b) Consider a diver in sea pointing a flashlight towards the surface of the water. What is the critical angle for the light beam to be reflected from the water surface?

1.7 Phase changes on TIR Consider a light wave of wavelength 870 nm traveling in a semiconductor medium (GaAs) of refractive index 3.6. It is incident on a different semiconductor medium (AlGaAs) of refractive index 3.4, and the angle of incidence is 80°. Will this result in total internal reflection? Calculate the phase change in the parallel and perpendicular components of the reflected electric field.

1.8 Thin film coating and multiple reflections Consider a thin film coating on an object as in Figure 1.32. If the incident wave has an amplitude of A_0, then there are various transmitted and reflected waves, as shown in Figure 1.32. We then have the following amplitudes based on the definitions of the reflection and transmission coefficients,

$$\begin{array}{llll}
A_1 = A_0 r_{12} & B_1 = A_0 t_{12} & B_2 = A_0 t_{12} r_{23} & C_1 = A_0 t_{12} t_{23} \\
A_2 = A_0 t_{12} r_{23} t_{21} & B_3 = A_0 t_{12} r_{23} r_{21} & B_4 = A_0 t_{12} r_{23} r_{21} r_{23} & C_2 = A_0 t_{12} r_{23} r_{21} t_{23} \\
A_3 = A_0 t_{12} r_{23} r_{21} r_{23} t_{21} & B_5 = A_0 t_{12} r_{23} r_{21} r_{23} r_{21} & B_6 = A_0 t_{12} r_{23} r_{21} r_{23} r_{21} r_{23} & C_3 = A_0 t_{12} r_{23} r_{21} r_{23} r_{21} t_{23}
\end{array}$$

and so on.

Assume that $n_1 < n_2 < n_3$ and that the thickness of the coating is d. For simplicity, we will assume normal incidence. The phase change in traversing the coating thickness d is $\phi = (2\pi/\lambda)n_2 d$ in which λ is the free space wavelength. The wave has to be multiplied by $\exp(-j\phi)$ to account for this phase difference.

The reflection and transmission coefficients are given by,

$$r_1 = r_{12} = \frac{n_1 - n_2}{n_1 + n_2} = -r_{21}, \qquad r_2 = r_{23} = \frac{n_2 - n_3}{n_2 + n_3}$$

and

$$t_1 = t_{12} = \frac{2n_1}{n_1 + n_2}, \qquad t_2 = t_{21} = \frac{2n_2}{n_1 + n_2},$$

(a) Show that

$$1 - t_1 t_2 = r_1^2$$

(b) Show that the reflection coefficient is

$$r = \frac{A_{\text{reflected}}}{A_0} = r_1 - \frac{t_1 t_2}{r_1} \sum_{k=1}^{\infty} (-r_1 r_2 e^{-j2\phi})^k$$

which can be summed to

$$r = \frac{r_1 - r_2 e^{-j2\phi}}{1 + r_1 r_2 e^{-j2\phi}}$$

(c) Show that the transmission coefficient is

$$t = \frac{C_{\text{transmitted}}}{A_0} = -\frac{t_1 t_2 e^{j\phi}}{r_1 r_2} \sum_{k=1}^{\infty} (-r_1 r_2 e^{-j2\phi})^k = \left(\frac{t_1 t_2 e^{j\phi}}{r_1 r_2}\right) \frac{r_1 r_2 e^{-j2\phi}}{1 + r_1 r_2 e^{-j2\phi}}$$

which can be summed to

$$t = \frac{t_1 t_2 e^{-j\phi}}{1 + r_1 r_2 e^{-j2\phi}}$$

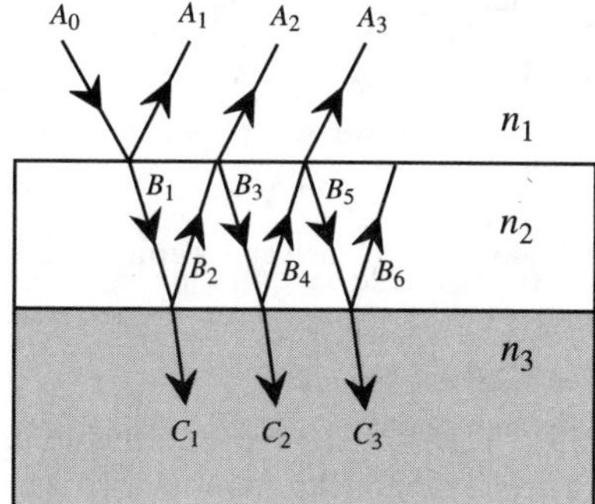

FIGURE 1.32 Thin film coating of refractive index n_2 on a semiconductor device

1.9 Antireflection coating Consider a thin film coating on an object, as in Figure 1.32. Using the transmission coefficient in Question 1.8, show that a normally incident light beam has maximum transmission into medium 3 when

$$d = \frac{m\lambda}{4n_2}$$

in which m is an odd-integer and λ is the free space wavelength. Show, in addition, that we need $r_1 = r_2$, which requires choosing $n_2 = (n_1 n_3)^{1/2}$. Derive the same result using the reflection coefficient in Question 1.8.

1.10 Fresnel's equation Fresnel's equations are sometimes given as follows:

$$r_\perp = \frac{E_{r0,\perp}}{E_{i0,\perp}} = \frac{n_1 \cos\theta_i - n_2 \cos\theta_t}{n_1 \cos\theta_i + n_2 \cos\theta_t}$$

$$r_{//} = \frac{E_{r0,//}}{E_{i0,//}} = \frac{n_1 \cos\theta_t - n_2 \cos\theta_i}{n_1 \cos\theta_t + n_2 \cos\theta_i}$$

$$t_\perp = \frac{E_{t0,\perp}}{E_{i0,\perp}} = \frac{2n_1 \cos\theta_i}{n_1 \cos\theta_i + n_2 \cos\theta_t}$$

and

$$t_{//} = \frac{E_{t0,//}}{E_{i0,//}} = \frac{2n_1 \cos\theta_i}{n_1 \cos\theta_t + n_2 \cos\theta_i}$$

Show that these reduce to Fresnel's equation given in Section 1.6.

Using Fresnel's equations, find the reflection and transmission coefficients for normal incidence and show that

$$r_\perp + 1 = t_\perp \quad \text{and} \quad r_{//} + nt_{//} = 1$$

in which $n = n_2/n_1$.

1.11 Fabry-Perot interferometer Figure 1.33 shows a Fabry-Perot etalon and a light beam of wavevector $k = 2\pi/\lambda$ incident at an angle θ. Show that the allowed cavity modes and hence the transmitted intensities are given by

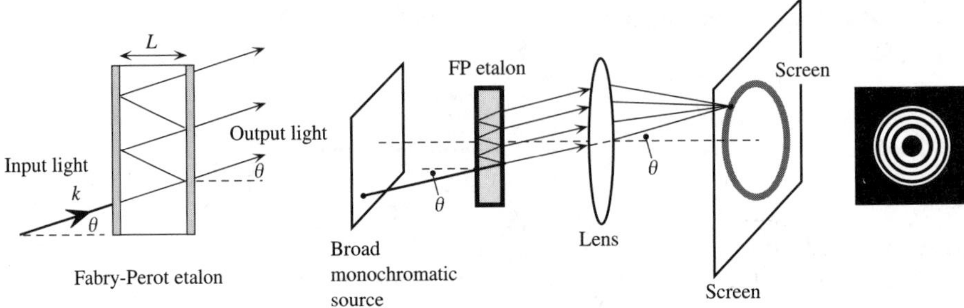

FIGURE 1.33 Fabry-Perot optical resonator and the Fabry-Perot interferometer (schematic)

$$2L\cos\theta = m\lambda.$$

in which m is an integer $(1, 2, 3 \ldots)$ (*Hint:* Resolve k normal to the etalon face).

Suppose that a light from a broad mononchromatic source is incident on a Fabry-Perot etalon and the transmitted waves are focused onto a screen as in Figure 1.33. The interference pattern is a ring of bright and dark rings. Explain why this is so. Suppose that L was adjustable, perhaps by a piezoelectric device. How would you adjust L to measure the wavelength of incident light?

1.12 Goos-Haenchen phase shift A ray of light that is traveling in a glass medium of refractive index $n_1 = 1.460$ becomes incident on a less dense glass medium of refractive index $n_2 = 1.430$. Suppose that the free space wavelength of the light ray is 850 nm. The angle of incidence $\theta_i = 85°$. Estimate the lateral Goos-Haenchen shift in the reflected wave for the perpendicular field component. Recalculate the Goos-Haenchen shift if the second medium has $n_2 = 1$ (air). What is your conclusion?

1.13 TIR and FTIR

(a) By considering the electric field component in medium B in Figure 1.20 (b), explain how you can adjust the amount of transmitted light.

(b) What is the critical angle at the hypotenuse face of a beam splitter cube made of glass with $n_1 = 1.6$ and having a thin film of liquid with $n_2 = 1.3$. Can you use 45° prisms with normal incidence?

(c) Explain how a light beam can propagate along a layer of material between two different media as shown in Figure 1.34 (a). Explain what the requirements are for the indices n_1, n_2, n_3. Will there be any losses at the reflections?

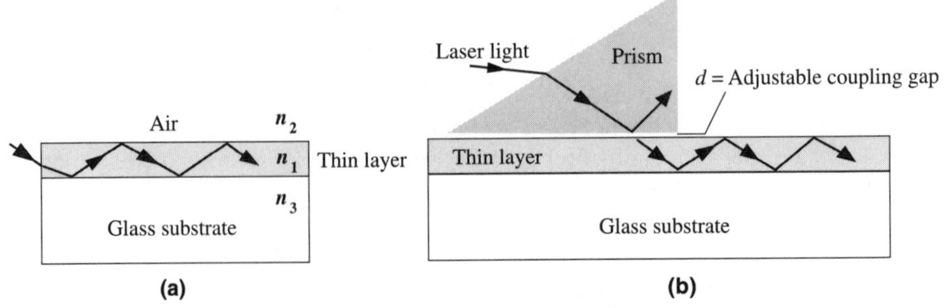

FIGURE 1.34 (a) Light propagation along an optical guide. (b) Coupling of laser light into a thin layer - optical guide - using a prism. The light propagates along the thin layer.

(d) Consider the prism coupler arrangement in Figure 1.34 (b). Explain how this arrangement works for coupling an external light beam from a laser into a thin layer on the surface of a glass substrate. Light is then propagated inside the thin layer along the surface of the substrate. What is the purpose of the adjustable coupling gap?

1.14 Coherence length A particular laser is operating in single mode and emitting a continuous wave lasing emission whose spectral width is 1 MHz. What is the coherence time and coherence length?

1.15 Spectral widths

(a) Suppose that frequency spectrum of a radiation emitted from a source has a central frequency v_o and a spectral width Δv. The spectrum of this radiation in terms of wavelength will have a central wavelength λ_o and a spectral width $\Delta\lambda$. Clearly, $\lambda_o = c/v_o$. Since $\Delta\lambda \ll \lambda_o$ and $\Delta v \ll v_o$, using $\lambda = c/v$, show that the line width $\Delta\lambda$ and hence the coherence length l_c are

$$\Delta\lambda = \Delta v\frac{\lambda_o}{v_o} = \Delta v\frac{\lambda_o^2}{c} \quad \text{and} \quad l_c = c\Delta t = \frac{\lambda_o^2}{\Delta\lambda}$$

(b) Calculate $\Delta\lambda$ for a lasing emission from a He-Ne laser that has $\lambda_o = 632.8$ nm and $\Delta v \approx 1.5$ GHz.

1.16 Diffraction by a lens Any lens in practice is an aperture and the image of a point is therefore a diffraction pattern. Suppose a lens with a diameter of 2 cm has a focal length of 40 cm. Suppose that it is illuminated with a plane wave, a collimated beam of light, of wavelength 590 nm. What is the diameter of the Airy disk at the focal point? What is your conclusion?

1.17 Bragg diffraction Suppose that parallel grooves are etched on the surface of a semiconductor to act as a reflection grating and that the periodicity (separation) of the grooves is 1 micron. If light of wavelength 1.3 μm is incident at an angle 89° to the normal, find the diffracted beams.

William Lawrence Bragg (1890–1971), Australian-born British physicist, won the Nobel prize with his father William Henry Bragg for his "famous equation" when he was only 25 years old *(Courtesy of Emilio Segrè Visual Archives, AIP, Weber Collection)*

"The important thing in science is not so much to obtain new facts as to discover new ways of thinking about them." —Sir William Lawrence Bragg

CHAPTER 2

Dielectric Waveguides and Optical Fibers

"The introduction of optical fiber systems will revolutionize the communications network. The low-transmission loss and the large bandwidth capability of the fiber systems allow signals to be transmitted for establishing communications contacts over large distances with few or no provisions of intermediate amplification."—Charles K. Kao[1]

Professor Charles Kao, who has been recognized as the inventor of fiber optics, left, receiving an IEE prize from Professor John Midwinter (1998 at IEE Savoy Place, London, UK; courtesy of IEE).

2.1 SYMMETRIC PLANAR DIELECTRIC SLAB WAVEGUIDE

A. Waveguide Condition

To understand the general nature of light wave propagation in optical waveguides, we first consider the planar dielectric slab waveguide shown in Figure 2.1, which is the simplest waveguide in terms of tractable analysis. A slab of dielectric of thickness $2a$ and

[1] Charles K. Kao (one of the pioneers of glass fibers for optical communications) *Optical Fiber Systems: Technology, Design, and Applications* (McGraw-Hill Book Company, New York, USA, 1982), p. 1

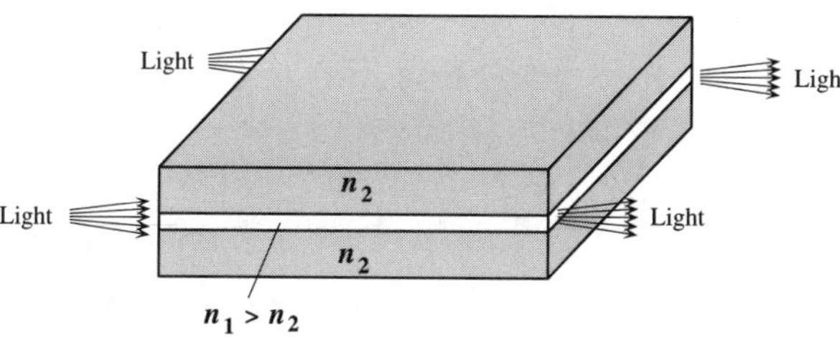

FIGURE 2.1 A planar dielectric waveguide has a central rectangular region of higher refractive index n_1 than the surrounding region, which has a refractive index n_2. It is assumed that the waveguide is infinitely wide and the central region is of thickness $2a$. It is illuminated at one end by a monochromatic light source.

refractive index n_1 is sandwiched between two semi-infinite regions both of refractive index $n_2(n_2 < n_1)$. The region of higher refractive index (n_1) is called the **core** and the region of lower refractive index (n_2) sandwiching the core is called the **cladding**.

A light ray can readily propagate along such a waveguide, in a zigzag fashion, provided it can undergo total internal reflection (TIR) at the dielectric boundaries. It seems that any light wave that has an angle of incidence (θ) greater than the critical angle (θ_c) for TIR, will be propagated. This, however, is true only for a very thin light beam with a diameter much less than the slab thickness, $2a$. We consider the realistic case when the whole end of the waveguide is illuminated, as depicted in Figure 2.1. To simplify the analysis, we will assume that light is launched from a line source in a medium of refractive index n_1. In general, the refractive index of the launching medium will be different than n_1, but this will affect only the amount of light coupled into the guide.

Consider a plane wave type of light ray propagating in the dielectric slab waveguide as shown in Figure 2.2. We will take the electric field E to be along x, parallel to the interface and perpendicular to z. The ray is guided in a zigzag fashion along the guide axis z by reflections from the core-cladding (n_1/n_2) boundaries. The result is the effective propagation of the electric field E along z. The figure also shows the constant phase wavefronts, normal to direction of propagation, on this ray. This particular ray is reflected at B and then at C. Just after the reflection at C, the wavefront at C overlaps the wavefront at A on the original ray. The wave interferes with itself. Unless these wavefronts at A and C are in phase, the two will interfere destructively and destroy each other. Only certain reflection angles θ give rise to the constructive interference and hence only certain waves can exist in the guide.

The phase difference between the points A and C corresponds to an optical path length $AB + BC$. In addition, there are two total internal reflections (TIR) at B and C and each introduces a further phase change of ϕ. Suppose that k_1 is the wavevector in

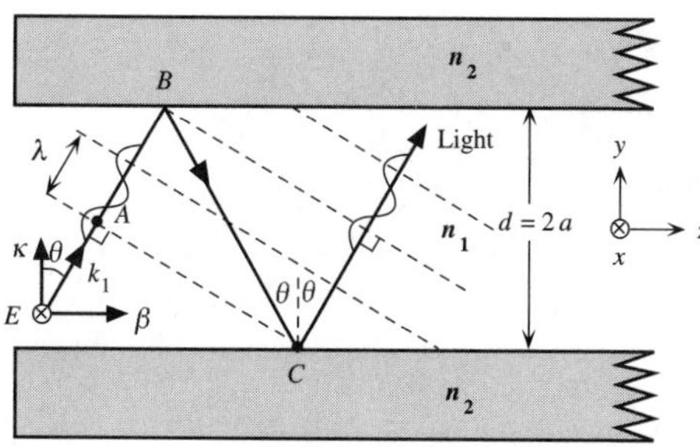

FIGURE 2.2 A light ray traveling in the guide must interfere constructively with itself to propagate successfully. Otherwise destructive interference will destroy the wave. The x-axis is into the paper.

n_1, *i.e.* $k_1 = kn_1 = 2\pi n_1/\lambda$, in which k and λ are the free space wavevector and wavelength. For constructive interference, the phase difference between A and C must be a multiple of 2π.

$$\Delta\phi(AC) = k_1(AB + BC) - 2\phi = m(2\pi) \tag{1}$$

in which $m = 0, 1, 2, \ldots$ is an integer.

We can easily evaluate the $AB + BC$ from geometrical considerations. $BC = d/\cos\theta$ and $AB = BC\cos(2\theta)$. Thus

$$AB + BC = BC\cos(2\theta) + BC = BC[(2\cos^2\theta - 1) + 1] = 2d\cos\theta$$

Thus, for wave propagation along the guide we need

Constructive interference

$$k_1[2d\cos\theta] - 2\phi = m(2\pi) \tag{2}$$

It is apparent that only certain θ and ϕ values can satisfy this equation for a given integer m. But ϕ depends on θ and also on the polarization state of the wave (direction of the electric field. So for each m, there will be one allowed angle θ_m and one corresponding ϕ_m. Dividing Eq. (2) by 2 we obtain the **waveguide condition**

Waveguide condition

$$\left[\frac{2\pi n_1(2a)}{\lambda}\right]\cos\theta_m - \phi_m = m\pi \tag{3}$$

in which ϕ_m indicates that ϕ is a function of the incidence angle θ_m.

It may be thought that the treatment above is somewhat artificial as we took a narrow angle for θ. It turns out that Eq. (3) can be derived as a general waveguide condition for guided waves whether we use a narrow or a wider angle, one or multiple rays. We can derive the same condition if we take two arbitrary parallel rays entering the guide as in Figure 2.3. The rays 1 and 2 are initially in phase, and represent the same "plane-wave". Ray 1 then suffers two reflections at A and B, and is then again traveling parallel to ray 2. Unless the wavefront on ray 1 just after reflection at B is in phase with the wavefront at B' on ray 2, the two would destroy each other. Both rays initially start in phase; ray 1 at A just before reflection and ray 2 at A'. Ray 1 at B, just after two reflections, has a phase $k_1AB - 2\phi$. Ray 2 at B' has a phase $k_1(A'B')$. The difference between the two phases must be $m(2\pi)$ and leads to the waveguide condition in Eq. (3).

We can resolve the wavevector k_1 into two propagation constants, β and κ, along and perpendicular to the guide axis z as shown Figure 2.2. Given θ_m that satisfies the waveguide condition, we define:

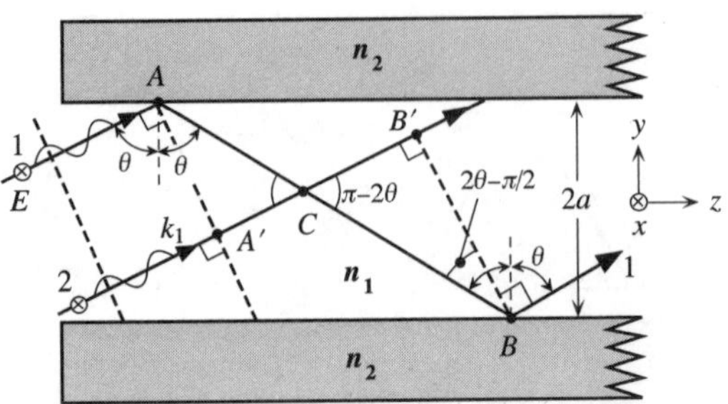

FIGURE 2.3 Two arbitrary waves 1 and 2 that are initially in phase must remain in phase after reflections. Otherwise the two will interfere destructively and cancel each other.

$$\beta_m = k_1 \sin\theta_m = \left(\frac{2\pi n_1}{\lambda}\right)\sin\theta_m \tag{4}$$

Propagation constant along the guide

and

$$\kappa_m = k_1 \cos\theta_m = \left(\frac{2\pi n_1}{\lambda}\right)\cos\theta_m \tag{5}$$

Transverse Propagation constant

The simplistic analysis as embedded in the waveguide condition in Eq. (3) shows quite clearly that only certain reflection angles are allowed within the guide corresponding to $m = 0, 1, 2, \ldots$. We note that higher m values yield lower θ_m. Each choice of m leads to a different reflection angle θ_m. Each different m value leads to a different propagation constant along the guide given by Eq. (4).

If we were to consider the interference of many rays, as in Figure 2.3, we would find that the resultant wave has a stationary electric field pattern along the y-direction, and this field pattern travels along the guide, z-axis, with a propagation constant β_m. We can show this by considering the resultant of the two parallel rays in Figure 2.3 that have incidence angles θ_m satisfying the waveguide condition. As shown in Figure 2.4, ray 1, after reflection at A, is traveling downward, whereas ray 2 is still traveling upward. The two meet at C, distance y above the guide center. The optical path difference between the two rays is $A'C - AC$, plus the phase change ϕ_m for ray 1 on TIR at A. Using the geometry shown in Figure 2.4, the phase difference between rays 1 and 2 is[2]

$$\Phi_m = (k_1 AC - \phi_m) - k_1 A'C = 2k_1(a - y)\cos\theta_m - \phi_m$$

Substituting the waveguide condition in Eq. (3) and simplifying we obtain the phase difference Φ_m as a function of y for a given m,

$$\Phi_m = \Phi_m(y) = m\pi - \frac{y}{a}(m\pi + \phi_m) \tag{6}$$

Just before C, rays 1 and 2 have opposite κy terms in their phases as they are traveling in opposite y-directions. The electric fields of rays 1 and 2 at C are:

$$E_1(y, z, t) = E_o \cos(\omega t - \beta_m z + \kappa_m y + \Phi_m)$$

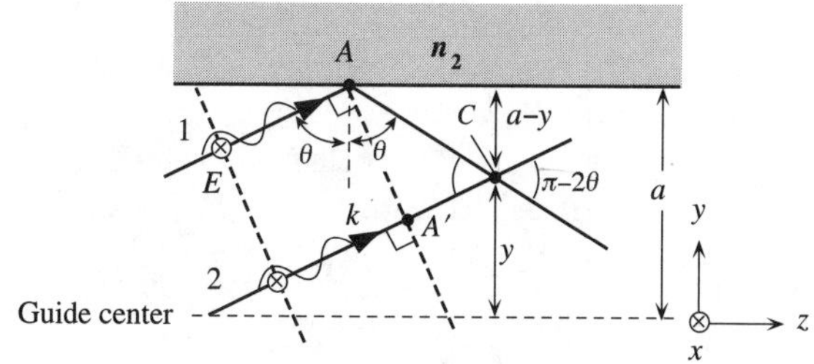

FIGURE 2.4 Interference of waves such as 1 and 2 leads to a standing wave pattern along the y direction which propagates along z

[2] The algebraic and trigonometric steps in the derivation are left as an exercise, and do not affect the logical flow.

and

$$E_2(y, z, t) = E_o \cos(\omega t - \beta_m z - \kappa_m y)$$

These two waves interfere to give[3]

$$E(y, z, t) = 2E_o \cos(\kappa_m y + \tfrac{1}{2}\Phi_m) \cos(\omega t - \beta_m z + \tfrac{1}{2}\Phi_m) \quad \textbf{(7)}$$

Equation (7) is a traveling wave along z, due to the $\cos(\omega t - \beta_m z)$ term, whose amplitude along y is modulated by the $\cos(\kappa_m y + \frac{1}{2}\Phi_m)$ term. The latter has no time dependence and corresponds to a standing wave pattern along y. Since each m value gives a different κ_m and Φ_m, for each m we obtain a distinct field pattern along y. Thus a light wave propagating along the guide is of the form

Possible waves in the guide

$$E(y, z, t) = 2E_m(y) \cos(\omega t - \beta_m z) \quad \textbf{(8)}$$

in which $E_m(y)$ is the field distribution along y for a given m. The distribution $E_m(y)$ *across* the guide is *traveling down the guide along z*.

Figure 2.5 shows the field pattern for the lowest mode, $m = 0$, which has a maximum intensity at the center. The whole field distribution is moving along z with a propagation vector, β_0. Also notice that the field penetrates into the cladding, which is due to a propagating evanescent wave in the cladding near the boundary. The field pattern in the core exhibits a harmonic variation across the guide, or along y, of the type in Eq. (7), whereas the field in the cladding is that of the evanescent wave and it decays exponentially with y. Figure 2.6 illustrates the field patterns for the first three modes, $m = 0$ to 2.

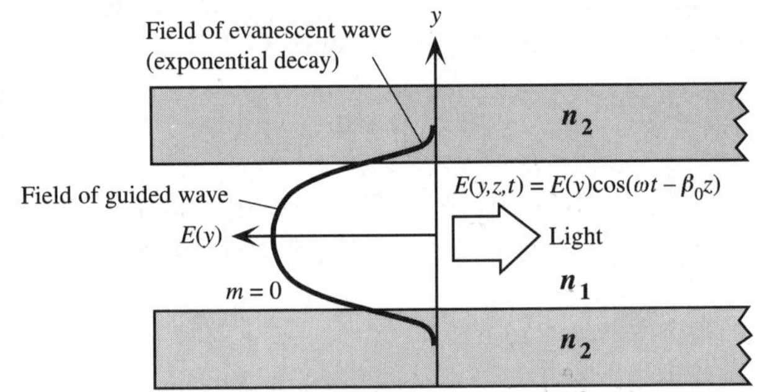

FIGURE 2.5 The electric field pattern of the lowest mode traveling wave along the guide. This mode has $m = 0$ and the lowest θ. It is often referred to as the glazing incidence ray. It has the highest phase velocity along the guide.

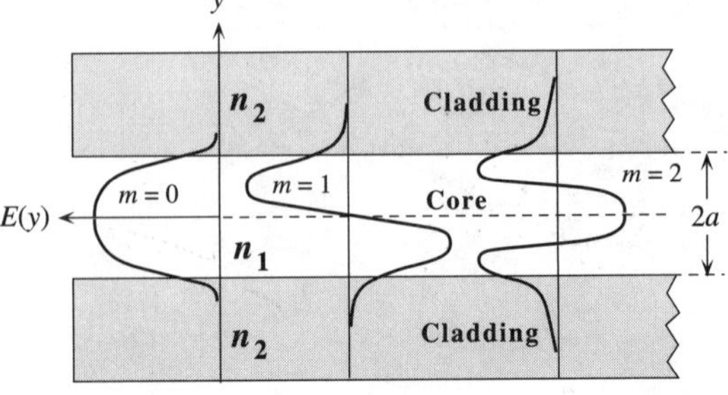

FIGURE 2.6 The electric field patterns of the first three modes ($m = 0, 1, 2$) traveling wave along the guide. Notice different extents of field penetration into the cladding.

[3] We use the trigonometric identity, $\sin A + \sin B = 2\sin[\frac{1}{2}(A + B)]\cos[\frac{1}{2}(A - B)]$

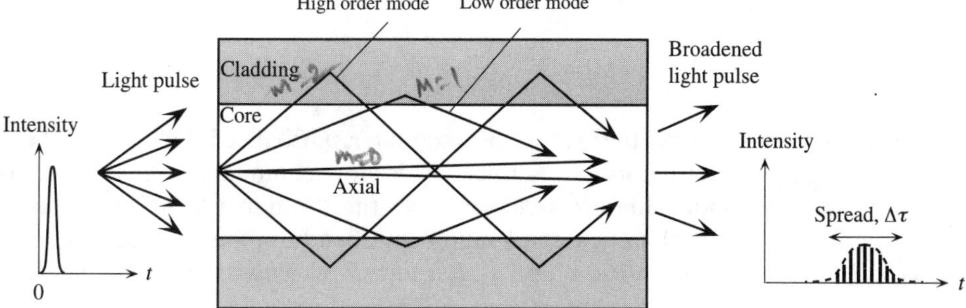

FIGURE 2.7 Schematic illustration of light propagation in a slab dielectric waveguide. Light pulse entering the waveguide breaks up into various modes that then propagate at different group velocities down the guide. At the end of the guide, the modes combine to constitute the output light pulse which is broader than the input light pulse.

We have seen that each m leads to an allowed θ_m value that corresponds to a particular traveling wave in the z-direction as described by Eq. (8) with a particular wavevector β_m as defined by Eq. (4). Each of these traveling waves, with a distinct field pattern, $E_m(y)$, constitutes a **mode of propagation**. The integer m identifies these modes and is called the **mode number**. The light energy can be transported only along the guide via one or more of these possible modes of propagation as depicted in Figure 2.7. Notice that the rays have been shown to penetrate the cladding, as in Figure 2.6, and reflected from an apparent plane in the cladding. Since θ_m is smaller for larger m, higher modes exhibit more reflections but they also penetrate much more into the cladding as schematically depicted in Figure 2.7. For the lowest mode, $m = 0$, which leads to θ_m being closest to 90° and the wave is said to travel *axially*. Light that is launched into the core of the waveguide can travel only along the guide in the allowed modes specified by Eq. (3). These modes will travel down the guide at different group velocities. When they reach the end of the guide they constitute the emerging light beam. If we launch a short-duration light pulse into the dielectric waveguide, the light emerging from the other end will be a broadened light pulse because light energy would have been propagated at different group velocities along the guide as depicted in Figure 2.7[4]. The light pulse therefore spreads as it travels along the guide.

B. Single and Multimode Waveguides

Although waveguide condition in Eq. (3) specifies the allowed θ_m values, θ_m must nonetheless satisfy TIR, that is, $\sin\theta_m > \sin\theta_c$. With the latter condition imposed in addition, we can only have up to a certain maximum number of modes being allowed in the waveguide. From Eq. (3) we can obtain an expression for $\sin\theta_m$ and then apply the TIR condition, $\sin\theta_m > \sin\theta_c$, to show that the mode number m must satisfy

$$m \leq (2V - \phi)/\pi \qquad \textbf{(9)}$$

Maximum value of m

[4] Although the spread in Figure 2.7 has been shown as "Gaussian" looking, this is not strictly so and the exact shape of the spread cannot be derived in a simple fashion.

in which V, called the **V-number**, is a quantity defined by

V-number definition

$$V = \frac{2\pi a}{\lambda}\left(n_1^2 - n_2^2\right)^{1/2} \tag{10}$$

The V-number also has other names, **V-parameter, normalized thickness**, and **normalized frequency**. The term normalized thickness is more common for the present planar guide, whereas in optical fibers (discussed later), the V-number term is more usual. For a given free space wavelength λ, the V-number depends on the waveguide geometry $(2a)$ and waveguide properties, n_1 and n_2. *It is therefore a characteristic parameter of the waveguide.*

The question arises whether there is a value of V that makes $m = 0$ the only possibility, that is, there is only one mode propagating. Suppose that for the lowest mode, the propagation is due to a glazing incidence at $\theta_m \rightarrow 90°$, then $\phi \rightarrow \pi$ and from Eq. (9), $V = (m\pi + \phi)/2$ or $\pi/2$. When $V < \pi/2$ there is only one mode propagating, which is the lowest mode with $m = 0$. From the expressions for V and ϕ we can show that $\phi \le 2V$ so that Eq. (9) never gives a negative m. *Thus, when $V < \pi/2$, $m = 0$ is the only possibility and only the fundamental mode ($m = 0$) propagates along the dielectric slab waveguide, which is then termed a* ***single mode*** *planar waveguide.* The free-space wavelength λ_c that leads to $V = \pi/2$ in Eq. (10) is the **cut-off wavelength**, and above this wavelength, only one-mode, the fundamental mode, will propagate.

C. TE and TM Modes

We have shown that, for a particular mode, the variation of the field intensity along y, $E_m(y)$, is harmonic. Figure 2.8 (a) and (b) consider two of the possibilities for the electric field direction of a wave traveling toward the core-cladding boundary.

(a) The electric field is perpendicular to the plane of incidence (plane of the paper) as indicated by $E_\perp$, and shown in Figure 2.8 (a). $E_\perp$ is along x, so that $E_\perp = E_x$.

(b) The magnetic field is perpendicular to the plane of incidence as indicated by $B_\perp$ and shown in Figure 2.8 (b). In this case the electric field is parallel to the plane of incidence and is denoted by $E_{//}$.

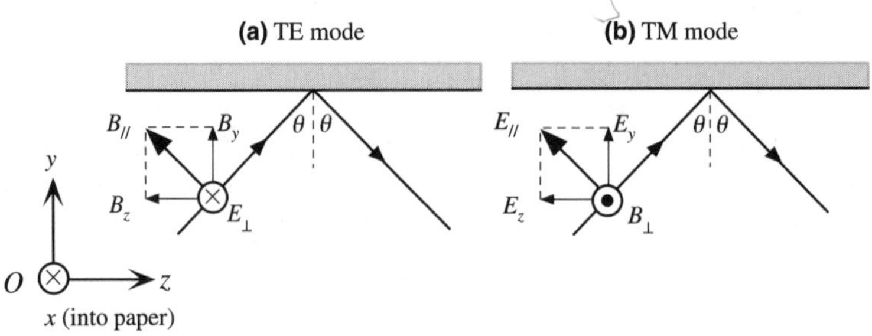

FIGURE 2.8 Possible modes can be classified in terms of (a) transverse eletric field (TE) and (b) transverse magnetic field (TM). Plane of incidence is the paper.

Any other field direction (perpendicular to the path of the ray) can be resolved to have electric field components along $E_{//}$ and $E_{\perp}$. These two fields experience different phase changes, $\phi_{//}$ and $\phi_{\perp}$, and consequently require different angles θ_m to propagate along the guide. We therefore have a different set of modes for $E_{//}$ and $E_{\perp}$. The modes associated with $E_{\perp}$ (or E_x) are termed **transverse electric field modes**, denoted by TE_m, because $E_{\perp}$ is actually perpendicular to the direction of propagation, z.

The modes associated with the $E_{//}$ field have a magnetic field $B_{\perp}$, which is perpendicular to the direction of propagation and are termed **transverse magnetic field modes**, denoted by TM_m. It is interesting that $E_{//}$ has a field component parallel to the z-axis, shown as E_z, which is along the direction of propagation. It is apparent that E_z is a propagating longitudinal electric field. In free space, it is *impossible* for such a longitudinal field to exist but within an optical guide, due to the interference phenomena, it is indeed possible to have a longitudinal field. Similarly, the TE modes, those with $B_{//}$, have a magnetic field along z that propagates along this direction as a longitudinal wave.

The phase change ϕ that accompanies TIR depends on the polarization of the field and is different for $E_{//}$ and $E_{\perp}$. The difference, however, is negligibly small for $n_1 - n_2 \ll 1$ and thus the waveguide condition and the cut-off condition can be taken to be identical for both TE and TM modes.

EXAMPLE 2.1.1 Waveguide modes

Consider a planar dielectric guide with a core thickness 20 μm, $n_1 = 1.455$, $n_2 = 1.440$, light wavelength of 900 nm. Given the waveguide condition in Eq. (3) and the expression for ϕ in TIR for the TE mode (Ch. 1),

$$\tan\left(\tfrac{1}{2}\phi_m\right) = \frac{\left[\sin^2\theta_m - \left(\dfrac{n_2}{n_1}\right)^2\right]^{1/2}}{\cos\theta_m}$$

using a graphical solution, find angles θ_m for all the modes . What is your conclusion?

Solution Consider the waveguide condition in Eq. (3), using $k_1\cos\theta_m = \kappa$, can be written as

$$(2a)k_1\cos\theta_m - m\pi = \phi_m$$

i.e.

$$\tan\left(ak_1\cos\theta_m - m\frac{\pi}{2}\right) = \frac{\left[\sin^2\theta_m - \left(\dfrac{n_2}{n_1}\right)^2\right]^{1/2}}{\cos\theta_m} = f(\theta_m) \qquad \textbf{(11)}$$

The left-hand side (LHS) simply reproduces itself whenever $m = 0, 2, 4, \ldots$ even. It becomes a cotangent function whenever $m = 1, 3, \ldots$ odd integer. The solutions therefore fall into odd and even m categories. Figure 2.9 shows the right-hand side (RHS), $f(\theta_m)$ vs. θ_m as well as the LHS, $\tan(ak_1\cos\theta_m - m\pi/2)$. Since the critical angle, $\theta_c = \arcsin(n_2/n_1) = 81.77°$, we can only find solutions in the range $\theta_m = 81.77° - 90°$. For example, the intersections for $m = 0$ and $m = 1$ are at 89.17°, 88.34°. We can also calculate the penetration δ_m of the field into the cladding using,

$$\frac{1}{\delta_m} = \alpha_m = \frac{2\pi n_2\left[\left(\dfrac{n_1}{n_2}\right)^2\sin^2\theta_m - 1\right]^{1/2}}{\lambda} \qquad \textbf{(12)}$$

Attenuation in cladding

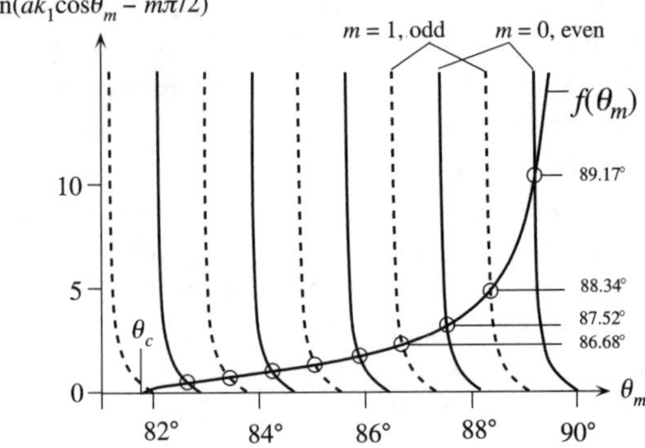

FIGURE 2.9 Modes in a planar dielectric waveguide can be determined by plotting the LHS and the RHS of Eq. (11).

Using Eqs. (11) and (12) we find the following mode angles θ_m and corresponding penetrations δ_m of the field into the cladding. The highest mode (highest m) has substantial field penetration into the cladding.

m	0	1	2	3	4	5	6	7	8	9
θ_m	89.2°	88.3°	87.5°	86.7°	85.9°	85.0°	84.2°	83.4°	82.6°	81.9°
δ_m (μm)	0.691	0.702	0.722	0.751	0.793	0.866	0.970	1.15	1.57	3.83

It is apparent that $f(\theta_m)$ intersects the LHS tangent function once for each choice of m until $\theta_m \leq \theta_c$. There are 10 modes.

An accurate solution of Eq. (11) would show that the fundamental mode angle for the TE mode is actually 89.172°. If we were to use the phase change ϕ_m for the TM mode from Chapter 1, we would find an angle 89.170°, which is almost identical to ϕ_m for the TE mode.

EXAMPLE 2.1.2 *V*-number and the number of modes

Using Eq. (9), *estimate* the number of modes that can be supported in a planar dielectric waveguide that is 100 μm wide and has $n_1 = 1.490$ and $n_2 = 1.470$ at the free-space source wavelength (λ), which is 1 μm. Compare your estimate with the formula

Number of modes

$$M = \text{Int}\left(\frac{2V}{\pi}\right) + 1 \tag{13}$$

in which $\text{Int}(x)$ is the integer function; it drops the decimal fraction of x.

Solution The phase change ϕ on TIR cannot be more than π so ϕ/π is less than 1. For a multimode guide ($V > 1$), we can write Eq. (9) as

$$m \leq \frac{2V - \phi}{\pi} \approx \frac{2V}{\pi}$$

We can calculate V since we are given $a = 50$ μm, $n_1 = 1.490$, $n_2 = 1.470$, and $\lambda = 1 \times 10^{-6}$ m or 1 μm,

$$V = (2\pi a/\lambda)(n_1^2 - n_2^2)^{1/2} = 76.44.$$

Then, $m \le 2(76.44)/\pi = 48.7$, or $m \le 48$. There are about 49 modes, because we must also include $m = 0$ as a mode. Using Eq. (13),

$$M = \text{Int}\left[\frac{2(76.44)}{\pi}\right] + 1 = 49$$

EXAMPLE 2.1.3 Mode field distance (MFD), $2w_o$

The field distribution along y penetrates into the cladding as depicted in Figure 2.5. The extent of the electric field across the guide is therefore more than $2a$. Within the core, the field distribution is harmonic whereas from the boundary into the cladding, the field is due to the evanescent wave and it decays exponentially according to

$$E_{\text{cladding}}(y') = E_{\text{cladding}}(0)\exp(-\alpha_{\text{cladding}}y')$$

in which $E_{\text{cladding}}(y')$ is the field in the cladding at a position y' measured from the boundary and α_{cladding} is the decay constant (or attenuation) for the evanescent wave in medium 2, which is given by[5]

$$\alpha_{\text{cladding}} = \frac{2\pi n_2}{\lambda}\left[\left(\frac{n_1}{n_2}\right)^2 \sin^2\theta_i - 1\right]^{1/2}$$

in which λ is the wavelength in free space. For the axial mode we can take the approximation $\theta_i \to 90°$

$$\alpha_{\text{cladding}} = \frac{2\pi n_2}{\lambda}\left[\left(\frac{n_1}{n_2}\right)^2 \sin^2\theta_i - 1\right]^{1/2} \approx \frac{2\pi}{\lambda}(n_1^2 - n_2^2)^{1/2} = \frac{V}{a}$$

The field in the cladding decays by a factor of e^{-1} when $y' = \delta = 1/\alpha_{\text{cladding}} = a/V$. The extent of the field in the cladding is about δ. The total extent of the field across the whole guide is therefore $2a + 2\delta$ which is called the **mode field distance** (MFD) and denoted by $2w_o$. Thus

$$2w_o \approx 2a + 2\frac{a}{V}$$

i.e.

$$2w_o \approx 2a\frac{(V + 1)}{V} \qquad \textbf{(14)}$$

Mode field distance

We note that as V increases, MFD becomes the same as the core thickness, $2a$. In single mode operation $V < \pi/2$ and MFD is considerably larger than $2a$. In fact at $V = \pi/2$, MFD is 1.6 times $2a$. In the case of cylindrical dielectric waveguides, e.g. optical fibers, MFD is called the **mode field diameter**.

[5] See Chapter 1.

2.2 MODAL AND WAVEGUIDE DISPERSION IN THE PLANAR WAVEGUIDE

A. Waveguide Dispersion Diagram

The propagating modes that exist in a slab waveguide are determined by the waveguide condition. Each choice of m from 0 to its maximum value results in one distinct solution and one possible propagation constant β_m. We see that each mode propagates with a different propagation constant even if illumination is by monochromatic radiation. Examination of Figure 2.7 gives the impression that the axial ray has the least reflections and therefore seems to arrive more quickly than a higher-mode ray. Higher modes zigzag more along the guide and appear to have longer ray paths. However, there are two important wrong impressions with this view. First, what is important is the group velocity v_g along the guide, the velocity at which the energy or information is transported. Second is that the higher modes penetrate more into the cladding where the refractive index is smaller and the waves travel faster.

We know from Chapter 1 that the group velocity v_g depends on $d\omega/d\beta$ in which ω is the frequency[6] and β is the propagation constant. For each mode m, the mode angle θ_m is determined by the *waveguide condition* that depends on the wavelength, hence on the frequency ω, and the waveguide properties (n_1, n_2 and a) as apparent from Eq. (3) in §2.1. Thus, $\theta_m = \theta_m(\omega)$ and consequently, $\beta_m = k_1 \sin\theta_m = \beta_m(\omega)$, a function of ω. The group velocity of a given mode is therefore a function of the light frequency and the waveguide properties.

We see that even if the refractive index were constant (independent of frequency or wavelength), the group velocity v_g of a given mode would still depend on the frequency by virtue of the guiding properties of the waveguide structure.

From the waveguide condition [Eq. (3) in §2.1], given the refractive indices (n_1 and n_2) and the guide dimension (a), we can calculate β_m for each ω and for each mode m to obtain the ω vs. β_m characteristics, which is called the **dispersion diagram**, as shown schematically in Figure 2.10. The *slope* $d\omega/d\beta_m$ at any frequency ω is the group velocity v_g. All allowed propagation modes are contained within the lines with slopes c/n_1 and c/n_2. The cut-off frequency $\omega_{\text{cut-off}}$ corresponds to the cut-off condition $(\lambda = \lambda_c)$ when $V = \pi/2$, and $\omega > \omega_{\text{cut-off}}$ leads to more than one mode. Examination of Figure 2.10 shows two immediate consequences. First, the group velocity at one frequency changes from one mode to another. Second, for a given mode it changes with the frequency.

B. Intermodal dispersion

In multimode operation, well above $\omega_{\text{cut-off}}$, the lowest mode ($m = 0$) has the slowest group velocity, close to c/n_1, and the highest mode has the highest group velocity. The reason is that a good portion of the field in higher modes is carried by the cladding where the refractive index is smaller (examine Figure 2.6). The lowest mode is contained substantially in the core. The modes therefore take different times to travel the length

[6] The term frequency implies angular frequency as it refers to ω.

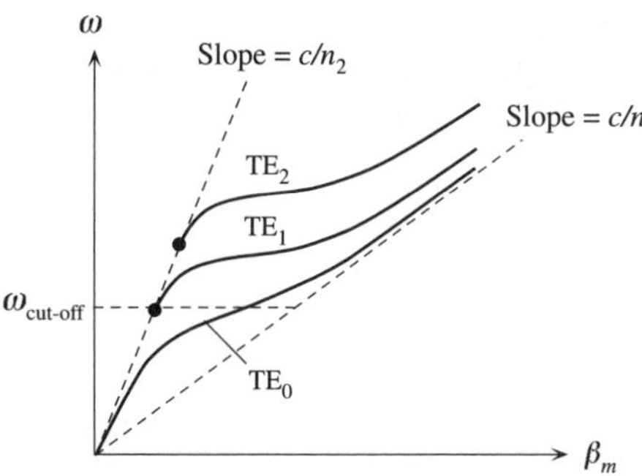

FIGURE 2.10 Schematic dispersion diagram, ω vs. β for the slab waveguide for various TE_m modes. $\omega_{\text{cut-off}}$ corresponds to $V = \pi/2$. The group velocity v_g at any ω is the slope of the ω vs. β curve at that frequency.

of the fiber. This phenomenon is called **modal dispersion** (*or* **intermodal dispersion**). A direct consequence of modal dispersion is that, if a short duration light pulse signal is coupled into the dielectric waveguide, then this pulse will travel along the guide via the excitation of various allowed modes. These modes will travel at different group velocities. The reconstruction of the light pulse at the receiving end from the various modes will result in a *broadened signal* as depicted in Figure 2.7. It is clear that in a single mode waveguide in which only one mode can propagate (the $m = 0$ mode), there will be no modal dispersion.

To evaluate the modal dispersion of a signal along the waveguide, we need to consider the shortest and longest times required for the signal to traverse the waveguide, which is tantamount to identifying the slowest and fastest modes excited in terms of their group velocities. If $\Delta\tau$ is the propagation time difference between the fastest and slowest modes over a distance L, then modal dispersion is defined by

$$\Delta\tau = \frac{L}{v_{g\min}} - \frac{L}{v_{g\max}} \qquad \textbf{(1)}$$

Modal dispersion

in which $v_{g\min}$ is the minimum group velocity of the slowest mode and $v_{g\max}$ is the maximum group velocity of the fastest mode.

The lowest order mode ($m = 0$), when $\omega > \omega_{\text{cut-off}}$, as apparent in Figure 2.10, has a group velocity $v_{g\min} \approx c/n_1$. The fastest propagation corresponds to the highest order mode, which has a group velocity very roughly c/n_2. Thus, approximately,

$$\frac{\Delta\tau}{L} \approx \frac{n_1 - n_2}{c} \qquad \textbf{(2)}$$

Modal dispersion

Equation (2) considers only two extreme modes, the lowest and the highest, and does not consider whether some intermediate modes can have group velocities falling outside the range c/n_1 to c/n_2 (consider the slope $d\omega/d\beta_m$ in Figure 2.10). Neither does it consider how the light energy is proportioned between various modes. Taking $n_1 = 1.48$ (core) and $n_2 = 1.46$ (cladding), we find $\Delta\tau/L \approx 6.7 \times 10^{-11}$ s m^{-1} or 67 ns km^{-1}. In general, intermodal dispersion is not as high as indicated by this estimate due to an "intermode coupling" the discussion of which is beyond the scope of this book.

The spread $\Delta\tau$ in Eq. (2) is between the two extremes of the broadened output light pulse. In optoelectronics we are frequently interested in the spread $\Delta\tau_{1/2}$ between the half intensity points that is smaller than the full width. The determination of $\Delta\tau_{1/2}$ depends on the temporal shape of the output light pulse, but as a first order approximation, when many modes are present, we can take $\Delta\tau_{1/2} \approx \Delta\tau$.

C. Intramodal Dispersion

Figure 2.10 shows that the lowest mode ($m = 0$) has a group velocity that depends on the frequency ω or wavelength λ. Thus, even if we use the guide in single mode operation, as long as the excitation source has a finite spectrum, it will contain various frequencies (there is no such thing as a perfect monochromatic light wave). These frequencies will then travel with different group velocities, as apparent in Figure 2.10, and hence arrive at different times. The higher the wavelength (lower the frequency), the greater the penetration of the field into the cladding, as depicted in Figure 2.11. Thus, a greater portion of the light energy is carried by the cladding in which the phase velocity is higher. Longer wavelengths propagate faster, even though by the same mode. This is called **waveguide dispersion** inasmuch as it results from the guiding properties of the dielectric structure and it has nothing to do with the frequency (or wavelength) dependence of the refractive index. Since increasing the wavelength decreases the V-number, a characteristic property of the guide, the dispersion can also be stated as due to the wavelength dependence of the V-number. There is no simple calculation for the waveguide dispersion because we have to incorporate the spectrum of the input light and calculate the group velocity from the dispersion diagram. This is discussed for optical fibers in §2.3.

The refractive index of the guide material will also depend on the wavelength and thus modify the ω–β_m behavior in Figure 2.10. The change in the group velocity of a given mode due to the n–λ dependence also gives rise to the broadening of a propagating light pulse. This is called **material dispersion**. Thus both waveguide and material dispersion act together to broaden a light pulse propagating within a given mode. Combined dispersion is called **intramode dispersion**.

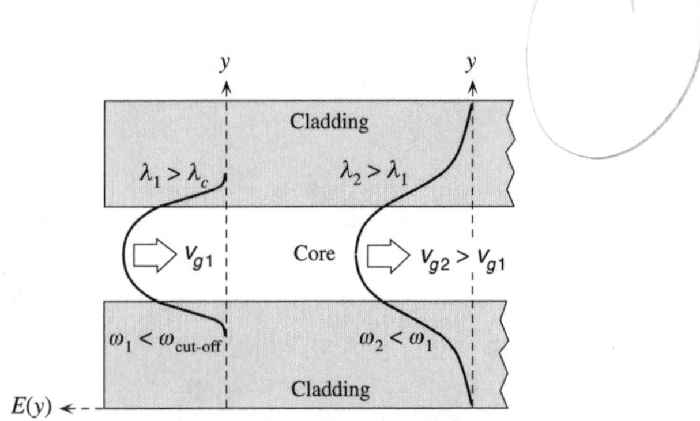

FIGURE 2.11 The electric field of TE_0 mode extends more into the cladding as the wavelength increases. As more of the field is carried by the cladding, the group velocity increases.

2.3 STEP INDEX FIBER

Reeled optical fibers on a drum. *(Courtesy of Corning.)*

The general ideas for guided wave propagation in a planar dielectric waveguide can be readily extended, with certain modifications, to the step indexed optical fiber shown in Figure 2.12. This is essentially a cylindrical dielectric waveguide with the inner core dielectric having a refractive index n_1 greater than n_2 of the outer dielectric, cladding. The **normalized index difference** Δ is defined by

$$\Delta = (n_1 - n_2)/n_1 \quad (1)$$

Normalized index difference

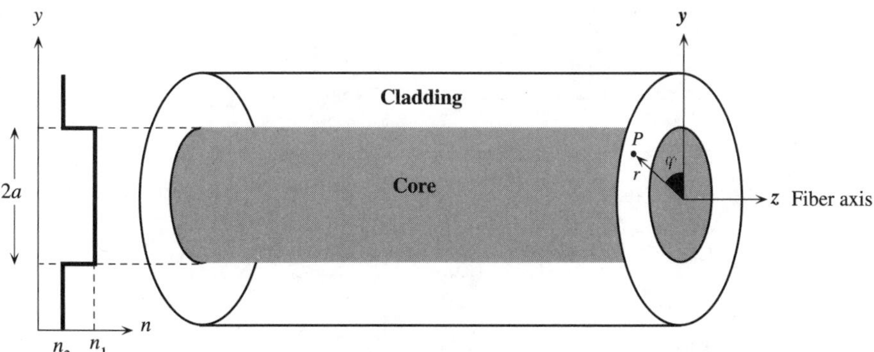

FIGURE 2.12 The step index optical fiber (schematic). The central region, the core, has greater refractive index than the outer region, the cladding. The fiber has cylindrical symmetry. The coordinates r, φ, z are used to represent any point P in the fiber. Cladding is normally much thicker than shown.

For all practical fibers used in optical communications, n_1 and n_2 differ only by a small amount, typically less than a few percent so that $\Delta \ll 1$.

We recall that the planar waveguide is bounded only in one dimension so that reflections occur only in the y-direction. The requirement of constructive interference of waves then leads to the existence of distinct modes each labeled by m. The cylindrical guide in Figure 2.12 is bounded in two dimensions and the reflections occur from all the surfaces, *i.e.* from a surface encountered along any radial direction r; along a radial direction at any angle φ to the y-axis in Figure 2.12. Since any radial direction can be represented in terms of x and y, reflections in both x and y directions are involved in constructive interference of waves and we therefore need two integers, l and m, to label all the possible traveling waves or guided modes that can exist in the guide.

We also recall that in the case of a planar waveguide we visualized a guided propagating wave in terms of rays that were zigzagging down the guide and all these rays necessarily passed through the axial plane of the guide. Further, all waves were either TE (transverse electric) or TM (transverse magnetic). A distinctly different feature of the step index fiber from the planar waveguide is existence of rays that zigzag down the fiber without necessarily crossing the fiber axis, so called skew rays. A **meridional ray** enters the fiber through the fiber axis and hence also crosses the fiber axis on each reflection as it zigzags down the fiber. It travels in a plane that contains the fiber axis as illustrated in Figure 2.13 (a). On the other hand, a **skew ray** enters the fiber off the fiber axis and zigzags down the fiber without crossing the axis. When viewed looking down the fiber (its projection in a plane normal to the fiber axis), it traces out a polygon around the fiber axis as shown in Figure 2.13 (b). A skew ray therefore has a *helical path* around the fiber axis. In a step index fiber both meridional and skew rays give rise to guided modes (propagating waves) along the fiber, each with a propagation constant β along z. Guided modes resulting from meridional rays are either TE or TM type as in the case

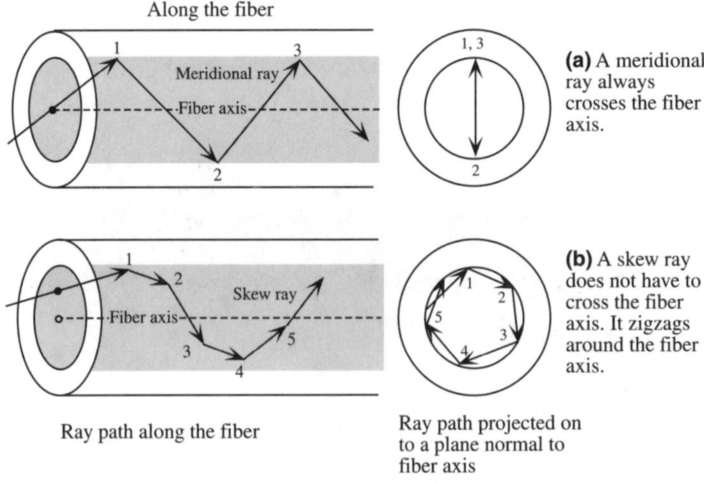

FIGURE 2.13 Illustration of the difference between a meridional ray and a skew ray. Numbers represent reflections of the ray.

of the planar waveguide. Skew rays, on the other hand, give rise to modes that have both E_z and B_z (or H_z) components and are therefore not TE or TM waves. They are called HE or EH modes as both electric and magnetic field can have components along z. They are called *hybrid modes*. It is apparent that guided modes in a step index fiber cannot be as easily described as those in the planar guide.

Guided modes in a step index fiber with $\Delta \ll 1$ (called *weakly guiding fibers*) are generally visualized by traveling waves that are almost plane polarized. They have transverse electric and magnetic fields (**E** and **B** are perpendicular to each other and also to z), analogous to field directions in a plane wave but the field magnitudes are not constant in the plane. These waves are called **linearly polarized** (LP) and have transverse electric and magnetic field characteristics. A guided LP mode along the fiber can be represented by the propagation of an electric field distribution $E(r, \varphi)$ along z. This field distribution, or pattern, is in the plane normal to the fiber axis and hence depends on r and φ but not on z. Further, because of the presence of two boundaries, it is characterized by two integers, l and m. The propagating field distribution in an LP mode is therefore given by $E_{lm}(r, \varphi)$ and we represent the mode as LP_{lm}. Thus, an LP_{lm} mode can be described by a traveling wave along z of the form,

$$E_{LP} = E_{lm}(r, \varphi) \exp j(\omega t - \beta_{lm} z) \tag{1}$$

in which E_{LP} is the field of the LP mode and β_{lm} is its propagation constant along z. It is apparent that for a given l and m, $E_{lm}(r, \varphi)$ represents a particular field pattern at a position z that is propagated along the fiber with an effective wavevector β_{lm}.

Figure 2.14 displays the electric field pattern (E_{01}) in the **fundamental mode** of the step index fiber that corresponds to $l = 0$ and $m = 1$, the LP_{01} mode. The field is maximum at the center of the core (or fiber axis) and penetrates somewhat into the cladding due to the accompanying evanescent wave. The extent of penetration depends on the V-number of the fiber (and hence on the wavelength). The light intensity in a mode is proportional to E^2, which means that the intensity distribution in the LP_{01} mode

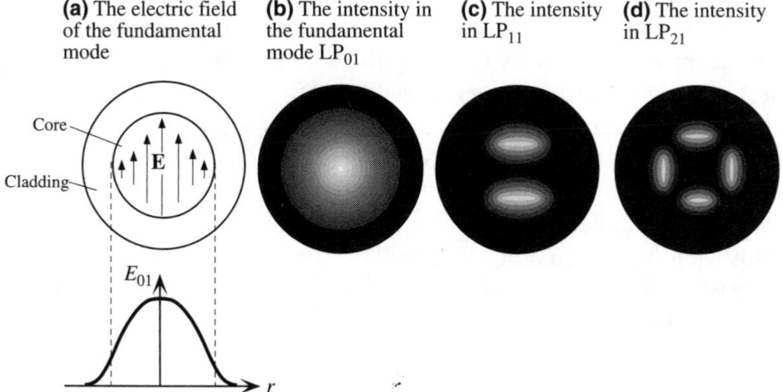

FIGURE 2.14 The electric field distribution of the fundamental mode in the transverse plane to the fiber axis z. The light intensity is greatest at the center of the fiber. Intensity patterns in LP_{01}, LP_{11} and LP_{21} modes.

has a maximum along the fiber axis as visualized in Figure 2.14; brightest zone at the center and brightness decreasing towards the cladding. Figure 2.14 also shows the intensity distributions in the LP_{11} and LP_{21} modes. The integers l and m are related to the intensity pattern in a LP_{lm} mode. There are m number of maxima along r starting from the core center and $2l$ number of maxima around a circumference as apparent in Figure 2.14. Further, within the ray picture, l represents the extent of helical propagation, or the amount of skew ray contribution to the mode. In the fundamental mode this is zero. Moreover, m is directly associated with the reflection angle θ of the rays as in the planar guide.

We see from the discussions above that light travels along the fiber through various modes of propagation with each mode having its own propagation vector β_{lm} and its own electric field pattern $E_{lm}(r, \varphi)$. Each mode has its own group velocity $v_g(l, m)$ that depends on the ω vs. β_{lm} dispersion behavior. When a light pulse is fed into the fiber, it travels down the fiber through various modes. These modes propagate with different group velocities and therefore emerge at the end of the fiber with a spread of arrival times, which means that the output pulse is a broadened version of the input pulse. As in the case of planar guide this broadening of the light pulse is an **intermodal dispersion** phenomenon. We can, however, design a suitable fiber that allows only the fundamental mode to propagate and hence exhibits no modal dispersion.

For a step index fiber, we define a ***V*-number**, or a **normalized frequency**, in a similar fashion to the planar waveguide,

V-number definition

$$V = \frac{2\pi a}{\lambda}(n_1^2 - n_2^2)^{1/2} = \frac{2\pi a}{\lambda}(2n_1 n\Delta)^{1/2} \quad \textbf{(2)}$$

in which a is the radius of the fiber core, λ is the *free space wavelength*, n is the average refractive index of the core and cladding, *i.e.* $n = (n_1 + n_2)/2$, and Δ is the **normalized index difference**, that is,

Normalized index difference

$$\Delta = (n_1 - n_2)/n_1 \approx (n_1^2 - n_2^2)/2n_1^2. \quad \textbf{(3)}$$

When the V-number is less than 2.405, it can be shown that only one mode, the fundamental mode (LP_{01}) shown in Figure 2.14 (a) and (b), can propagate through the fiber core. As V is decreased further by reducing the core size, the fiber can still support the $LP_{0,1}$ mode, but the mode extends increasingly into the cladding. The finite cladding size may then result in some of the power in the wave being lost. A fiber that is designed (by the choice of a and Δ) to allow only the fundamental mode to propagate at the required wavelength is called a **single mode fiber**. Typically, single mode fibers have a much smaller core radius and a smaller Δ than multimode fibers. If the wavelength λ of the source is reduced sufficiently, a single mode fiber will become multimode as V will exceed 2.405; higher modes will also contribute to propagation. The cut-off wavelength λ_c above which the fiber becomes single mode is given by

Single mode cut-off wavelength

$$V_{\text{cut-off}} = \frac{2\pi a}{\lambda_c}(n_1^2 - n_2^2)^{1/2} = 2.405 \quad \textbf{(4)}$$

When the V-parameter increases above 2.405, the number of modes rises sharply. A good approximation to the number of modes M in a step index multimode fiber is given by

$$M \approx \frac{V^2}{2} \qquad (5)$$

Number of modes when $V \gg 2.405$

The effect of varying the various physical parameters of a step-index optical fiber on the number of propagating modes can be readily deduced from the V number in Eq. (2). For example, increasing the core radius (a) or its refractive index (n_1) increases the number of modes. On the other hand, increasing the wavelength or the cladding refractive index (n_2) decreases the number of modes. The cladding diameter does not enter the V-number equation and so it may be thought that it plays no significant role in wave propagation. In multimode fibers, light propagates through many modes and these are mainly all confined to the core. In a step index fiber, the field of the fundamental mode penetrates into the cladding as an evanescent wave traveling along the boundary. If the cladding is not sufficiently thick, this field will reach the end of the cladding and escape, resulting in intensity loss. For example, typically, the cladding diameter for a single-mode step-index fiber is at least 10 times the core diameter.

Since the propagation constant β_{lm} of an LP mode depends on the waveguide properties and the source wavelength, it is convenient to describe light propagation in terms of a "normalized" propagation constant that depends only on the V-number. Given $k = 2\pi/\lambda$ and guide indices n_1 and n_2, the **normalized propagation constant** b is related to $\beta = \beta_{lm}$ by the definition,

$$b = \frac{(\beta/k)^2 - n_2^2}{n_1^2 - n_2^2} \qquad (6)$$

Normalized propagation constant

According to the definition above, the lower limit $b = 0$ corresponds to $\beta = kn_2$, propagation in the cladding material, and the upper limit $b = 1$ corresponds to $\beta = kn_1$, propagation in the core material. The dependence of b on the V-number has been calculated in the literature for various modes as shown in Figure 2.15 for a few lower order LP modes. Notice that the fundamental mode (LP_{01}) exists for all V-numbers and LP_{11} is cut-off at $V = 2.405$. There is a cut-off V and a corresponding cut-off wavelength for each particular LP mode higher than the fundamental mode. Given the V-parameter of the fiber we can easily find b, and hence β, for the allowed LP modes from Figure 2.15.

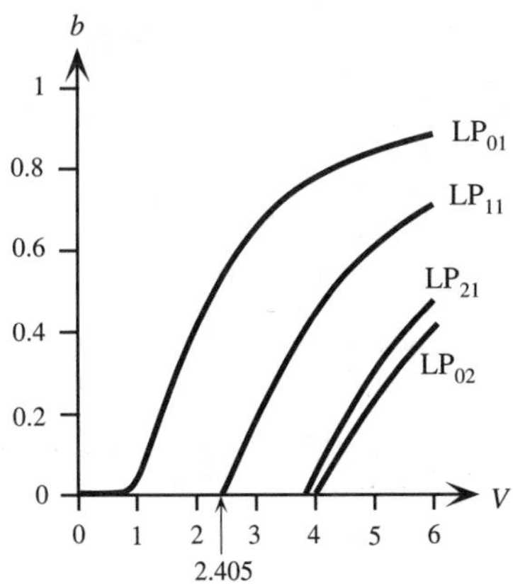

FIGURE 2.15 Normalized propagation constant b vs. V-number for a step index fiber for various LP modes.

EXAMPLE 2.3.1 A multimode fiber

Calculate the number of allowed modes in a multimode step index fiber that has a core of refractive index of 1.468 and diameter 100 μm, and a cladding of refractive index of 1.447 if the source wavelength is 850 nm.

Solution The V-number for this fiber can be calculated by substituting $a = 50$ μm, $\lambda = 0.850$ μm, $n_1 = 1.468$, and $n_2 = 1.447$ into the expression for the V-number,

$$V = (2\pi a/\lambda)(n_1^2 - n_2^2)^{1/2} = (2\pi 50/0.850)(1.468^2 - 1.447^2)^{1/2} = 91.44.$$

Since $V \gg 2.405$, the number of modes is

$$M \approx V^2/2 = (91.44)^2/2 = 4181$$

which is large. There are numerous modes.

EXAMPLE 2.3.2 A single mode fiber

What should be the core radius of a single mode fiber that has a core of $n_1 = 1.468$, cladding of $n_2 = 1.447$, and is to be used with a source wavelength of 1.3 μm?

Solution Single mode propagation is achieved when $V \leq 2.405$. Thus we need,

$$V = (2\pi a/\lambda)(n_1^2 - n_2^2)^{1/2} \leq 2.405,$$

or

$$[2\pi a/(1.3\ \mu\text{m})](1.468^2 - 1.447^2)^{1/2} \leq 2.405$$

which gives $a \leq 2.01$ μm. As suspected, this is rather thin for easy coupling of the fiber to a light source or to another fiber and special coupling techniques must be used. We also note that the size of a is comparable in magnitude to the wavelength, which means that the geometric ray picture, strictly, cannot be used to describe light propagation.

EXAMPLE 2.3.3 Single mode cut-off wavelength

Calculate the cut-off wavelength for single mode operation for a fiber that has a core with diameter of 7 μm, a refractive index of 1.458, and a cladding of refractive index of 1.452. What is the V-number and the mode field diameter (MFD) when operating at $\lambda = 1.3$ μm?

Solution For single mode operation,

$$V = (2\pi a/\lambda)(n_1^2 - n_2^2)^{1/2} \leq 2.405$$

Substituting for a, n_1 and n_2 and rearranging we get,

$$\lambda > [2\pi(3.5\ \mu\text{m})(1.458^2 - 1.452^2)^{1/2}]/2.405 = 1.208\ \mu\text{m}$$

Wavelengths shorter than 1.028 μm will result in a multimode propagation.
At $\lambda = 1.3$ μm,

$$V = 2\pi[(3.5\ \mu\text{m})/(1.3\ \mu\text{m})](1.458^2 - 1.452^2)^{1/2} = 2.235$$

The mode field diameter MFD is then,

$$2w_o \approx 2a(V + 1)/V = (7\ \mu\text{m})(2.235 + 1)/2.235 = 10.13\ \mu\text{m}.$$

EXAMPLE 2.3.4 Group velocity and delay

Consider a single mode fiber with core and cladding indices of 1.448 and 1.440, core radius of 3 μm, operating at 1.5 μm. Given that we can approximate the fundamental mode normalized propagation constant b by[7]

$$b \approx \left(1.1428 - \frac{0.996}{V}\right)^2 \qquad (1.5 < V < 2.5) \tag{7}$$

calculate the propagation constant β. Change the operating wavelength to λ' by a small amount, say 0.01%, and then recalculate the new propagation constant β'. Then determine the group velocity v_g of the fundamental mode at 1.5 μm, and the group delay τ_g over 1 km of fiber.

Solution Equation (6) for a weakly guiding fiber can also be written as

$$b = \frac{(\beta/k) - n_2}{n_1 - n_2} \qquad i.e. \qquad \beta = n_2 k[1 + b\Delta] \tag{8}$$

We can calculate V from the fiber properties that are given, and then use Eq. (7) to calculate b. From b, using $k = 2\pi/\lambda$ in Eq. (8), we can calculate β. In addition, $\omega = 2\pi c/\lambda$. Thus, $V = (2\pi a/\lambda)(n_1^2 - n_2^2)^{1/2} = 1.910088$. From Eq. (7) $b = 0.3860859$, and from Eq. (8), $\beta = 6.044796 \times 10^6$ m^{-1}. Calculations are summarized in Table 2.1. The group velocity is

$$v_g = \frac{\omega' - \omega}{\beta' - \beta} = \frac{(1.256511 - 1.256624) \times 10^{15}}{(6.044189 - 6.044796) \times 10^6} = 2.0714 \times 10^8 \text{ m s}^{-1}.$$

The group delay τ_g over 1 km is 4.83 μs.

TABLE 2.1

Calculation →	V	k (m^{-1})	ω (rad s^{-1})	b	β (m^{-1})
$\lambda = 1.500000$ μm	1.910088	4188790	1.256624×10^{15}	0.3860859	6.044796×10^6
$\lambda' = 1.50015$ μm	1.909897	4188371	1.256511×10^{15}	0.3860211	6.044189×10^6

2.4 NUMERICAL APERTURE

Not all source radiation can be guided along an optical fiber. Only rays falling within a certain cone at the input of the fiber can normally be propagated through the fiber. Figure 2.16 shows the path of a light ray launched from the outside medium of refractive index n_o (not necessarily air) into the fiber core. Suppose that the incidence angle at the end of the fiber core is α, and inside the waveguide the ray makes an angle θ with the normal to the fiber axis. Then unless the angle θ is greater than the critical angle θ_c

[7] See J. Senior, *Optical Fiber Communications, Principles and Practice, Second Ed.* (Prentice-Hall, New York, 1992.), p. 65

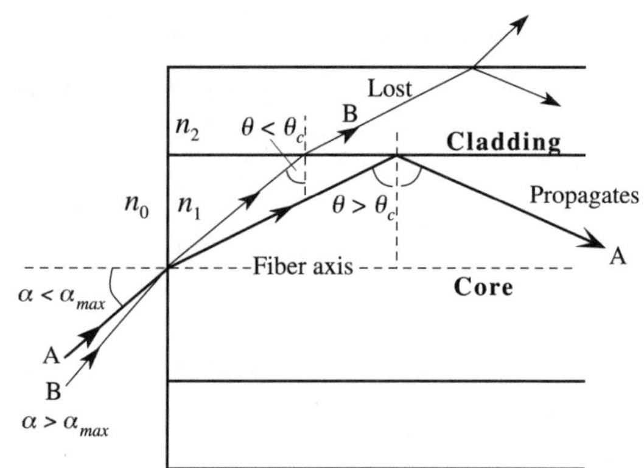

FIGURE 2.16 Maximum acceptance angle α_{max} is that which just gives total internal reflection at the core-cladding interface, i.e. when $\alpha = \alpha_{max}$ then $\theta = \theta_c$. Rays with $\alpha > \alpha_{max}$ (e.g. ray B) become refracted and penetrate the cladding and are eventually lost.

for TIR, the ray will escape into the cladding. Thus for light propagation, the launching angle α has to be such that TIR is supported within the fiber. From Figure 2.16, it should be apparent that the maximum value of α is that which results in $\theta = \theta_c$. At the n_o/n_1 interface, Snell's law gives

$$\sin \alpha_{max}/\sin(90° - \theta_c) = n_1/n_o$$

in which θ_c is determined by the onset of TIR, that is $\sin \theta_c = n_2/n_1$. We can now eliminate θ_c to obtain,

$$\sin \alpha_{max} = \frac{(n_1^2 - n_2^2)^{1/2}}{n_o}$$

The **numerical aperture** NA is a characteristic parameter of an optical fiber defined by

Definition of numerical aperture, NA

$$NA = (n_1^2 - n_2^2)^{1/2} \quad \textbf{(1)}$$

so that in terms of NA, the **maximum acceptance angle**, α_{max}, becomes,

Maximum acceptance angle

$$\sin \alpha_{max} = \frac{NA}{n_o} \quad \textbf{(2)}$$

Meridional rays only

The angle $2\alpha_{max}$ is called the **total acceptance angle** and depends on the NA of the fiber and the refractive index of the launching medium. NA is an important factor in light launching designs into the optical fiber. We should note that Eq. (2) is strictly applicable for meridional rays. Skew rays have a wider acceptance angle. Since NA is defined in terms of the refractive indices, we can obtain a relationship between the V-number and NA,

V-number and NA

$$V = \frac{2\pi a}{\lambda} NA \quad \textbf{(3)}$$

Multi-mode propagation in a fiber involves many modes, the majority of which are usually skew rays. The acceptance angle is larger for these skew rays than that predicted by Eq. (2), which is for meridional rays.

EXAMPLE 2.4.1 A multimode fiber and total acceptance angle

A step index fiber has a core diameter of 100 μm and a refractive index of 1.480. The cladding has a refractive index of 1.460. Calculate the numerical aperture of the fiber, acceptance angle from air, and the number of modes sustained when the source wavelength is 850 nm.

Solution The numerical aperture is

$$NA = \left(n_1^2 - n_2^2\right)^{1/2} = \left(1.480^2 - 1.460^2\right)^{1/2} = 0.2425.$$

From $\sin\alpha_{\max} = NA/n_o = 0.2425/1$, the acceptance angle is $\alpha_{\max} = 14°$ and the total acceptance angle is 28°.

The V-number in terms of the numerical aperture can be written as,

$$V = (2\pi a/\lambda)NA = \left[(2\pi 50)/(0.85)\right](0.2425) = 89.62.$$

The number of modes, $M \approx V^2/2 = 4016$.

EXAMPLE 2.4.2 A single mode fiber

A typical single mode optical fiber has a core of diameter 8 μm and a refractive index of 1.46. The normalized index difference is 0.3%. The cladding diameter is 125 μm. Calculate the numerical aperture and the acceptance angle of the fiber. What is the single mode cut-off wavelength λ_c of the fiber?

Solution The numerical aperture is

$$NA = \left(n_1^2 - n_2^2\right)^{1/2} = \left[(n_1 + n_2)(n_1 - n_2)\right]^{1/2}$$

Substituting $\left(n_1 - n_2\right) = n_1\Delta$ and $\left(n_1 + n_2\right) \approx 2n_1$, we get

$$NA \approx \left[(2n_1)(n_1\Delta)\right]^{1/2} = n_1(2\Delta)^{1/2} = 1.46(2 \times 0.003)^{1/2} = 0.113$$

The acceptance angle is given by

$$\sin\alpha_{\max} = NA/n_o = 0.113/1 \text{ or } \alpha_{\max} = 6.5°.$$

The condition for single mode propagation is $V \leq 2.405$, which corresponds to a minimum wavelength λ_c given by

$$\lambda_c = [2\pi a NA]/2.405 = \left[(2\pi 4\ \mu\text{m})(0.113)\right]/2.405 = 1.18\ \mu\text{m}.$$

Illumination wavelengths shorter than 1.18 μm will result in a multimode operation.

2.5 DISPERSION IN SINGLE MODE FIBERS

A. Material Dispersion

Since only one mode propagates in a single mode, step-index fiber, it should be apparent that there is no intermodal dispersion of an input light pulse. This is the advantage of the single mode fiber. However, even if light propagation occurs by a single mode, there will still be dispersion due to the variation of the refractive index n_1 of the core glass with the wavelength of light coupled into the fiber. The propagation velocity of the guided wave along the fiber core depends on n_1, which in turn depends on the wavelength. This type of dispersion that results from the wavelength dependence of the material

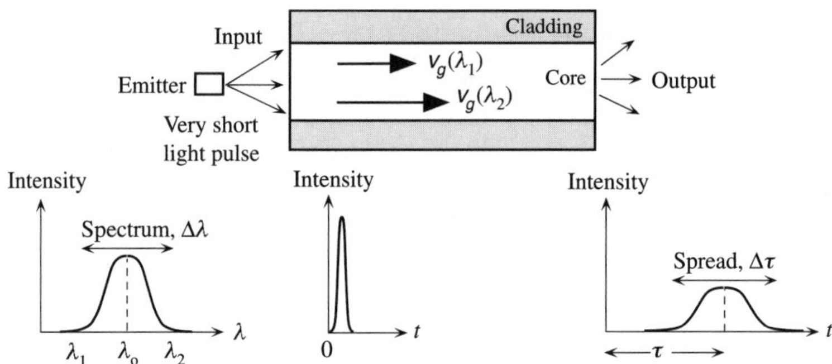

FIGURE 2.17 All excitation sources are inherently non-monochromatic and emit within a spectrum, $\Delta\lambda$, of wavelengths. Waves in the guide with different free space wavelengths travel at different group velocities due to the wavelength dependence of n_1. The waves arrive at the end of the fiber at different times and hence result in a broadened output pulse.

properties of the guide is called **material dispersion.** Its manifestation is the result of no practical light source being perfectly monochromatic so that there are waves in the guide with various free space wavelengths, that is, a range of λ values. For example, even excitation from a laser will result in a spectrum of source wavelengths fed into the fiber as schematically illustrated in Figure 2.17. The wavelength width $\Delta\lambda$ of the spectrum will depend on the nature of the light source but it is never zero. Each radiation with a different λ will propagate, via the fundametal mode, at a different group velocity v_g because the group index N_g of the medium (related to n_1) depends on the wavelength. The waves will therefore arrive at different times at the end of the fiber and hence will give rise to a broadened output light pulse as depicted in Figure 2.17. The group index N_g, for the silica glass (Chapter 1) is almost constant at wavelengths around 1.3 μm, which implies zero material dispersion around this wavelength.

Material dispersion can be evaluated by considering the dependence of the group velocity v_g on the wavelength through the N_g vs. λ behavior, and the spectrum of the light source that defines the range $\Delta\lambda$ of emitted wavelengths from the light source. We consider a very short-duration light pulse as the input signal with a spectrum of wavelengths over $\Delta\lambda$ as in Figure 2.17. Then the output is a pulse that is broadened by $\Delta\tau$ due to the spread in the arrival times τ of the waves. $\Delta\tau$ increases with the fiber length L because slower waves fall further behind the faster waves over a longer distance. Dispersion is therefore normally expressed as spread per unit length and is given by

Material dispersion

$$\frac{\Delta\tau}{L} = |D_m|\Delta\lambda \tag{1}$$

in which D_m is a coefficient called the **material dispersion coefficient**. It is approximately given by the second derivative of the refractive index,

Material dispersion coefficient

$$D_m \approx -\frac{\lambda}{c}\left(\frac{d^2n}{d\lambda^2}\right) \tag{2}$$

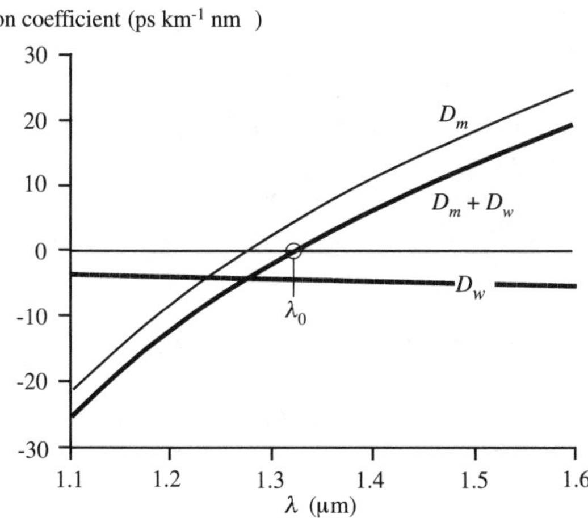

FIGURE 2.18 Material dispersion coefficient (D_m) for the core material (taken as SiO_2), waveguide dispersion coefficient (D_w) (a = 4.2 μm) and the total or chromatic dispersion coefficient $D_{ch}(= D_m + D_w)$ as a function of free space wavelength, λ.

Note that although D_m may be negative or positive, $\Delta\tau$ and $\Delta\lambda$ are defined as positive quantities, which is the reason for the magnitude sign in Eq. (1). To find $\Delta\tau/L$ in Eq. (1) we must evaluate D_m at the center wavelength λ_o of the spectrum. The dependence of D_m on λ for silica as a typical fiber core glass material is displayed in Figure 2.18. Notice that the curve passes through zero at $\lambda \approx 1.27$ μm. When silica (SiO_2) is doped with germania (GeO_2) to increase the refractive index for the core, the D_m vs. λ curve shifts slightly to higher wavelengths.

The transit time τ of a light pulse represents a delay of information between the output and the input. The signal delay time per unit distance, τ/L, is called the **group delay** (τ_g) and is determined by the group velocity v_g as it refers to the transit time of signals (energy). In Eq. (1), $\Delta\tau/L$ is therefore the spread in group delay times due to a finite input spectrum. If β_{01} is the propagation constant of the fundamental mode then by definition

$$\tau_g = \frac{1}{v_g} = \frac{d\beta_{01}}{d\omega} \quad (3)$$

Fundamental mode group delay time

which will be a function of wavelength. Material dispersion in Eq. (1) is the spread in τ_g in Eq. (3) due to the dependence of β_{01} on wavelength through N_g.

B. Waveguide Dispersion

Another dispersion mechanism, called the **waveguide dispersion**, is due to the dependence of the group velocity $v_g(01)$ of the fundamental mode on the V-number, which depends on the source wavelength λ, even if n_1 and n_2 were constant. It should be emphasized that waveguide dispersion is quite separate from material dispersion. Even if n_1 and n_2 were wavelength independent (no material dispersion), we would still have waveguide dispersion by virtue of $v_g(01)$ depending on V and V depending inversely on λ. Waveguide dispersion arises as a result of the guiding properties of the waveguide, which impose a nonlinear ω–β_{lm} relationship, as discussed in §2.2. Therefore, a spectrum of source wavelengths will result in different V-numbers for each source wavelength and hence different propagation velocities. The result is that there will be a spread in the group delay times of the fundamental-mode waves with different λ and hence a broadening of the output light pulse as depicted in Figure 2.17.

If we use a light pulse of very short duration as input, as in Figure 2.17, with a wavelength spectrum of $\Delta\lambda$, then the broadening or dispersion per unit length, $\Delta\tau/L$, in the output light pulse due to waveguide dispersion can be found from

Waveguide dispersion

$$\frac{\Delta\tau}{L} = |D_w|\Delta\lambda \tag{4}$$

in which D_w is a coefficient called the **waveguide dispersion coefficient**. It depends on the waveguide properties (in a non-trivial way) and, over the range $1.5 < V < 2.4$, it is given approximately by

Waveguide dispersion coefficient

$$D_w \approx \frac{1.984 N_{g2}}{(2\pi a)^2 2cn_2^2} \tag{5}$$

in which N_{g2} *and* n_2 are the group and refractive indices of the cladding (medium 2). It is apparent that D_w depends on the guide geometry through the core radius a. Figure 2.18 also shows the dependence of D_w on the wavelength for a core radius of $a = 4.2$ µm. We note that D_m and D_w have opposite tendencies.

C. Chromatic Dispersion or Total Dispersion

In single mode fibers the dispersion of a propagating pulse arises because of the finite width $\Delta\lambda$ of the source spectrum; it is not perfectly monochromatic. Dispersion mechanism is based on the fundamental mode velocity depending on the source wavelength, λ. This type of dispersion caused by a range of source wavelengths is generally termed **chromatic dispersion** and includes both material and waveguide dispersion since both depend on $\Delta\lambda$. As a first approximation, the two dispersion effects can be simply added so that the overall dispersion per unit length becomes,

Chromatic dispersion

$$\frac{\Delta\tau}{L} = |D_m + D_w|\Delta\lambda \tag{6}$$

This is shown in Figure 2.18 where $D_m + D_w$, defined as the **chromatic dispersion coefficient**, D_{ch}, passes through zero at a certain wavelength, λ_0. For the example in Figure 2.18, chromatic dispersion is zero at around 1300 nm.

Since D_w depends on the guide geometry, as apparent in Eq. (5), it is therefore possible to shift the zero dispersion wavelength λ_0 by suitably designing the guide. For example, by reducing the core radius and increasing the core doping, λ_0 can be shifted to 1550 nm where light attenuation in the fiber is minimal. Such fibers are called **dispersion shifted fibers**.

Although chromatic dispersion, $D_m + D_w$, passes through zero, this does not mean that there would be no dispersion at all. First, we should note that $D_m + D_w$ can be made zero for only one wavelength, λ_0, not at every wavelength within the spectrum, $\Delta\lambda$, of the source. Further, other second order effects also contribute to dispersion.

D. Profile and Polarization Dispersion Effects

Although material and waveguide dispersion are the chief sources of broadening of a propagating light pulse there are other dispersion effects. There is an additional dispersion mechanism called **profile dispersion** that arises because the group velocity, $v_g(01)$, of the fundamental mode also depends on the refractive index difference Δ, *i.e.* $\Delta = \Delta(\lambda)$. If Δ changes with wavelength, then different wavelengths from the source would have different group velocities and experience different group delays leading to pulse broadening. It is part of chromatic dispersion because it depends on the input spectrum $\Delta\lambda$:

$$\frac{\Delta\tau}{L} = |D_p|\Delta\lambda \quad \textbf{(7)}$$

Profile dispersion

in which D_p is the **profile dispersion coefficient**, which can be calculated (though not in any simple fashion; see Question 2.13). Typically, D_p is less than 1 ps nm^{-1} km^{-1} and thus negligible compared with D_w. The overall chromatic dispersion coefficient then becomes $D_{ch} = D_m + D_w + D_p$. It should be mentioned that the reason Δ exhibits a wavelength dependence is due to material dispersion characteristics, *i.e.* n_1 vs. λ and n_2 vs. λ behavior, so that, in reality, profile dispersion originates from material dispersion.

Polarization dispersion arises when the fiber is not perfectly symmetric and homogeneous, that is, the refractive index is *not isotropic*. When the refractive index depends on the direction of the electric field, the propagation constant of a given mode depends on its polarization. Due to various variations in the fabrication process, such as small changes in the glass composition, geometry and induced local strains (either during fiber drawing or cabling), the refractive indices n_1 and n_2 may not be isotropic. Suppose that n_1 has the values n_{1x} and n_{1y} when the electric field is parallel to the x and y axes respectively as shown in Figure 2.19. Then the propagation constant for fields along x and y would be different, $\beta_x(01)$ and $\beta_y(01)$, leading to different group delays and hence

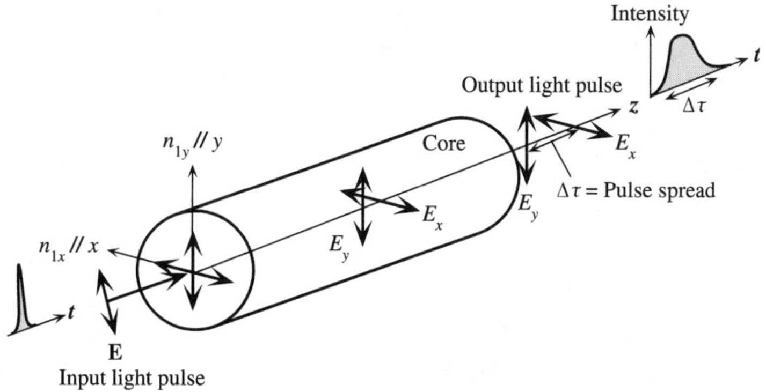

FIGURE 2.19 Suppose that the core refractive index has different values along two orthogonal directions corresponding to electric field oscillation directions (polarizations). We can take x and y axes along these directions. An input light will travel along the fiber with E_x and E_y polarizations having different group velocities and hence arrive at the output at different times

dispersion even if the source wavelength is monochromatic. The situation in reality is more complicated because n_{1x} and n_{1y} would vary along the fiber length and there will be interchange of energy between these modes as well. The final dispersion nonetheless depends on the extent of anisotropy, $n_{1x} - n_{1y}$, which is kept minimum by various manufacture procedures (such as rotating the fiber during drawing, *etc.*). Typically, polarization dispersion is less than a fraction of a picosecond per kilometer of fiber; dispersion does not scale linearly with fiber length L (in fact, it scales roughly as L^2).

E. Dispersion Flattened Fibers

The doping of the core material to shift material dispersion (represented by D_m) and hence overall dispersion to longer wavelength results in an increased attenuation of the signal (see §2.8 and 2.9). Further, it may be desirable to have minimal dispersion over a range of wavelengths not just at the zero-crossing wavelength λ_0 as in Figure 2.18. Waveguide dispersion, represented by D_w, can be adjusted by changing the waveguide geometry. Recall that waveguide dispersion arises from the dependence of the group velocity v_g on the wavelength λ. As the wavelength increases, the field penetrates more into the cladding, which changes the proportion of light energy carried by the core and the cladding and hence changes v_g. We can thus alter the waveguide geometry, that is the refractive index profile, and thereby control D_w to yield a total chromatic dispersion that is *flattened* between two wavelengths λ_1 and λ_2 as in the **dispersion flattened fiber** shown in Figure 2.20. The refractive index profile of such a fiber looks like a W in which the cladding is a thin layer with a depressed refractive index; the fiber is called *doubly clad*. The simple step index fiber is *singly clad*. Greater control on waveguide dispersion can be obtained by using multiply clad fibers. Such fibers are more difficult to manufacture but can exhibit excellent chromatic dispersion $1 - 3$ ps km^{-1} nm^{-1} over the wavelength range $1.3 - 1.6$ μm. Low dispersion over a wavelength range, of course, allows *wavelength multiplexing*, *e.g.* using a number of wavelengths (*e.g.* 1.3, 1.55 μm) as communication channels.

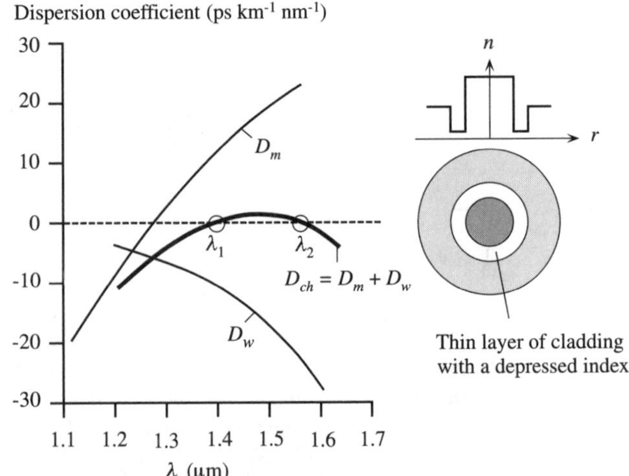

FIGURE 2.20 Dispersion flattened fiber example. The material dispersion coefficient (D_m) for the core material and waveguide dispersion coefficient (D_w) for the doubly clad fiber result in a flattened small chromatic dispersion between λ_1 and λ_2.

EXAMPLE 2.5.1 Material dispersion

By convention, the width $\Delta\lambda$ of the wavelength spectrum of the source and the dispersion $\Delta\tau$ refer to half-power widths and not widths from one extreme end to the other. Suppose that $\Delta\lambda_{1/2}$, called the **linewidth**, is the width of intensity vs. wavelength spectrum between the half intensity points and $\Delta\tau_{1/2}$ is the width of the output light intensity vs. time signal between the half-intensity points.

Estimate the material dispersion effect per km of silica fiber operated from a light emitting diode (LED) emitting at 1.55 μm with a linewidth of 100 nm. What is the material dispersion effect per km of silica fiber operated from a laser diode emitting at the same wavelength with a linewidth of 2 nm?

Solution From Figure 2.18, at 1.55 μm, the material dispersion coefficient $D_m = 22$ ps km^{-1} nm^{-1}. This is given as per km of length and per nm of spectral linewidth. For the LED, $\Delta\lambda_{1/2} = 100$ nm.

$$\Delta\tau_{1/2} \approx L|D_m|\Delta\lambda_{1/2} \approx (1 \text{ km})(22 \text{ ps km}^{-1} \text{ nm}^{-1})(100 \text{ nm}) = 2200 \text{ ps, or } 2.2 \text{ ns.}$$

For the laser diode, $\Delta\lambda_{1/2} = 2$ nm and

$$\Delta\tau_{1/2} \approx L|D_m|\Delta\lambda_{1/2} \approx (1 \text{ km})(22 \text{ ps km}^{-1} \text{ nm}^{-1})(2 \text{ nm}) = 44 \text{ ps, or } 0.044 \text{ ns.}$$

There is clearly a big difference between the dispersion effects of the two sources. The total dispersion, however, as indicated by Figure 2.18, will be less. Indeed, if the fiber is properly dispersion shifted so that $D_m + D_w = 0$ at 1.55 μm, then dispersion due to excitation from a typical laser diode will be a few picoseconds per km (but not zero!).

EXAMPLE 2.5.2 Material, waveguide, and chromatic dispersion

Consider a single mode optical fiber with a core of SiO_2-13.5%GeO_2 for which the material and waveguide dispersion coefficients are shown in Figure 2.21. Suppose that the fiber is excited from a 1.5 μm laser source with a linewidth $\Delta\lambda_{1/2}$ of 2 nm. What is the dispersion per km of fiber if the core diameter $2a$ is 8 μm? What should be the core diameter for zero chromatic dispersion at $\lambda = 1.5$ μm?

Solution In Figure 2.21, at $\lambda = 1.5$ μm, $D_m = +10$ ps km^{-1} nm^{-1} and, with $a = 4$ μm, $D_w = -6$ ps km^{-1} nm^{-1} so that chromatic dispersion is

$$D_{ch} = D_m + D_w = 10 - 6 = 4 \text{ ps km}^{-1} \text{ nm}^{-1}.$$

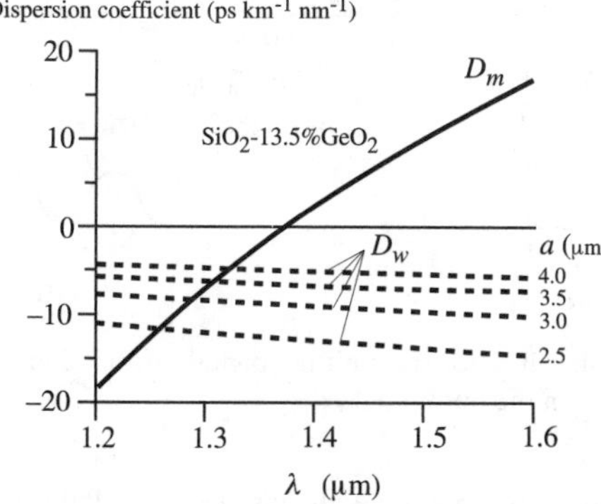

FIGURE 2.21 Material and waveguide dispersion coefficients in an optical fiber with a core SiO_2-13.5%GeO_2 for $a = 2.5$ to 4 μm.

Total dispersion or chromatic dispersion per km of fiber length is then,

$$\Delta\tau_{1/2}/L = |D_{ch}|\Delta\lambda_{1/2} = (4 \text{ ps km}^{-1} \text{ nm}^{-1})(2 \text{ nm}) = 8 \text{ ps km}^{-1}.$$

Dispersion will be zero at 1.5 μm when $D_w = -D_m$ or when $D_w = -10$ ps km^{-1} nm^{-1}. Examination of D_w vs. λ for $a = 2.5$ to 4 μm shows that a should be about 3 μm. It should be emphasized that although $D_m + D_w = 0$ at 1.5 μm, this is only at one wavelength, whereas the input radiation is over a range $\Delta\lambda$ of wavelengths so that in practice chromatic dispersion is never actually zero. It will be minimum at 1.5 μm when, in the present case, the core radius $a = 3$ μm.

2.6 BIT RATE, DISPERSION, ELECTRICAL, AND OPTICAL BANDWIDTH

A. Bit Rate and Dispersion

In digital communications, signals are generally sent as light pulses along an optical fiber. Information is first converted to an electrical signal that is in the form of pulses as depicted in Figure 2.22. The pulses represent bits of information that are in digital form. For simplicity we have taken the pulses to be very short but generally there is a well-defined pulse duration. The electrical signal drives a light emitter such as a laser diode whose light output is appropriately coupled into a fiber for transmission to the destination. The light output at the destination end of the fiber is coupled to a photodetector that converts the light signal back to an electric signal. The information is then decoded from this electrical signal. Digital communications engineers are interested in the maximum rate at which the digital data can be transmitted along the fiber. This rate is called the **bit rate capacity** B (bits per second) of the fiber and is directly related to the dispersion characteristics.[8]

Suppose that we feed a light pulse of very short duration into the fiber. The output pulse will be delayed by the transit time τ it takes for the light pulse to travel down the fiber. Due to various dispersion mechanisms there will be a spread $\Delta\tau$ in the arrival times of different guided waves, for example, different fundamental modes for different source wavelengths. This dispersion is typically measured between half-power (or in-

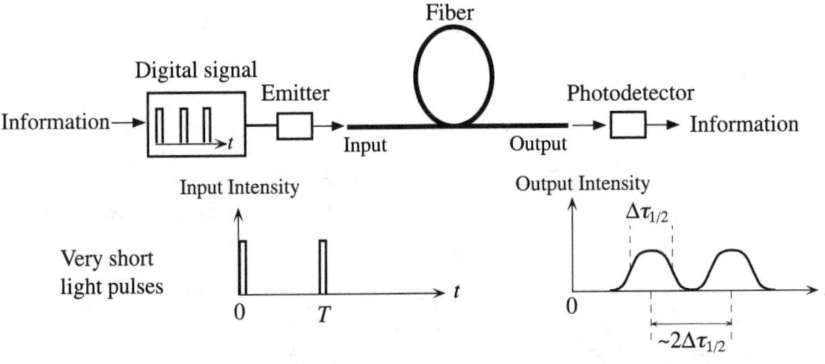

FIGURE 2.22 An optical fiber link for transmitting digital information and the effect of dispersion in the fiber on the output pulses.

[8] The discussion below in this section is more intuitive than rigorous. Mathematical derivations may be found in textbooks on Optical Communications.

tensity) points and is called **full width at half power (FWHP)**, or **full width at half maximum (FWHM)**. To clarify the definition of $\Delta\tau$ we can use $\Delta\tau_{1/2}$ to represent the extent of dispersion based on FWHP. Intuitively, as apparent from Figure 2.22, clear distinguishability between two consecutive output pulses, that is no **intersymbol interference**, requires that they be time-separated from peak to peak by at least $2\,\Delta\tau_{1/2}$. Thus, we can only feed in pulses at the input, at best, at every $2\,\Delta\tau_{1/2}$ seconds, which then defines the period (T) of the input pulses. Thus the maximum bit rate, or simply the bit rate B, at which pulses can be transmitted is very roughly $1/(2\,\Delta\tau_{1/2})$,

$$B \approx \frac{0.5}{\Delta\tau_{1/2}} \qquad (1)$$

Intuitive RZ bit rate and dispersion

The maximum bit rate B in Eq. (1) assumes a pulse representing the binary information 1 must return to zero for a duration before the next binary information. Two consecutive pulses for two consecutive 1's must have a zero in between as in the output pulses shown in Figure 2.22 and Figure 2.23. This bit rate is called the **return-to-zero (RZ) bit rate or data rate**. On the other hand, it is also possible to send the two consecutive binary 1 pulses without having to return to zero at the end of each 1-pulse. That is, two 1-pulses are immediately next to each other. The two pulses in Figure 2.23 can be brought closer until the repetition period $T \approx \Delta\tau_{1/2}$ and the signal is nearly uniform over the length of these two consecutive 1's. Such a maximum data rate is called **nonreturn to zero (NRZ) bit rate**. The NRZ bit rate is twice the RZ bit rate. The maximum bit rate and dispersion discussions below refer to the RZ data rate.

In a more rigorous analysis we need to know the temporal shape of the output signal and the criterion for discerning the information; for example, what is the extent of allowed overlap in the output light pulses. There is a general approximate relationship between the bit rate and the *root-mean-square (rms) dispersion* (or the mean standard deviation from the mean[9]) σ which is shown in Figure 2.23. Notice that the

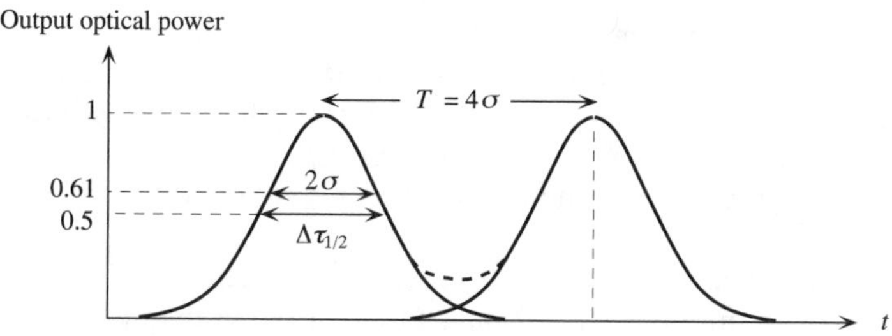

FIGURE 2.23 A Gaussian output light pulse and some tolerable intersymbol interference between two consecutive output light pulses (y-axis in relative units). At time $t = \sigma$ from the pulse center, the relative magnitude is $e^{-1/2} = 0.607$ and full width root mean square (rms) spread in $\Delta\tau_{\text{rms}} = 2\sigma$.

[9]The shape $h(t)$ of a perfectly Gaussian output light pulse centered at $t = 0$ is $h(t) = [1/(2\pi\sigma^2)^{1/2}]\exp[-t^2/(2\sigma^2)]$ where σ is the root mean square deviation; $\Delta t_{\text{rms}} = \sigma$. When $t = \sigma$, $h(t) = e^{-1}$. In practice, Gaussian is only an approximation to the actual light output pulse.

full-width rms time spread $\Delta\tau_{rms}$ of the pulse between its rms points is 2σ. For a Gaussian output light pulse, σ and $\Delta\tau_{1/2}$ are related by $\sigma = 0.425\Delta\tau_{1/2}$. The relationship is different for other light pulse shapes. The bit rate B in terms of σ requires that two consecutive light output pulses are separated by 4σ between their peaks as shown in Figure 2.23. Thus,

Maximum RZ bit rate and dispersion

$$B \approx \frac{0.25}{\sigma} \tag{2}$$

For a Gaussian pulse, $\sigma = 0.425\Delta\tau_{1/2}$ so that $B = 0.59/\Delta\tau_{1/2}$, and the bit rate is ~18% greater than the intuitive estimate in Eq. (1). In general, the input pulses are not infinitely short and have a certain time-width that must also be considered in the discernibility of the output information along with the method of information decoding. All these factors typically modify the numerical factor in Eq. (2).[10] Dispersion increases with fiber length L and also with the range of source wavelengths $\Delta\lambda_{1/2}$, measured between the half-intensity points of the source spectrum (intensity vs. wavelength). This means that the bit rate decreases with increasing L and $\Delta\lambda_{1/2}$. It is therefore customary to specify the product of the bit rate B with the fiber length L at the operating wavelength for a given emitter (LED, laser diode, *etc.*). Suppose that the rms spread of wavelengths in the light output spectrum of the emitter is σ_λ. For a Gaussian output spectrum, for example, $\sigma_\lambda = 0.425\Delta\lambda_{1/2}$. If D_{ch} is the chromatic dispersion coefficient, then the rms dispersion of the output light pulse is $LD_{ch}\sigma_\lambda$. Then the BL product, called the bit-rate × distance product, is given by

Maximum bit rate × distance

$$BL \approx \frac{0.25L}{\sigma} = \frac{0.25}{|D_{ch}|\sigma_\lambda} \tag{3}$$

It is clear that BL is a characteristic of the fiber, through D_{ch}, and also of the range of source wavelengths. In specifications, the fiber length is taken as 1 km. For example, BL for a step-index single mode fiber operating at a wavelength of 1,300 nm and excited by a laser diode source is several Gb s^{-1} km. Dividing this by the actual length of the fiber gives the operating bit rate for that length.

When both chromatic (or intramodal) and intermodal dispersion are present, as in a graded index fiber, we have to combine the two effects taking into account that their origins are different. Both material and waveguide dispersion arise from a range of input wavelength (that is $\Delta\lambda$) and the net effect is simply the linear addition of the two dispersion coefficients, $D_{ch} = D_w + D_m$. The same is *not* true for combining intermodal and intramodal dispersion simply because their origins are different. Overall dispersion in terms of an rms dispersion σ can be found from individual rms dispersions by

Total rms dispersion

$$\sigma^2 = \sigma^2_{\text{intermodal}} + \sigma^2_{\text{intramodal}} \tag{4}$$

and σ can then be used in Eq. (2) to approximately find B. Equation (4) is generally valid for finding the resultant rms deviation from individual rms deviations whenever two independent processes are superimposed.

---

[10] The maximum bit rate depends on the input pulse shape, fiber dispersion characteristics and hence the output pulse shape and, not least, the modulation scheme of encoding the information. The treatment is covered in optical communications courses and obviously beyond the scope of an elementary book.

Equations (2) and (3) seem innocuously simple due to the fact that they have been written in terms of the rms dispersion. To determine B from $\Delta\tau_{1/2}$ we need to know the pulse shape. For example, for a rectangular pulse, the *full width*, ΔT, is the same as $\Delta\tau_{1/2}$. The mathematics shows that $\sigma = 0.29\Delta\tau_{1/2} = 0.29\Delta T$, which is quite different than the Gaussian pulse case. For the rectangular pulse, therefore, the bandwidth $B = 0.25/\sigma = 0.87/\Delta T = 0.87/\Delta\tau_{1/2}$. On the other hand, for an ideal **Gaussian pulse**, $\sigma = 0.425\Delta\tau_{1/2}$ and $B = 0.25/\sigma = 0.59/\Delta\tau_{1/2}$; ΔT, full width, has no meaning.

B. Optical and Electrical Bandwidth

The emitter in the simple optical fiber link in Figure 2.22 can also be driven, or modulated, by an analog signal that varies continuously with time. We can, for example, drive the emitter using a sinusoidal signal as shown in Figure 2.24. The input light intensity into the fiber then becomes modulated to be a sinusoidal with time at the same frequency f as the modulating signal. The light output intensity at the fiber destination should also be a sinusoidal with only a shift in phase due to the time it takes for waves to travel along the fiber. We can determine the transfer characteristics of the fiber by feeding in sinusoidal light intensity signals, which have the same intensity but different modulation frequencies f. Ideally, the light output should have the same intensity for the various modulation frequencies. Figure 2.24 shows the observed optical transfer characteristic of the fiber, which is defined as the output light power per unit input light power (P_o/P_i), as a function of modulation frequency f. The response is flat and then falls with frequency. The reason is that the frequency becomes too fast so that dispersion effects smear out the light at the output. The frequency f_{op} in which the output intensity is 50% below the flat region defines the optical bandwidth f_{op} of the fiber and hence the useful frequency range in which modulated optical signals can be transferred long the fiber. Intuitively, the optical cut-off frequency f_{op} should correspond roughly to the bit rate, that is, $f_{op} = B$. This is not entirely true because B can tolerate some pulse overlap depending on the shape of the output pulse and the discernibility criterion. If the fiber dispersion characteristics are Gaussian, then

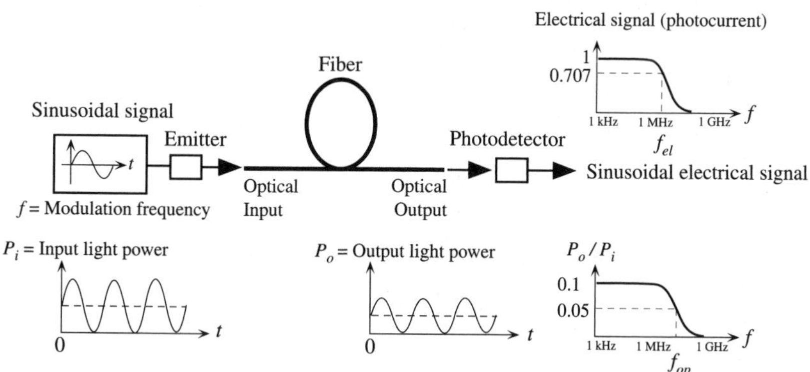

FIGURE 2.24 An optical fiber link for transmitting analog signals and the effect of dispersion in the fiber on the bandwidth, f_{op}.

TABLE 2.2 Relationships between dispersion parameters, maximum bit rates, and bandwidths. RZ = Return to zero pulses. NRZ = Nonreturn to zero pulses. B' is the maximum bit rate for NRZ pulses.

Dispersed pulse shape	$\Delta\tau_{1/2}$ = FWHM width	B (RZ)	B' (NRZ)	f_{op}	f_{el}
Gaussian with rms deviation σ	$\sigma = 0.425\Delta\tau_{1/2}$	$0.25/\sigma$	$0.5/\sigma$	$0.75B = 0.19/\sigma$	$0.71f_{op} = 0.13/\sigma$
Rectangular with full width ΔT	$\sigma = 0.29\Delta T = 0.29\Delta\tau_{1/2}$	$0.25/\sigma$	$0.5/\sigma$	$0.69B = 0.17/\sigma$	$0.73f_{op} = 0.13/\sigma$

Optical bandwidth for Gaussian dispersion

$$f_{op} \approx 0.75B \approx \frac{0.19}{\sigma} \tag{5}$$

in which σ is the total rms dispersion through the fiber. The optical bandwidth and fiber length product $f_{op}L$ is then approximately 25% smaller than the BL product.

It is important to realize that the electrical signal from the photodetector (photocurrent or voltage) does not exhibit the same bandwidth. This is because bandwidth f_{el} for electrical signals is measured where the signal is 70.7% of its low frequency value as indicated in Figure 2.24. The electrical signal (photocurrent) from the photodetector is proportional to the fiber-output light power and when this is 50% below the flat region at f_{op}, so is the electrical signal. This f_{op} is therefore greater than f_{el}. It is, of course, f_{el} that is of interest in optical receiver system design. The relationship between f_{el} and f_{op} depends on the dispersion through the fiber. For Gaussian dispersion, $f_{el} \approx 0.71f_{op}$. Table 2.2 summarizes the relationships between various dispersion parameters, maximum bit rates and bandwidths for Gaussian, and rectangular dispersed pulse shapes. Notice that digital information transmission using a nonreturn-to-zero (NRZ) scheme has twice the bit rate of the RZ scheme.

EXAMPLE 2.6.1 Bit rate and dispersion

Consider an optical fiber with a chromatic dispersion coefficient 8 ps km^{-1} nm^{-1} at an operating wavelength of 1.5 μm. Calculate the bit rate × distance product (BL), and the optical and electrical bandwidths for a 10 km fiber if a laser diode source with a FWHP linewidth $\Delta\lambda_{1/2}$ of 2 nm is used.

Solution For FWHP dispersion,

$$\Delta\tau_{1/2}/L = |D_{ch}|\Delta\lambda_{1/2} = (8\text{ ps km}^{-1}\text{ nm}^{-1})(2\text{ nm}) = 16\text{ ps km}^{-1}$$

Assuming a Gaussian light pulse shape, the RZ bit rate × distance product (BL) is

$$BL = 0.59L/\Delta t_{1/2} = 0.59/(16\text{ ps km}^{-1}) = 36.9\text{ Gb s}^{-1}\text{ km}.$$

The optical and electrical bandwidths for a 10 km distance is

$$f_{op} = 0.75B = 0.75(36.9\text{ Gb s}^{-1}\text{ km})/(10\text{ km}) = 2.8\text{ GHz}.$$

and

$$f_{el} = 0.71f_{op} = 2.0\text{ GHz}.$$

2.7 THE GRADED INDEX (GRIN) OPTICAL FIBER

The main drawback of the single mode step index fiber is the relatively small numerical aperture NA and hence the difficulty in the amount of light that can be coupled into it. Increasing the NA means only increasing the V-number, which must be less than 2.405. Multimode fibers have a relatively larger NA and hence accept more light from broader angles. The greater NA and wider core diameter of the multimode fiber allows not only easier light coupling but also more optical power to be launched into the fiber. Multimode fibers suffer from intermodal dispersion. In terms of visualizing rays this means that the rays representing the modes travel along different path lengths and hence arrive at different times at the end of the fiber as shown in Figure 2.25 (a). The velocity along the path (not along z) is c/n_1 so that those trajectories that are longer, those that have many reflections, take longer, which means that the axial ray 1 arrives first, then ray 2 and then 3, and so on. (This is the intuitive ray picture.)

In the **graded index (GRIN) fiber** the refractive index is not constant within the core but decreases from n_1 at the center, as a power law, to n_2 at the cladding as shown in Figure 2.25 (b). The refractive index profile across the core is close to being parabolic. Such a refractive index profile is capable of minimizing intermodal dispersion to a virtually inoccuous level. All the rays (*e.g.* marked 1, 2, and 3 etc.) in the graded index fiber arrive at the same time as depicted in Figure 2.25 (b). The intuitive reason for this is that the velocity along the ray path, c/n, is not constant and increases as the ray is farther away from the center. A ray such as 2 that has a longer path than ray 1 then experiences a faster velocity during a part of its journey to enable it to catch up with ray 1. Similarly, ray 3 experiences a faster velocity than 2 during part of its propagation to catch up with ray 2, and so on.

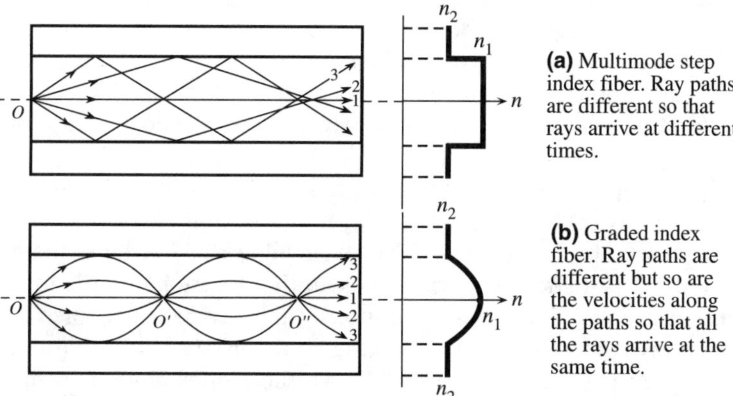

n_2
n_1
n
O
3
2
1
(a) Multimode step index fiber. Ray paths are different so that rays arrive at different times.
n_2
O
O'
O''
n_1
n
n_2
(b) Graded index fiber. Ray paths are different but so are the velocities along the paths so that all the rays arrive at the same time.

FIGURE 2.25 (a) Multimode step index fiber. Ray paths are different so that rays arrive at different times. (b) Graded index fiber. Ray paths are different but so are the velocities along the paths so that all the rays arrive at the same time.

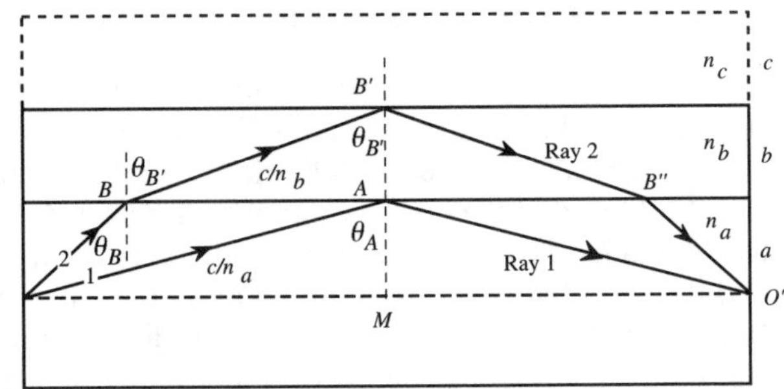

FIGURE 2.26 We can visualize a graded index fiber by imagining a stratified medium with the layers of refractive indices $n_a > n_b > n_c \ldots$ Consider two close rays 1 and 2 launched from O at the same time but with slightly different launching angles. Ray 1 just suffers total internal reflection. Ray 2 becomes refracted at B and reflected at B'.

We can intuitively understand the reason for the absence of modal dispersion by considering the graded index fiber to consist of many thin concentric layers each of constant refractive index, n_a, n_b, n_c, etc., as depicted in Figure 2.26. We view the core as a stratified region. The refractive index decreases from layer to layer towards the cladding. All rays are launched from O at the core axis. Can two rays such as 1 and 2 launched at different angles but at the same time arrive at O' at the same time? Can we make ray 2 arrive at O' at the same time as ray 1? The launching angle of ray 1 is such that its angle of incidence, θ_A, at the *ab* interface is just the critical angle $\theta_c(ab) = \arcsin(n_b/n_a)$ between layers n_a and n_b so that it suffers TIR at A and then arrives at O'. Ray 2 has a slightly wider launching angle that leads to an angle of incidence, θ_B, which is less than $\theta_c(ab)$. Ray 2 therefore becomes refracted and enters layer b where its angle is $\theta_{B'}$. If we choose n_b to be appropriately lower than n_a we can direct $\theta_{B'}$ so that ray 2 impinges on layer c at B', exactly above A, halfway between O and O'. Because of symmetry, B' must be midway between O and O' otherwise ray 2 cannot arrive at O'. Further, if we choose n_c to be appropriately lower than n_b then we can arrange for ray 2 to suffer TIR at B', *i.e.* $\theta_{B'} > \theta_c(bc) = \arcsin(n_c/n_b)$. Thus by appropriate choices of n_a, n_b, and n_c we can ensure rays 1 and 2 pass through O'. Do the rays 1 and 2 arrive at the same time? In layer B, ray 2 moves faster because the refractive index is less than that in A. The path OBB' for ray 2 is longer than OA for 1 but ray 2 moves faster along BB'. If the refractive indices have been appropriately chosen, then ray 2 traveling faster in layer b manages to catch up with ray 1 and both rays arrive nearly at the same time. The appropriate choices for the refractive indices follow an approximately parabolic decrease of n from the core axis.

In the graded index fiber, of course, the index changes continuously, which is analogous to having a ray travel from layer to layer almost immediately so that there is an immediate sequence of refractions as visualized in Figure 2.27. After so many refractions the angle eventually satisfies the critical angle at that location and the ray suffers TIR. In a medium in which n decreases continuously, the ray path becomes bent continuously until the ray suffers TIR. The ray paths in the graded index core are therefore curved trajectories as shown in Figure 2.27. All the modes have their maxima at the same location on the z axis and at about the same time. These ideas apply essentially to meridional rays, those rays that cross the core axis. In addition, there are also continuously bent skew rays, which are **helical rays** that result from rays entering the fiber core off the axis. When we examine intermodal dispersion we have to also consider these he-

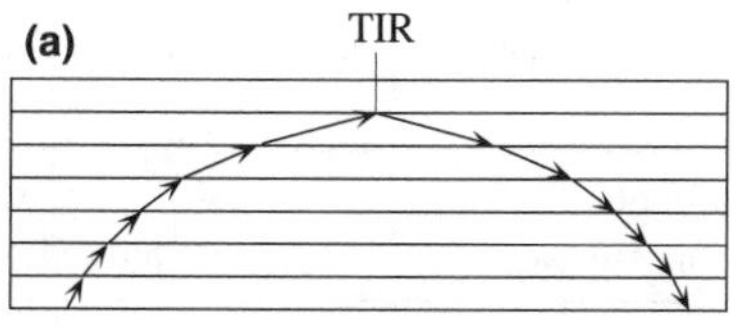

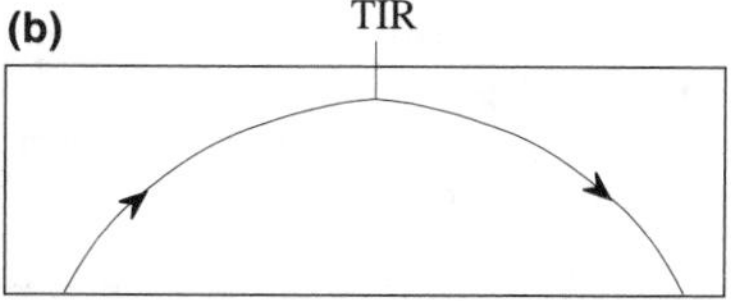

n decreases step by step from one layer to next upper layer; very thin layers.

Continuous decrease in *n* gives a ray path changing continuously.

FIGURE 2.27 (a) A ray in a thinly stratified medium becomes refracted as it passes from one layer to the next upper layer with lower *n* and eventually its angle satisfies TIR. (b) In a medium where *n* decreases continuously the path of the ray bends continuously.

lical rays. When we consider the propagation of all the modes through the graded index fiber, intermodal dispersion is not totally absent though it is reduced by orders of magnitude from the multimode step index fiber.

The refractive index profile can generally be described by a power law with an index γ called the **profile index** (or **the coefficient of index grating**) so that,

$$n = n_1[1 - 2\Delta(r/a)^{\gamma}]^{1/2} \quad ; r < a, \tag{1a}$$

$$n = n_2 \quad ; r = a \tag{1b}$$

which looks like the schematic refractive index profile shown in Figure 2.25 (b). The intermodal dispersion is then minimum when γ is

$$\gamma = \frac{4 + 2\Delta}{2 + 3\Delta} \approx 2(1 - \Delta) \tag{2}$$

in which Δ is small so that γ is close to 2 (parabolic). This is the **optimal profile index**.

With this optimal profile index, the rms dispersion $\sigma_{\text{intermode}}$ in the output light pulse per unit length is given by[11]

$$\frac{\sigma_{\text{intermode}}}{L} \approx \frac{n_1}{20\sqrt{3}c}\Delta^2 \tag{3}$$

Dispersion in graded index fiber

Table 2.3 compares the typical characteristics of the three general types of fibers, multimode step index, single mode step index, and graded index fibers, and lists some of their advantages and disadvantages.

EXAMPLE 2.7.1 Dispersion in a graded-index fiber and bit rate

Consider a graded index fiber whose core has a diameter of 50 μm and a refractive index of $n_1 = 1.480$. The cladding has $n_2 = 1.460$. If this fiber is used at 1.30 μm with a laser diode that has very a narrow linewidth, what will the bit rate × distance product be? Evaluate the *BL* product

[11] See J. Senior, *Optical Fiber Communications, Principles and Practice, Second Ed.* (Prentice-Hall, New York, 1992.) Ch. 3.

TABLE 2.3 Comparison of typical characteristics of multimode step-index, single-mode step-index, and graded-index fibers. (Typical values combined from various sources.)

Property	Multimode step-index fiber	Single-mode step-index fiber	Graded index fiber
$\Delta = (n_1 - n_2)/n_1$	0.02	0.003	0.015
Core diameter (μm)	100	8.3 (MFD = 9.3 μm)	62.5
Cladding diameter (μm)	140	125	125
NA	0.3	0.1	0.26
Bandwidth × distance or Dispersion	20 − 100 MHz km.	<3.5 ps km^{-1} nm^{-1} at 1.3 μm >100 Gb s^{-1} km in common use	300 MHz km − 3 GHz km at 1.3 μm at 1.3 μm
Attenuation of light	4 − 6 dB km^{-1} at 850 nm 0.7 − 1 dB km^{-1} at 1.3 μm	1.8 dB km^{-1} at 850 nm 0.34 dB km^{-1} at 1.3 μm 0.2 dB km^{-1} at 1.55 μm	3 dB km^{-1} at 850 nm 0.6 − 1 dB km^{-1} at 1.3 μm 0.3 dB km^{-1} at 1.55 μm
Typical light source	Light emitting diode (LED)	Lasers, single mode injection lasers	LED, lasers
Typical applications	Short haul or subscriber local network communications	Long haul communications	Local and wide-area networks. Medium haul communications

if this were a multimode step index fiber given that the light output is nearly rectangular so that $\sigma \approx 0.29\Delta\tau$ in which $\Delta\tau$ is the full spread.

Solution The normalized refractive index difference $\Delta = (n_1 - n_2)/n_1 = (1.48 - 1.46)/1.48 = 0.0135$.

Dispersion for 1 km of fiber is

$$\frac{\sigma_{\text{intermode}}}{L} \approx \frac{n_1}{20\sqrt{3}c}\Delta^2 = \frac{1.480}{20\sqrt{3}(3 \times 10^8)}(0.0135)^2$$

$$= 2.6 \times 10^{-14}\ \text{s m}^{-1} \text{ or } 0.026\ \text{ns km}^{-1}.$$

so that

$$BL \approx \frac{0.25L}{\sigma_{\text{intermode}}} = \frac{0.25}{(2.6 \times 10^{-11}\ \text{s km}^{-1})} = 9.6\ \text{Gb s}^{-1}\ \text{km}$$

We have ignored any material dispersion and, further, we assumed the index variation to perfectly follow the optimal profile, which means that in practice BL will be worse. (For example, a 15% variation in γ from the optimal value can result in $\sigma_{\text{intermode}}$ and hence BL that are more than 10 times worse.)

If this were a multimode step-index fiber with the same n_1 and n_2, then the full dispersion (total spread) would roughly be

$$\frac{\Delta\tau}{L} \approx \frac{n_1 - n_2}{c} = \frac{1.480 - 1.460}{3 \times 10^8} = 6.67 \times 10^{-11}\ \text{s m}^{-1} \text{ or } 66.7\ \text{ns km}^{-1}.$$

To calculate the BL we need $\sigma_{\text{intermode}} \approx 0.29\Delta\tau$

$$BL \approx \frac{0.25L}{\sigma_{\text{intermode}}} = \frac{0.25}{(0.29 \times 66.7 \times 10^{-9}\ \text{s km}^{-1})} = 12.9\ \text{Mb s}^{-1}\ \text{km}$$

which is nearly 1,000 times smaller.

Note: Since the dispersion $\Delta\tau$ increases linearly with the fiber length, the bit rate $\times$ distance (BL) product for a multimode fiber appears to be constant. It implies $B \propto L^{-1}$. Although this is approximately true over short distances (perhaps a few kilometers), over long distances, the bit rate $\times$ distance product is not constant and typically $B \propto L^{-\gamma}$ in which γ is an index between 0.5 and 1. The reason is that, due to various fiber imperfections, there is mode mixing that reduces the extent of pulse spreading.

2.8 LIGHT ABSORPTION AND SCATTERING

In general, when light propagates through a material it becomes *attenuated* in the direction of propagation as illustrated in Figure 2.28. We distinguish between *absorption* and *scattering*, both of which give rise to a loss of intensity in the regular direction of propagation. In addition, extrinsic factors such as *fiber bending* can also lead to light attenuation.

A. Absorption

In absorption, some of the energy from the propagating wave is converted to other forms of energy, for example, to heat by the generation of lattice vibrations. There are a number of absorption processes that dissipate the energy from the wave. An example of one mechanism, called **lattice absorption**, is displayed in Figure 2.29. The solid in this example is made of ions and as an EM wave propagates it displaces the oppositely charged ions in opposite directions and forces them to vibrate at the frequency of the wave. In other words, the medium experiences *ionic polarization*. It is the displacements of these ions that give rise to ionic polarization and its contribution to the relative permittivity, ε_r. As the ions and hence the lattice is made to vibrate by the passing EM wave,

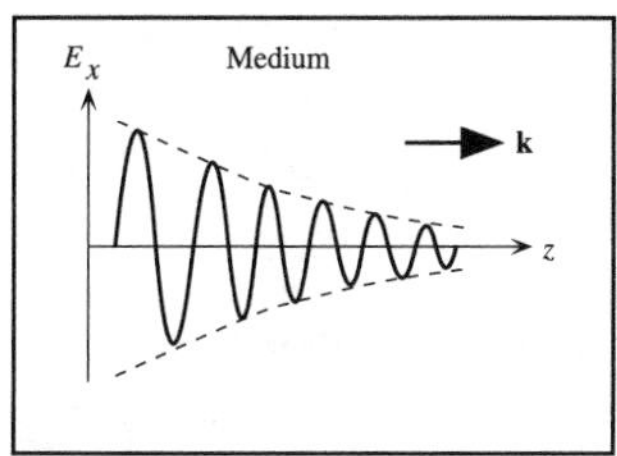

FIGURE 2.28 Attenuation of light in the direction of propagation.

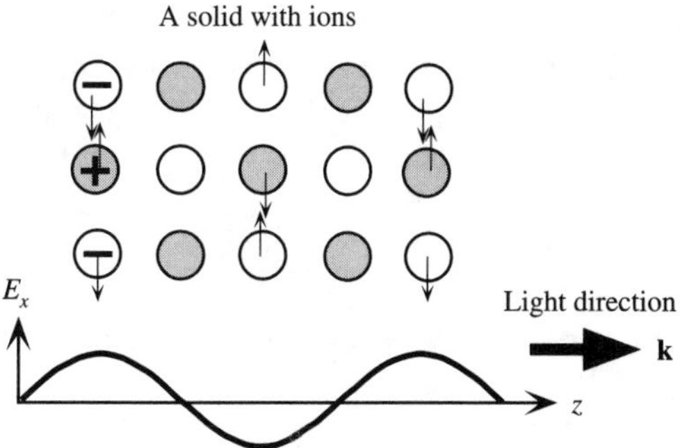

FIGURE 2.29 Lattice absorption through a crystal. The field in the wave oscillates the ions which consequently generate "mechanical" waves in the crystal; energy is thereby tranferred from the wave to lattice vibrations.

some energy is coupled into the lattice vibrations of the solid. This energy peaks when the frequency of the wave is close to the natural lattice vibrational frequencies. Typically, these frequencies are in the *infrared region*. Most of the energy is then absorbed from the EM wave and converted to lattice vibrational energy (heat). We associate this absorption with the resonance peak or relaxation peak of ionic polarization loss (imaginary part of the relative permittivity,[12] ε_r'').

Although Figure 2.29 depicts an ionic solid to visualize absorption due to lattice waves, energy from a passing EM wave can also be absorbed by various ionic impurities in a medium, as these charges can couple to the electric field and oscillate . Bonding between an oscillating ion and the neighboring atoms causes the mechanical oscillations of the ion to be coupled to neighboring atoms. This leads to a generation of lattice waves that take away energy from the EM wave.

B. Scattering

Scattering of an EM wave implies that a portion of the energy in a light beam is directed away from the original direction of propagation as illustrated for a small dielectric particle scattering a light beam in Figure 2.30. There are various types of scattering processes.

Consider what happens when a propagating wave encounters a molecule, or a small dielectric particle (or region) that is smaller than the wavelength . The electric field in the wave polarizes the particle by displacing the lighter electrons with respect to the heavier positive nuclei. The electrons in the molecule couple and oscillate with the electric field in the wave (ac electronic polarization). The oscillation of charge "up" and "down," or the oscillation of the induced electric dipole, radiates EM waves all around the molecule as depicted in Figure 2.30. We should remember that an oscillating charge is like an alternating current that always radiates EM waves (like an antenna). The net effect is that the incident wave becomes partially reradiated in different directions and hence loses intensity in its original direction of propagation. We may think of the process as the particle absorbing some of the energy via electronic polarization and reradiating it in different directions. It may be thought that the scattered waves constitute a spherical wave emanating from the scattering molecule, but this is not generally the case as the re-emitted radiation depends on the shape and polarizability of the molecule in different directions. We assumed a small particle so that at any time the field has no spatial variation through the particle, whose polarization then oscillates with the electric field oscillations. Whenever the size of a scattering region, whether an inhomogeneity, a small particle, or a molecule, is much smaller than the wavelength λ of the incident wave, the scattering process is generally termed **Rayleigh scattering**. In this type of scattering, the particle size is typically smaller than one-tenth of the wavelength.

[12] S.O. Kasap, *Principles of Electronic Materials and Devices, Second Edition* (McGraw-Hill, 2001.) Ch. 7.

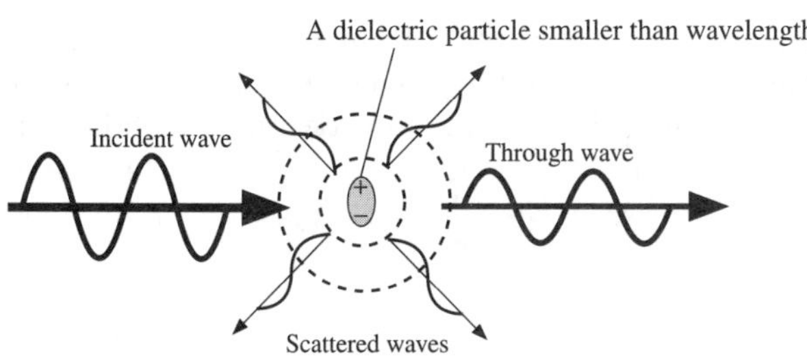

FIGURE 2.30 Rayleigh scattering involves the polarization of a small dielectric particle or a region that is much smaller than the light wavelength. The field forces dipole oscillations in the particle (by polarizing it) which leads to the emission of EM waves in "many" directions so that a portion of the light energy is directed away from the incident beam.

Lord Rayleigh (John William Strutt) was an English physicist (1877–1919) and a Nobel Laureate (1904) who made a number of contributions to wave physics of sound and optics. *(Courtesy of AIP Emilio Segrè Visual Archives, Physics Today Collection)*

Rayleigh scattering of waves in a medium arises whenever there are small inhomogeneous regions in which the refractive index is different than the medium (which has some average refractive index). This means a local change in the relative permittivity and polarizability. The result is that the small inhomogeneous region acts like a small dielectric particle and scatters the propagating wave in different directions. In the case of optical fibers, dielectric inhomogeneities arise from fluctuations in the relative permittivity that is part of the intrinsic glass structure. As the fiber is drawn by freezing a liquid-like flow, random thermodynamic fluctuations in the composition and structure that occur in the liquid-state become frozen into the solid structure. Consequently, the glass fiber has small fluctuations in the relative permittivity that lead to Rayleigh scattering. Nothing can be done to eliminate Rayleigh scattering in glasses as it is part of their intrinsic structure.

It is apparent that the scattering process involves electronic polarization of the molecule or the dielectric particle. We know that this process couples most of the energy at ultraviolet frequencies where the dielectric loss due to electronic polarization is maximum and the loss is due to EM wave radiation. Therefore, as the frequency of light increases, the scattering becomes more severe. In other words, *scattering decreases with increasing wavelength.* For example, blue light that has a shorter wavelength than red light is scattered more strongly by air molecules. When we look at the sun directly it appears yellow because the blue light has been scattered in the direct light more than the red light. When we look at the sky in any direction but the sun our eyes receive scattered light, which appears blue; hence the sky is blue. At sunrise and sunset, the rays from the sun have to traverse the longest distance through the atmosphere and have the most blue light scattered, which gives the sun its red color at these times.

2.9 ATTENUATION IN OPTICAL FIBERS

As light propagates through an optical fiber it becomes attenuated by a number of processes that depend on the wavelength of light. Suppose that the input optical power into a fiber of length L is P_{in} and the output optical power at the destination end is P_{out} and intensity anywhere in the fiber at a distance x from the input is P. The **attenuation coefficient** α is defined as the *fractional decrease in the optical power per unit distance, i.e.*

Definition of attenuation coefficient

$$\alpha = -\frac{1}{P}\frac{dP}{dx} \tag{1}$$

We can integrate this over the length L of the fiber to relate α to P_{out} and P_{in} by

$$\alpha = \frac{1}{L}\ln\left(\frac{P_{in}}{P_{out}}\right) \tag{2}$$

We can just as well define attenuation in terms of light intensity but we have used power to follow convention since attenuation tests on optical fibers measure the optical power. If we know α then we can always find P_{out} from P_{in} through,

$$P_{out} = P_{in}\exp(-\alpha L)$$

In general, optical power attenuation in an optical fiber is expressed in terms of decibels per unit length of fiber, typically as dB per km. The attenuation of the signal in decibels per unit length is defined in terms of the logarithm to base 10 by

Attenuation coefficient in dB/length

$$\alpha_{\text{dB}} = \frac{1}{L}10\log\left(\frac{P_{in}}{P_{out}}\right) \tag{3}$$

Substituting for P_{in}/P_{out} from above we obtain

$$\alpha_{\text{dB}} = \frac{10}{\ln(10)}\alpha = 4.34\alpha \tag{4}$$

Figure 2.31 shows the attenuation coefficient, as dB per km, of a typical silica glass-based optical fiber as a function of wavelength. The sharp increase in the attenuation at wavelengths beyond 1.6 μm in the *infrared* region is due to energy absorption by "lat-

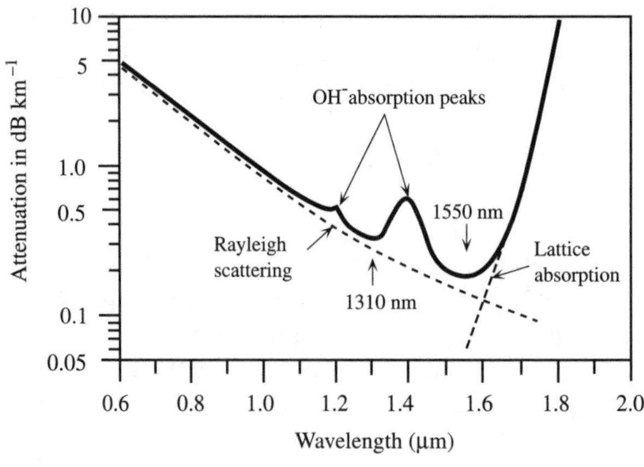

FIGURE 2.31 Illustration of a typical attenuation vs. wavelength characteristics of a silica based optical fiber. There are two communications channels at 1310 nm and 1550 nm.

tice vibrations" of the constituent ions of the glass material. Fundamentally, energy absorption in this region corresponds to the stretching of the Si-O bonds in ionic polarization induced by the EM wave. Absorption increases with wavelength as we approach the resonance wavelength of the Si-O bond, which is around 9 μm. In the case of Ge-O glasses, this is further away, around 11 μm. There is another intrinsic material absorption in the region below 500 nm, not shown in the figure, which is due to photons exciting electrons from the valence band to the conduction band of glass.

There is a marked attenuation peak centered at 1.4 μm, and a barely discernible minor peak at about 1.24 μm. These attenuation regions arise from the presence of hydroxyl ions as impurities in the glass structure inasmuch as it is difficult to remove all traces of hydroxyl (water) products during fiber production. Further, hydrogen atoms can easily diffuse into the glass structure at high temperatures during production, which leads to the formation of hydrogen bonds in the silica structure and OH ions. Absorbed energy is mainly by the stretching vibrations of the OH bonds within the silica structure, which has a fundamental resonance in the infrared region (beyond 2.7 μm) but overtones or harmonics at lower wavelengths (or higher frequencies). The first overtone at around 1.4 μm is the most significant as can be seen in the figure. The second overtone is around 1 μm and in high-quality fibers this is negligible. A combination of first overtone of the OH vibration and the fundamental vibrational frequency of SiO_2 gives rise to a minor loss peak at around 1.24 μm. There are two important windows in the attenuation vs. wavelength behavior in which the attenuation exhibits minima. The window at around 1.3 μm is the region between two neighboring OH^- absorption peaks. This window is widely used in optical communications at 1310 nm. The window at around 1.55 μm is between the first harmonic absorption of OH^- and the infrared lattice absorption tail and represents the lowest attenuation. Current technological drive is to use this window for long-haul communications. It can be seen that it is important to keep the hydroxyl content in the fiber within tolerable levels.

There is a background attenuation process that decreases with wavelength and is due to the Rayleigh scattering of light by the local variations in the refractive index. Glass has a non-crystalline or an amorphous structure, which means that there is no long-range order to the arrangement of the atoms but only a short range order, typically a few bond lengths. The glass structure is as if the structure of the melt has been suddenly frozen. We can only define the number of bonds a given atom in the structure will have. Random variations in the bond angle from atom to atom lead to a disordered structure. There is therefore a random local variation in the density over a few bond lengths that leads to fluctuations in the refractive index over a few atomic lengths. These random fluctuations in the refractive index give rise to light scattering and hence light attenuation along the fiber. It should be apparent that since a degree of structural randomness is an intrinsic property of the glass structure this scattering process is unavoidable and represents the lowest attenuation possible through a glass medium. As one may surmise, attenuation by scattering in a medium is minimum for light propagating through a "perfect" crystal. In this case the only scattering mechanisms will be due to thermodynamic defects (vacancies) and the random thermal vibrations of the lattice atoms.

As mentioned above, the Rayleigh scattering process decreases with wavelength and, according to Rayleigh, it is inversely proportional to λ^4. The expression for the attenuation α_R in a single component glass due to Rayleigh scattering is approximately given by,

Rayleigh scattering in silica

$$\alpha_R \approx \frac{8\pi^3}{3\lambda^4}(n^2 - 1)^2\beta_T k_B T_f \qquad \textbf{(5)}$$

in which λ is the free space wavelength, n is the refractive index at the wavelength of interest, β_T is the isothermal compressibility (at T_f) of the glass, k_B is the Boltzmann constant, and T_f is a quantity called the *fictive temperature* (roughly the *softening temperature of glass*) at which the liquid structure during the cooling of the fiber is frozen to become the glass structure. Fiber is drawn at high temperatures and as the fiber cools eventually the temperature drops sufficiently for the atomic motions to be so sluggish that the structure becomes essentially "frozen-in" and remains like this even at room temperature. Thus, T_f marks the temperature below which the liquid structure is frozen and hence the density fluctuations are also frozen into the glass structure. It is apparent that Rayleigh scattering represents the lowest attenuation one can achieve using a glass structure. By proper design, the attenuation window at 1.5 μm may be lowered to approach the Rayleigh scattering limit.

External factors can also lead to attenuation in the optical fiber. The most important are microbending and macrobending losses. **Microbending loss** is due to a "sharp" local bending of the fiber that changes the guide geometry and refractive index profile locally, which leads to some of the light energy radiating away from the guiding direction. A sharp bend, as illustrated intuitively in Figure 2.32, will change the local waveguide geometry in such a way that a zigzagging ray suddenly finds itself with an incidence angle θ', narrower than its normal angle θ ($\theta' < \theta$), which gives rise to either a transmitted wave (a refracted wave into the cladding) or to a greater cladding penetration. If $\theta' < \theta_c$, the critical angle, then there will be no total internal reflection and substantial light power will be radiated into the cladding and eventually to the outside medium (polymer coating, *etc.*).

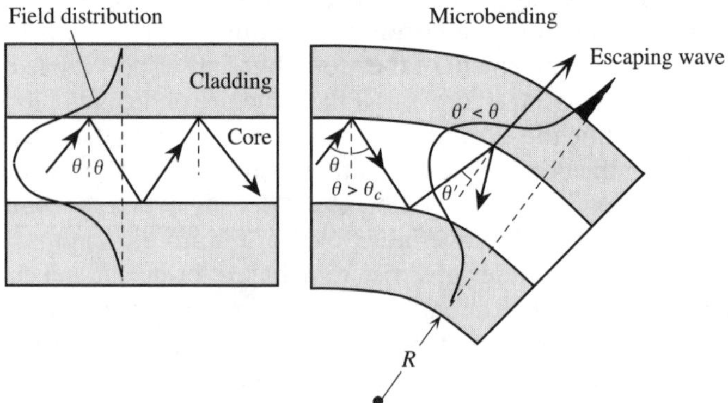

FIGURE 2.32 Sharp bends change the local waveguide geometry that can lead to waves escaping. The zigzagging ray suddenly finds itself with an incidence angle θ' that gives rise to either a transmitted wave, or to a greater cladding penetration; the field reaches the outside medium and some light energy is lost.

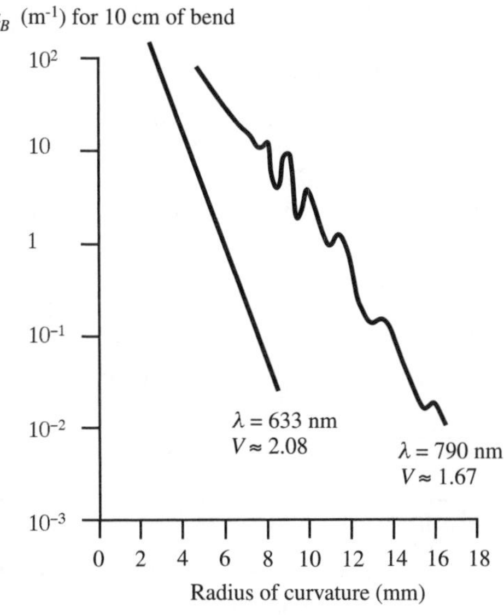

FIGURE 2.33 Measured microbending loss for a 10 cm fiber bent by different amounts of radius of curvature R. Single mode fiber with a core diameter of 3.9 μm, cladding radius 48 μm, $\Delta = 0.004$, $NA = 0.11, V \approx 1.67$ and 2.08 (Data extracted and replotted with Δ correction from, A.J. Harris and P.F. Castle, *IEEE J. Light Wave Technology*, Vol. LT14, pp. 34–40, 1986; see original article for discussion of peaks in α_B vs. R at 790 nm.)

Greater penetration can lead to the optical field reaching the cladding-outer boundary and hence to some of the light being lost away into the outer coating. Attenuation increases sharply with the extent of bending; as θ' gets narrow and TIR is lost, substantially more energy is transferred into the cladding. Further, highest modes propagate with incidence angles θ close to θ_c, which means that these modes are most severely affected. Multimode fibers therefore suffer more from bending losses then single mode fibers.

Microbending loss α_B increases rapidly with increasing bend "sharpness," *i.e.* with decreasing radius of bend curvature, R (which is defined in Figure 2.32). Figure 2.33 shows a typical microbending loss α_B dependence on the radius of curvature R for a single mode fiber for two different operating wavelengths. As apparent, α_B increases exponentially with R (this is a semilogarithmic plot), which depends on the wavelength and the fiber characteristics (*e.g.* V-number). Typically, bend radii less than ~10 mm can lead to appreciable microbending loss.

Macrobending loss is due to small changes in the refractive index of the fiber due to induced strains when it is bent during its use, *e.g.*, when it is cabled and laid. Induced strains change n_1 and n_2, and hence affect the mode field diameter, that is, the field penetration into the cladding. Some of this increased cladding field will reach the cladding boundary to become lost in the outer medium (radiation, absorption, *etc.*). Typically, macrobending loss crosses over into microbending loss when the radius of curvature becomes less than a few centimeters.

EXAMPLE 2.9.1 Rayleigh scattering limit

What is the attenuation due to Rayleigh scattering at around the $\lambda = 1.55$ μm window given that pure silica (SiO_2) has the following properties: $T_f = 1730$°C (softening temperature); $\beta_T = 7 \times 10^{-11}$ m^2 N^{-1} (at high temperatures); $n = 1.4446$ at 1.5 μm.

Solution We simply calculate the Rayleigh scattering attenuation using

$$\alpha_R \approx \frac{8\pi^3}{3\lambda^4}(n^2 - 1)^2\beta_T k_B T_f$$

so that

$$\alpha_R \approx \frac{8\pi^3}{3(1.55 \times 10^{-6})^4}(1.4446^2 - 1)^2(7 \times 10^{-11})(1.38 \times 10^{-23})(1730 + 273)$$

$$= 3.27 \times 10^{-5}\ \text{m}^{-1} \text{ or } 3.27 \times 10^{-2}\ \text{km}^{-1}.$$

Attenuation in dB per km is then

$$\alpha_{dB} = 4.34\alpha_R = (4.34)(3.27 \times 10^{-2}\ \text{km}^{-1}) = 0.142\ \text{dB km}^{-1}.$$

This represents the lowest possible attenuation for a silica glass core fiber at 1.55 μm.

EXAMPLE 2.9.2 Attenuation along an optical fiber

The optical power launched into a single-mode optical fiber from a laser diode is approximately 1 mW. The photodetector at the output requires a minimum power of 10 nW to provide a clear signal (above noise). The fiber operates at 1.3 μm and has an attenuation coefficient of 0.4 dB km^{-1}. What is the maximum length of fiber that can be used without inserting a repeater (to regenerate the signal)?

Solution We can use

$$\alpha_{dB} = \frac{1}{L}10\log\left(\frac{P_{in}}{P_{out}}\right)$$

so that

$$L = \frac{1}{\alpha_{dB}}10\log\left(\frac{P_{in}}{P_{out}}\right) = \frac{1}{0.4}10\log\left(\frac{10^{-3}}{10 \times 10^{-9}}\right) = 125\ \text{km}$$

There will be additional losses, such as fiber bending losses, which arise from the bending of the fiber that will reduce this length to below this limit. For long distance communications, the signal has to be amplified, using an optical ampfier, after a distance of about 50 − 100 km, and eventually regenerated by using a repeater.

2.10 FIBER MANUFACTURE

A. Fiber Drawing

There are a number of processes for producing optical fibers for various applications. We will consider the **outside vapor deposition** (OVD) technique, which is one of the widely-used processes and produces fiber with low loss. It is also known as the outside vapor phase oxidation process.

The first step is to prepare a **preform**, which is a glass rod that has the right refractive index profile across its cross section and the right glass properties (*e.g.* negligible amounts of impurities). This rod is typically 10 − 30 mm in diameter and about one to two meters in length. The optical fiber is drawn from this preform using special fiber-drawing equipment that is schematically illustrated in Figure 2.34.

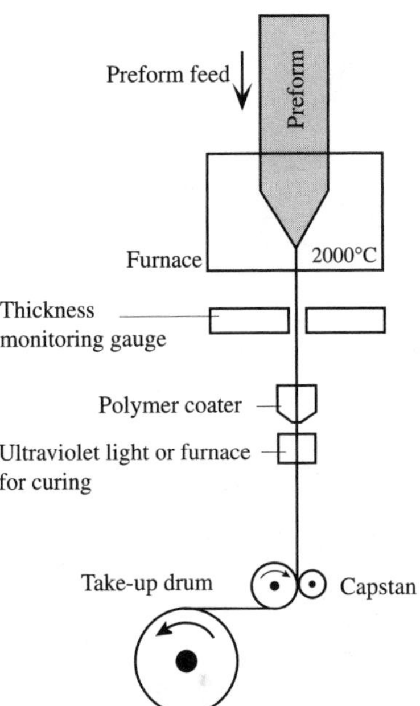

FIGURE 2.34 Schematic illustration of a fiber drawing tower.

The preform rod is slowly fed into a hot furnace that has a hot zone around 1900–2000 °C where the glass flows like a viscous melt (resembling honey). As the rod reaches the hot zone and its end begins to flow, its tip is pulled, with just the right tension, to come out as a fiber and is spooled on a rotating take-up drum. The diameter of the fiber must be highly controlled to achieve the required waveguide characteristics. An optical thickness monitor gauge provides information on the changes of the fiber diameter that is used (in an automatic feedback control system) to adjust the speed of the fiber winding mechanism and the speed of the preform feeder to maintain a constant fiber diameter, typically better than 0.1%. In some cases, the preform is hollow, that is, it has a thin central hole along the rod axis. The hollow simply collapses during the drawing and does not affect the final drawn fiber.

It is essential that, as soon as the fiber is drawn, it is coated with a polymeric layer (*e.g.* urethane acrylate) to mechanically and chemically protect the fiber surface. When a bare fiber glass surface is exposed to ambient conditions it quickly develops various microcracks on the surface and these dramatically reduce the mechanical strength (fracture strength) of the fiber. The applied polymeric coating is initially a viscous liquid and needs to be cured (hardened), which is done as the coated fiber passes through a curing oven, or with ultraviolet lamps if it is UV hardenable. Sometimes two layers of polymeric coating are applied. Cladding is typically 125–150 μm and the overall diameter with the polymeric coatings is 250–500 μm. A schematic diagram of the cross section of a typical single-mode optical fiber is shown in Figure 2.35. In this example, there is a thick polymeric buffer tube, or a buffer jacket, surrounding the fiber and its coating to cushion the fiber against mechanical pressure and microbending (sharp bending). Some fibers are buffered by having the fiber loose within a buffer tube. The tube may then contain a filling compound to increase the buffering ability. Single and multiple fibers are

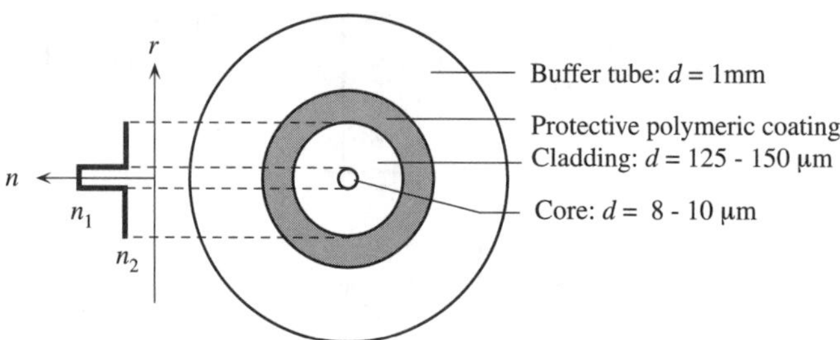

FIGURE 2.35 The cross section of a typical single-mode fiber with a tight buffer tube. (d = diameter)

invariably used in cable form and the structure of the cable depends on the application (*e.g.* long-haul communications), the number of fibers carried, and the cable environment (*e.g.* underground, underwater, overhead, *etc.*).

B. Outside Vapor Deposition (OVD)

Outside vapor deposition (OVD) is one of the vapor deposition techniques used to produce the rod preform used in fiber drawing. The OVD process is illustrated in Figure 2.36 and has two stages. The first **laydown** stage involves using a fused silica glass rod (or a ceramic rod such as alumina) as a target rod as shown Figure 2.36 (a). This acts as a mandrel and is rotated. The required glass material for the preform with the right composition is grown on the outside surface of this target rod by depositing glass soot particles. The deposition is achieved by burning various gases in an oxyhydrogen burner (torch) flame where glass soot is produced as reaction products.

Suppose that we need a preform with a core that has germania (GeO_2) in silica glass so that the core has a higher refractive index. The required gases, $SiCl_4$ (silicon tetrachloride), $GeCl_4$ (germanium tetrachloride), and fuel in the form of oxygen O_2 and

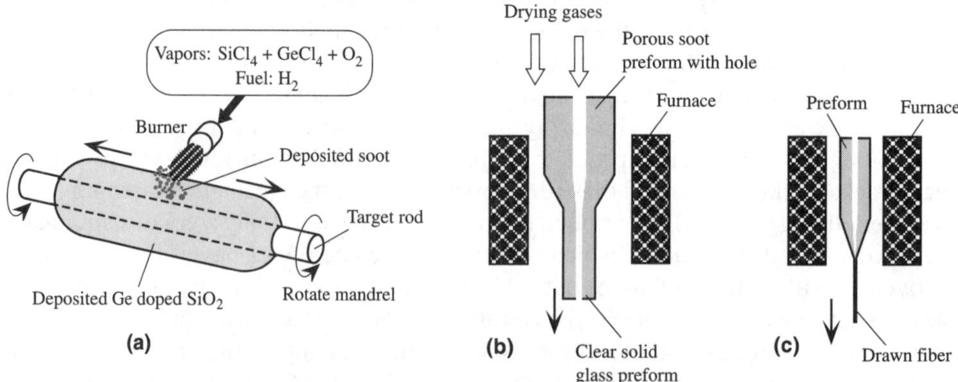

FIGURE 2.36 Schematic illustration of OVD and the preform preparation for fiber drawing. (a) Reaction of gases in the burner flame produces glass soot that deposits on to the outside surface of the mandrel. (b) The mandrel is removed and the hollow porous soot preform is consolidated; the soot particles are sintered, fused, together to form a clear glass rod. (c) The consolidated glass rod is used as s preform in fiber drawing.

hydrogen H_2 are burnt in a burner flame over the target rod surface as shown in Figure 2.36 (a). The important reactions of the gases in the flame are

$$SiCl_4(\text{gas}) + O_2(\text{gas}) \longrightarrow SiO_2(\text{solid}) + 2Cl_2(\text{gas})$$
$$GeCl_4(\text{gas}) + O_2(\text{gas}) \longrightarrow GeO_2(\text{solid}) + 2Cl_2(\text{gas})$$

These reactions produce fine glass particles, silica and germania, called "soot" that deposit on the outside surface of the target rod and form a porous glass layer as the burner travels along the mandrel. The glass preform is built layer by layer by slowly running the burner up and down along the length of the rotating mandrel either by actually moving the burner or moving the mandrel (the same result). First, the layers for the core region are deposited and then the gas composition is adjusted to deposit the layers for the cladding region. Typically, there may be about 200 layers in the final preform. The composition and hence the refractive index of each layer can of course be controlled by adjusting the relative amounts of $SiCl_4$ and $GeCl_4$ gases fed into the burner for the chemical reaction. Indeed, any desired refractive index profile, in principle, can be obtained by controlling the layer compositions.

Once all the necessary glass layers have been deposited, the central target mandrel is removed, which leaves a hollow porous glass preform rod; a porous opaque glass tube. The second **consolidation** stage involves sintering this porous glass rod as illustrated in Figure 2.36 (b). The porous preform is fed through a consolidation furnace (1400 – 1600 °C) in which the high temperature sinters (fuses) the fine glass particles into a dense, clear solid, the **glass preform**. At the same time, drying gases (such as chlorine or thionyl chloride) are forced through to remove water vapors and hydroxyl impurities that otherwise would result in unacceptably high attenuation. This clear glass preform is then fed into a **draw furnace** to draw the fiber as shown in Figure 2.36 (c) and described above. The central hollow simply collapses and fuses at the high temperatures of the draw process. Typically, a preform takes several hours to make and the subsequent drawing of a fiber from the preform takes a few more hours. The manufacturing cost reflects in the fiber cost, which is in excess of $25 per km (1999).

Left: The soot rod fed into the consolidation furnace for sintering. Right: Glass preform fed into the fiber drawing furnace. *(Courtesy of Corning.)*

EXAMPLE 2.10.1 Fiber drawing

In a certain fiber production process a preform of length 110 cm and diameter 20 mm is used to draw a fiber. Suppose that the fiber drawing rate is 5 m s^{-1}. What is the maximum length of fiber that can be drawn from this preform if the last 10 cm of the preform is not drawn and the fiber diameter is 125 μm? How long does it take to draw the fiber?

Solution We assume that a length L_p of the preform is drawn as fiber. Since the density is the same and mass is conserved, the volume must also be conserved. If d_f and d_p are the fiber and preform diameters, and L_f and L_p are the lengths of the fiber and drawn preform respectively, then,

$$L_f d_f^2 = L_p d_p^2$$

i.e.

$$L_f = \frac{(1.1 - 0.1\text{ m})(20 \times 10^{-3}\text{ m})^2}{(125 \times 10^{-6}\text{ m})^2} = 25{,}600\text{ m or }25.6\text{ km.}$$

Since the rate is 5 m/s, the time it takes to draw the fiber in minutes is

$$\text{Time (hrs)} = \frac{\text{Length (km)}}{\text{Rate (km/hr)}} = \frac{25{,}600\text{ m}}{(5\text{ m/s})(60 \times 60\text{ s/hr})} = 1.4\text{ hrs.}$$

Typical drawing rates are in the range 5 − 20 m s^{-1} so that 1.4 hrs would be on the long-side.

QUESTIONS AND PROBLEMS

2.1 Dielectric slab waveguide

(a) Consider the rays 1 and 2 in Figure 2.3. Derive the waveguide condition.

(b) Consider the two rays 1 and 2 in Figure 2.4. Show that the phase difference when they meet at C at a distance y above the guide center is

$$\Phi_m = k_1 2(a - y)\cos\theta_m - \phi_m$$

(c) Using the waveguide condition, show that

$$\Phi_m = \Phi_m(y) = m\pi - \frac{y}{a}(m\pi + \phi_m)$$

2.2 TE field pattern in slab waveguide Consider two parallel rays 1 and 2 interfering in the guide as in Figure 2.4. Given the phase difference (as in Question 2.1)

$$\Phi_m = \Phi_m(y) = m\pi - \frac{y}{a}(m\pi + \phi_m)$$

between the waves at C, distance y above the guide center, find the electric field pattern $E(y)$ in the guide.

Plot the field pattern for the first three modes taking a planar dielectric guide with a core thickness 20 μm, $n_1 = 1.455$ $n_2 = 1.440$, light wavelength of 1.3 μm.

2.3 TE and TM Modes in dielectric slab waveguide Consider a planar dielectric guide with a core thickness 20 μm, $n_1 = 1.455$ $n_2 = 1.440$, light wavelength of 1.3 μm. Given the waveguide condition, Eq. (3) in §2.1, and the expressions for phase changes ϕ and ϕ' in TIR for the TE and TM modes respectively,

$$\tan\left(\tfrac{1}{2}\phi_m\right) = \frac{\left[\sin^2\theta_m - \left(\dfrac{n_2}{n_1}\right)^2\right]^{1/2}}{\cos\theta_m} \quad \text{and} \quad \tan\left(\tfrac{1}{2}\phi'_m\right) = \frac{\left[\sin^2\theta_m - \left(\dfrac{n_2}{n_1}\right)^2\right]^{1/2}}{\left(\dfrac{n_2}{n_1}\right)^2 \cos\theta_m}$$

using a graphical solution find the angle θ for the fundamental TE and TM modes and compare their propagation constants along the guide.

2.4 Group velocity We can calculate the group velocity of a given mode as a function of frequency ω using a convenient math software package. It is assumed that the math software package can carry out symbolic algebra such as partial differentiation (the author used Mathview by Maple, though others can also be used). The propagation constant of a given mode is $\beta = k_1 \sin\theta$ in which β and θ imply β_m and θ_m. The objective is to express β and ω in terms of θ. Since $k_1 = n_1\omega/c$, the waveguide condition in Eq. (11) in §2.1 is

$$\tan\left(a\frac{\beta}{\sin\theta}\cos\theta - m\frac{\pi}{2}\right) = \frac{\left[\sin^2\theta - \left(\dfrac{n_2}{n_1}\right)^2\right]^{1/2}}{\cos\theta}$$

so that

$$\beta = \frac{\tan\theta}{a}\left[\arctan\left(\sec\theta\sqrt{\sin^2\theta - \left(\frac{n_2}{n_1}\right)^2}\right) + m\frac{\pi}{2}\right] = F_m(\theta) \quad (1)$$

The frequency is given by

$$\omega = \frac{c\beta}{n_1\sin\theta} = \frac{c}{n_1\sin\theta}F_m(\theta) \quad (2)$$

Both β and ω are now functions of θ. Then the group velocity is

$$v_g = \frac{d\omega}{d\beta} = \left[\frac{d\omega}{d\theta}\right]\left[\frac{d\theta}{d\beta}\right] = \frac{c}{n_1}\left[\frac{F'_m(\theta)}{\sin\theta} - \frac{\cos^2\theta}{\sin\theta}F_m(\theta)\right]\left[\frac{1}{F'_m(\theta)}\right]$$

i.e.

$$v_g = \frac{c}{n_1\sin\theta}\left[1 - \cos^2\theta\frac{F_m(\theta)}{F'_m(\theta)}\right] \quad (3)$$

For a given m value, Eqs (2) and (3) can be plotted parametrically, that for each θ value we can calculate ω and v_g and plot v_g vs. ω. Figure 2.37 shows an example for a guide with the characteristics in the caption. Using a convenient math software package, or by other means, obtain the same v_g vs. ω behavior and discuss intermodal dispersion and whether the Equation (2) in §2.2 is appropriate.

2.5 Dielectric slab waveguide Consider a dielectric slab waveguide that has a thin GaAs layer of thickness 0.2 μm between two AlGaAs layers. The refractive index of GaAs is 3.66 and that of the AlGaAs layers is 3.40. What is the cut-off wavelength beyond which only a single mode can propagate in the waveguide, assuming that the refractive index does not vary greatly with the wavelength? If a radiation of wavelength 870 nm (corresponding to bandgap radiation) is propagating in the GaAs layer, what is the penetration of the evanescent wave into the AlGaAs layers? What is the mode field distance of this radiation?

2.6 Dielectric slab waveguide Consider a slab dielectric waveguide that has a core thickness ($2a$) of 10 μm, $n_1 = 3$, $n_2 = 1.5$. Solution of the waveguide condition in Eq. (11) in Example 2.1.1 gives the mode angles θ_0 and θ_1 for the TE_0 and TE_1 modes for selected wavelengths as summarized in the table below. For each wavelength calculate ω and β_m and

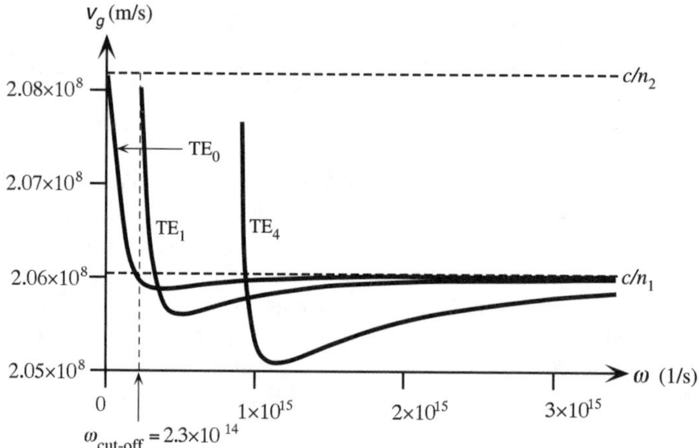

FIGURE 2.37 Group velocity vs. angular frequency for three modes for a planar dielectric waveguide which has $n_1 = 1.455$, $n_2 = 1.440$, $a = 10$ μm (Results from Mathview, Waterloo Maple math-software application). TE$_0$ is for $m = 0$ *etc.*

then plot ω vs. β_m. On the same plot show the lines with slopes c/n_1 and c/n_2. Compare your plot with the dispersion diagram in Figure 2.10.

λ, μm	15	20	25	30	40	45	50	70	100	150	200
θ_0°	77.8	74.52	71.5	68.7	63.9	61.7	59.74	53.2	46.4	39.9	36.45
θ_1°	65.2	58.15	51.6	45.5	35.5	32.02	30.17	–	–	–	–

2.7 Dielectric slab waveguide Consider a planar dielectric waveguide with a core thickness 10 μm, $n_1 = 1.4446$, $n_2 = 1.4440$. Calculate the V-number, the mode angle θ_m for $m = 0$ (use a graphical solution, if necessary), penetration depth, and mode field distance (MFD $= 2\alpha + 2\delta$), for light wavelengths of 1.0 μm and 5 μm. What is your conclusion? Compare your MFD calculation with $2w_o = 2a(V + 1)/V$.

2.8 A multimode fiber Consider a multimode fiber with a core diameter of 100 μm, core refractive index of 1.475, and a cladding refractive index of 1.455 both at 850 nm. Consider operating this fiber at $\lambda = 850$ nm.

(a) Calculate the V-number for the fiber and estimate the number of modes.

(b) Calculate the wavelength beyond which the fiber becomes single mode.

(c) Calculate the numerical aperture.

(d) Calculate the maximum acceptance angle.

(e) Calculate the modal dispersion $\Delta\tau$ and hence the bit rate × distance product given that rms dispersion $\sigma \approx 0.29\,\Delta\tau$ in which $\Delta\tau$ is the full spread.

2.9 A single mode fiber Consider a fiber with a SiO_2–13.5%GeO_2 core of diameter of 8 μm and refractive index of 1.468 and a cladding refractive index of 1.464 both refractive indices at 1300 nm where the fiber is to be operated using a laser source with a half maximum width (FWHM) of 2 nm.

(a) Calculate the V-number for the fiber. Is this a single mode fiber?

(b) Calculate the wavelength below which the fiber becomes multimode.

(c) Calculate the numerical aperture.

(d) Calculate the maximum acceptance angle.

(e) Obtain the material dispersion and waveguide dispersion and hence estimate the bit rate × distance product $(B \times L)$ of the fiber.

2.10 A single mode fiber design According to Question 1.3 (Ch. 1), the Sellmeier dispersion equation provides n vs. λ for pure SiO_2 and SiO_2–13.5 mol.%GeO_2. The refractive index increases linearly with the addition of GeO_2 to SiO_2 from 0 to 13.5 mol.%. A single mode step index fiber is required to have the following properties: $NA = 0.1$, core diameter of 9 μm, and a core of SiO_2–13.5% GeO_2. What should the cladding composition be?

2.11 Material dispersion If N_{g1} is the group refractive index of the core material of a step fiber, then the propagation time (group delay time) of the fundamental mode is

$$\tau = \frac{L}{v_g} = \frac{LN_{g1}}{c}$$

Since N_g will depend on the wavelength, show that the material dispersion coefficient D_m is given approximately by

$$D_m = \frac{d\tau}{Ld\lambda} \approx \frac{\lambda}{c}\frac{d^2n}{d\lambda^2}$$

Using the Sellmeier equation in Question 1.3 in Chapter 1, evaluate material dispersion at $\lambda = 1.55$ μm for pure silica (SiO_2) and SiO_2–13.5%GeO_2 glass.

2.12 Waveguide dispersion Waveguide dispersion arises as a result of the dependence of the propagation constant on the V-number, which depends on the wavelength. It is present even when the refractive index is constant; no material dispersion. Let us suppose that n_1 and n_2 are wavelength (or k) independent. Suppose that β is the propagation constant of mode lm and $k = 2\pi/\lambda$ in which λ is the free space wavelength. Then the normalized propagation constant b and propagation constant k are related by (see Example 2.3.4)

$$\beta = n_2k[1 + b\Delta] \tag{1}$$

The group velocity is defined and given by

$$v_g = \frac{d\omega}{d\beta} = c\frac{dk}{d\beta}$$

Show that the propagation time, or the group delay time, τ of the mode is

$$\tau = \frac{L}{v_g} = \frac{Ln_2}{c} + \frac{Ln_2\Delta}{c}\frac{d(kb)}{dk} \tag{2}$$

Given the definition of V,

$$V = ka\left[n_1^2 - n_2^2\right]^{1/2} \approx kan_2(2\Delta)^{1/2} \tag{3}$$

and

$$\frac{d(Vb)}{dV} = \frac{d}{dV}\left[bkan_2(2\Delta)^{1/2}\right] = an_2(2\Delta)^{1/2}\frac{d}{dV}(bk) \tag{4}$$

Show that

$$\frac{d\tau}{d\lambda} = -\frac{Ln_2\Delta}{c\lambda}V\frac{d^2(Vb)}{dV^2} \tag{5}$$

and that the waveguide dispersion coefficient is

$$D_w = \frac{d\tau}{Ld\lambda} = -\frac{n_2\Delta}{c\lambda}V\frac{d^2(Vb)}{dV^2} \tag{6}$$

FIGURE 2.38 $V\left[d^2(Vb)/dV^2\right]$ vs. V-number for a step index fiber (after W.A. Gambling *et al., The Radio and Electronics Engineer*, **51**, 313, 1981)

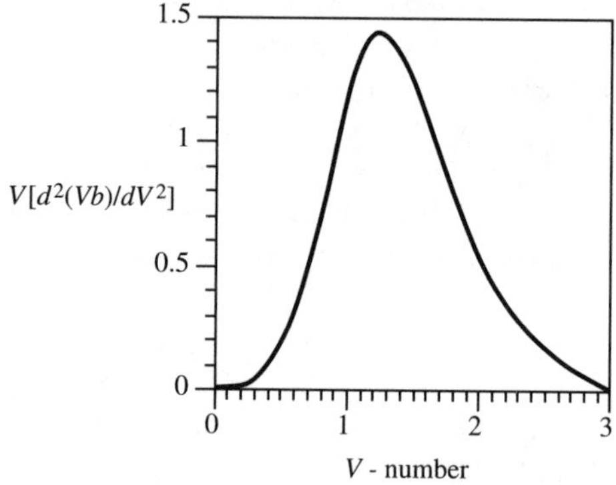

Figure 2.38[13] shows the dependence of $V\left[d^2(Vb)/dV^2\right]$ on the V-number. In the range $1.5 < V < 2.4$,

$$V\frac{d^2(Vb)}{dV^2} \approx \frac{1.984}{V^2}$$

Show that,

$$D_w \approx -\frac{n_2\Delta}{c\lambda}\frac{1.984}{V^2} = -\frac{(n_1 - n_2)}{c\lambda}\frac{1.984}{V^2} \qquad \textbf{(7)}$$

which simplifies to

$$D_w \approx -\frac{1.984}{c(2\pi a)^2 2n_2}\lambda \qquad \textbf{(8)}$$

Equation (2) should really have N_{g2} instead of n_2 in which case Eq. (8) would be

$$D_w \approx -\frac{1.984N_{g2}}{c(2\pi a)^2 2n_2^2}\lambda \qquad \textbf{(9)}$$

Consider a fiber with a core of diameter of 8 μm and refractive index of 1.468 and a cladding refractive index of 1.464, both refractive indices at 1300 nm. Suppose that a 1.3 μm laser diode with a spectral linewidth of 2 nm is used to provide the input light pulses. Estimate the waveguide dispersion per kilometer of fiber using Eqs. (6) and (9).

2.13 Profile dispersion Total dispersion in a single mode, step index fiber is primarily due to material dispersion and waveguide dispersion. However, there is an additional dispersion mechanism called *profile dispersion* that arises from the propagation constant β of the fundamental mode also depending on the refractive index difference Δ. Consider a light source with a range of wavelengths $\delta\lambda$ coupled into a step index fiber. We can view this as a change $\delta\lambda$ in the input wavelength λ. Suppose that n_1, n_2, hence Δ depend on the wavelength λ. The propagation time, or the group delay time, τ_g *per unit length* is

$$\tau_g = \frac{1}{v_g} = \frac{1}{c}\left(\frac{d\beta}{dk}\right) \qquad \textbf{(1)}$$

Since β depends on n_1, Δ and V, let us consider τ_g as a function of n_1, Δ (thus n_2), and V. A change $\delta\lambda$ in λ will change each of these quantities. Using the partial differential chain rule,

[13] Many books state the approximation $Vd^2(Vb)/dV^2 \approx 1.984/V^2$ and show a graph of $Vd^2(Vb)/dV^2$ vs. V (from Gloge, *Applied Optics*, **10**, 2442, 1971) that does not match this approximation over the range considered.

$$\frac{\delta\tau_g}{\delta\lambda} = \frac{\partial\tau_g}{\partial n_1}\frac{\partial n_1}{\partial\lambda} + \frac{\partial\tau_g}{\partial V}\frac{\partial V}{\partial\lambda} + \frac{\partial\tau_g}{\partial\Delta}\frac{\partial\Delta}{\partial\lambda} \tag{2}$$

The mathematics turns out to be complicated but the statement in Eq. (2) is equivalent to

Total dispersion = Material dispersion (due to $\partial n_1/\partial\lambda$)

+ Waveguide dispersion (due to $\partial V/\partial\lambda$)

+ Profile dispersion (due to $\partial\Delta/\partial\lambda$)

in which the last term is due to Δ depending on λ; although small, this is not zero. Even the statement in Eq. (2) above is over simplified but nonetheless provides an insight into the problem. The total intramode (chromatic) dispersion coefficient D_{ch} is then given by

$$D_{ch} = D_m + D_w + D_p \tag{3}$$

in which D_m, D_w, D_p are material, waveguide, and profile dispersion coefficients respectively. The waveguide dispersion is given by Eq. (8) in Question 2.12 and the profile dispersion coefficient is (very) approximately,

$$D_p \approx -\frac{N_{g1}}{c}\left(V\frac{d^2(Vb)}{dV^2}\right)\left(\frac{d\Delta}{d\lambda}\right) \tag{4}$$

in which b is the normalized propagation constant and $Vd^2(Vb)/dV^2$ vs. V is shown in Figure 2.38. The term $Vd^2(Vb)/dV^2 \approx 1.984/V^2$.

Consider a fiber with a core of diameter of 8 μm. The refractive and group indices of the core and cladding at $\lambda = 1.55$ μm are $n_1 = 1.4504$, $n_2 = 1.4450$, $N_{g1} = 1.4676$, $N_{g2} = 1.4625$, and $d\Delta/d\lambda = 161\ \text{m}^{-1}$. Estimate the waveguide and profile dispersion per km of fiber per nm of input light linewidth at this wavelength.

2.14 A graded index fiber

(a) Consider an optimal graded index fiber with a core diameter of 30 μm and a refractive index of 1.474 at the center of the core and a cladding refractive index of 1.453. Suppose that the fiber is coupled to a laser diode emitter at 1300 nm and a spectral linewidth (FWHM) of 3 nm. Suppose that the material dispersion coefficient at this wavelength is about -5 ps km^{-1} nm^{-1}. Calculate the total dispersion and estimate the bit rate × distance product of the fiber. How does this compare with the performance of a multimode fiber with same core radius, and n_1 and n_2? What would the total dispersion and maximum bit rate be if an LED source of spectral width (FWHM) $\Delta\lambda_{1/2} \approx 80$ nm is used?

(b) If $\sigma_{\text{intermode}}(\gamma)$ is the rms dispersion in a graded index fiber with a profile index γ, and if γ_o is the optimal profile index, then

$$\frac{\sigma_{\text{intermode}}(\gamma)}{\sigma_{\text{intermode}}(\gamma_o)} = \frac{2(\gamma - \gamma_o)}{\Delta(\gamma + 2)}$$

and so is given by Eq. (2) in §2.7. Calculate the new dispersion and bit rate × distance product if γ is 10% greater than the optimal value γ_o.

2.15 A planar waveguide with stratified medium (approximation to a graded index fiber) Figure 2.39 shows a planar dielectric waveguide in which the refractive index changes from n_1 to n_2 to n_3 and so on along y at $y = \delta/2, 3\,\delta/2, 5\,\delta/2, \ldots$. Thus, the refractive index decreases one step at a time from $y = 0$ along y as depicted in the figure.

Consider the paths of two rays, A and B, starting from O. The first ray suffers total internal refraction at A and travels to O'. The launching angle for the first ray is such that θ_A is the

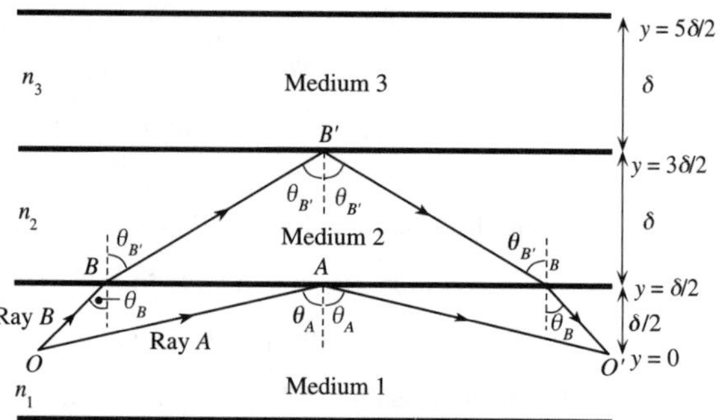

FIGURE 2.39 Step-graded-index dielectric waveguide. Two rays are launched from the center of the waveguide at O at angles θ_A and θ_B such that ray A suffers TIR at A and ray B suffers TIR at B'. Both TIRs are at critical angles.

critical angle for TIR. Ray B, launched at a narrower angle, penetrates into medium 2 at point B and then travels toward point B' at the boundary between media 2 and 3. Ray B is launched such that at point B' its angle $\theta_{B'}$ is the critical angle for TIR between media 2 and 3.

The problem is what the relationship between n_1, n_2, and n_3 should be to have both rays arrive at the same time at point O' so that we observe no time dispersion. By symmetry, if two rays arrive at the same time at O', they must also arrive at their respective TIR points at A and B' at the same time.

(a) Show that the time taken for the first ray to travel from O to A is

$$t_{OA} = \frac{\dfrac{\left(\frac{1}{2}\delta\right)}{\cos\theta_A}}{\dfrac{c}{n_1}} = \frac{\left(\frac{1}{2}\delta\right)n_1}{c\left[1 - \left(\dfrac{n_2}{n_1}\right)^2\right]^{1/2}} \tag{1}$$

(b) Show that the time taken for the second ray to travel from O to B' is

$$t_{OB'} = \frac{\left(\frac{1}{2}\delta\right)n_1}{c\left[1 - \left(\dfrac{n_3}{n_1}\right)^2\right]^{1/2}} + \frac{\delta n_2}{c\left[1 - \left(\dfrac{n_3}{n_2}\right)^2\right]^{1/2}} \tag{2}$$

Let the step variations of n at $y = \delta/2, 3\delta/2, \ldots$ obey

$$n^2 = n_1^2\left[1 - 2\Delta\left(\frac{y}{a}\right)^\gamma\right] \tag{3}$$

in which Δ is some constant (less than unity), and γ is an index that describes the profile of the refractive index along y. Obviously, at $y = 0, n = n_1$ as we expect. Show that

$$n_2^2 = n_1^2(1 - \varepsilon)$$

in which

$$\varepsilon = 2\Delta\left(\frac{\delta}{2a}\right)\gamma \tag{4}$$

and

$$n_3^2 = n_1^2[1 - \varepsilon(3^\gamma)] \tag{5}$$

(d) Consider the condition $t_{OA} - t_{OB'} = 0$ so that the two rays arrive at the same time at O'. Using the results in Eqs (1) and (2) and the refractive indices in Eqs (4) and (5), show that the two rays arrive at the same time if

$$\frac{2(1 - \varepsilon)}{(3^\gamma - 1)^{1/2}} + \frac{1}{3^{\gamma/2}} - 1 = 0 \tag{6}$$

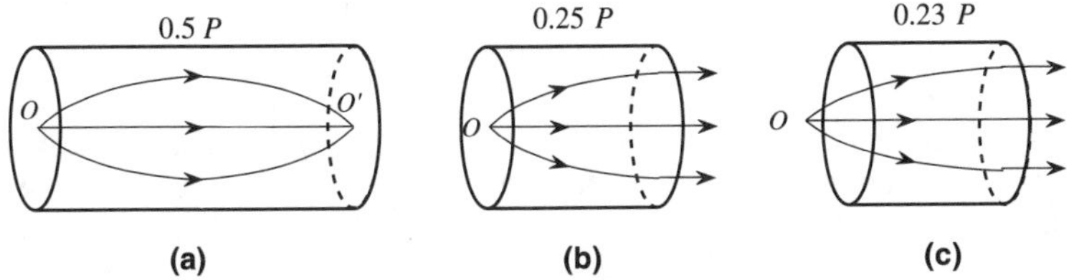

FIGURE 2.40 Graded index (GRIN) rod lenses of different pitches. (a) Point O is on the rod face center and the lens focuses the rays onto O' on to the center of the opposite face. (b) The rays from O on the rod face center are collimated out. (c) O is slightly away from the rod face and the rays are collimated out.

As the layer thickness δ becomes small we have $\varepsilon \rightarrow 0$. Show that as $\varepsilon \rightarrow 0, \gamma = 2.067$ is a solution of Eq. (6).

Do you have a conclusion from this exercise? What type of graded index fiber would you *recommend* for minimal intermodal dispersion? Sketch schematically its index profile along a radial direction.

What is the main theoretical limitation of the treatment above? Is $\varepsilon \rightarrow 0$ a valid assumption? What will happen to γ if κ was not zero but a small number?

2.16 GRIN rod lenses *Graded index* (GRIN) rod lens is a glass rod whose refractive index changes parabolically from its central axis where the index is maximum. It is like a very thick, short graded index fiber whose diameter is perhaps 0.5 − 5 mm. Such GRIN rod lenses of different lengths can be used to focus or collimate light rays, as illustrated in Figure 2.40. The principle of operation can be understood by considering ray trajectories in a stratified medium as in Figure 2.27 and Figure 2.25 (b), in which ray trajectories are sinusoidal looking paths. One pitch (P) is a full one period variation in the ray trajectory along the rod axis. Figure 2.40 (a), (b), and (c) show half-pitch ($0.5P$), quarter-pitch and ($0.25P$), and $0.23P$ GRIN rod lenses. The point O in (a) and (b) is on the rod face center where as in (c) it is slightly away from the rod face.

(a) How would you represent Figure 2.40 (a) using two conventional converging lenses. What are O and O'?

(b) How would you represent Figure 2.40 (b) using a conventional converging lens. What is O?

(c) Sketch ray paths for a GRIN rod with a pitch between $0.25P$ and $0.5P$ starting from O at the face center. Where is O'?

(d) What use is $0.23P$ GRIN rod lens in Figure 2.40 (c)?

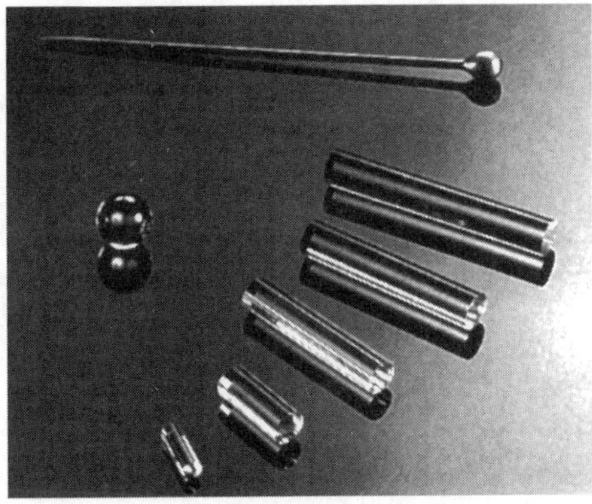

GRIN rod lenses and a spherical lens (a ball lens) used in coupling light into fibers. *(Courtesy of Melles Griot.)*

2.17 Optical Fibers Consider the manufacture of optical fibers and the materials used.

(a) What factors would reduce dispersion?

(b) What factors would reduce attenuation?

2.18 Microbending loss It is found that for a single mode fiber with a cut-off wavelength $\lambda_c = 1180$ nm, operating at 1300 nm, the microbending loss reaches 1 dB m^{-1} when the radius of curvature of the bend is roughly 6 mm for $\Delta = 0.00825$, 12 mm for $\Delta = 0.00550$, and 35 mm for $\Delta = 0.00275$. Explain these findings.

2.19 Microbending loss Microbending loss α_B depends on the fiber characteristics and wavelength. We will calculate α_B approximately given various fiber parameters using the single mode fiber microbending loss equation (D. Marcuse, *J. Op. Soc. Am.*, Vol. 66, pp. 216–220, 1976)

$$\alpha_B = \frac{\pi^{1/2}\kappa^2}{2\gamma^{3/2}V^2[K_1(\gamma a)]^2} R^{-1/2} \exp\left(-\frac{2\gamma^3}{3\beta^2} R\right)$$

in which R is the bend radius of curvature, a is the fiber radius, β is the propagation constant, determined by b, normalized propagation constant, which is related to V, $\beta = n_2 k[1 + b\Delta]$; $k = 2\pi/\lambda$ is the free-space wavevector; $\gamma = \sqrt{[\beta^2 - n_2^2 k^2]}$; $\kappa = \sqrt{[n_1^2 k^2 - \beta]}$, and $K_1(x)$ is a first-order modified Bessel function, readily available in math software packages. b can be found from Eq. (7) in Example 2.3.4, $b = (1.1428 - 0.996V^{-1})^2$. Consider a single mode fiber with $n_1 = 1.450$, $n_2 = 1.446$, $2a$ (diameter) $= 3.9$ μm. Plot α_B vs. R for $\lambda = 633$ nm and 790 nm from $R = 2$ mm to 15 mm. What is your conclusion? (You might wish to compare your calculations with the experiments of A.J. Harris and P.F. Castle, *IEEE J. Light Wave Technol.*, Vol. LT4, 34–41, 1986).

2.20 Self-phase modulation dispersion At sufficiently high light intensities, the refractive index of glass n' can be written as $n' = n + CI$ in which C is a constant and I is the light intensity. The intensity of light modulates its own phase. Consider what will happen if the input light is very intense, and given that it must have a finite spectrum $\Delta\lambda$. The intensity is maximum at λ_o and follows a Gaussian $I(\lambda)$ vs. λ shape around λ_o. Discuss how this can lead to an *additional* dispersion mechanism.

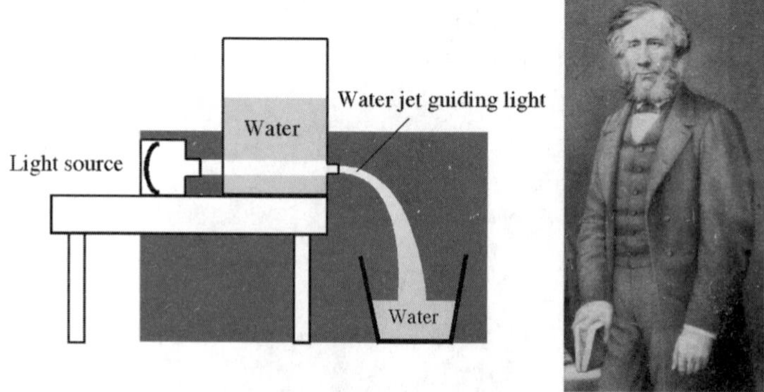

John Tyndall in 1854 demonstrated to the Royal Institution that a water jet can act as a light guide. *(Left: Author's imagination. Right: Courtesy of AIP Emilio Segrè Visual Archives, Zelzny Collection)*

C H A P T E R   3

Semiconductor Science and Light Emitting Diodes

"Although the hole and its negative counterpart, the excess electron, have been prominent in the theory of solids since the work of A.H. Wilson in 1931, the announcement of the transistor in 1948 has given holes and electrons new technological significance."

—William Shockley
from *Electrons and Holes in Semiconductors*
(D. van Nostrand Co. Inc., 1950)

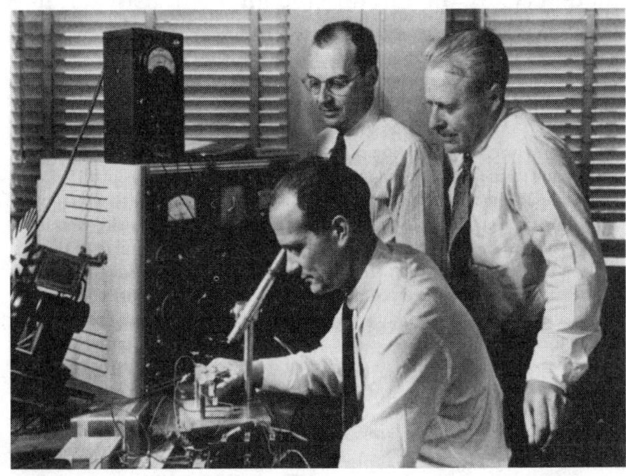

William Shockley (seated), John Bardeen (left), and Walter Brattain (right) invented the transistor at Bell Labs and thereby ushered in a new era of semiconductor devices. The three inventors shared the Nobel prize in 1956. *(Courtesy of Bell Laboratories)*

3.1 SEMICONDUCTOR CONCEPTS AND ENERGY BANDS

A. Energy Band Diagrams

We know from modern physics that the energy of the electron in an atom is quantized and can have only certain discrete values as visualized for the Li atom in Figure 3.1, which has two electrons in the 1*s* shell and one electron in the 2*s* subshell. The same

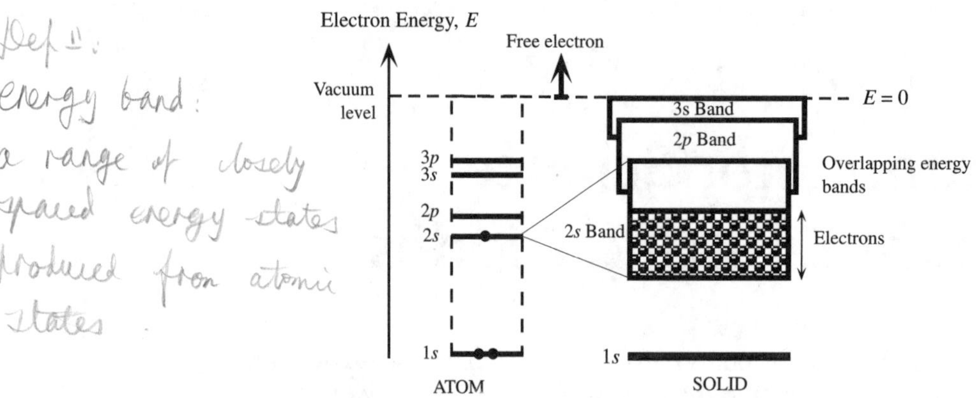

FIGURE 3.1 In a metal the various energy bands overlap to give a single band of energies that is only partially full of electrons. There are states with energies up to the vacuum level where the electron is free.

concept also applies to the electron energy in a molecule with several atoms. Again the electron energy is quantized. However, when we bring together something like 10^{23} Li atoms to form a metal crystal, the interatomic interactions result in the formation of electron energy bands. The 2*s* energy level splits into some 10^{23} closely spaced energy levels that effectively form an **energy band**, which is called the 2*s* band. Similarly, other higher energy levels also form bands as depicted in Figure 3.1. These energy bands overlap to form one continuous energy band that represents the energy band structure of a metal. The 2*s* energy level in the Li atom is half full (2*s* subshell needs 2 electrons), which means that the 2*s* band in the crystal will also be half full. Metals characteristically have partially filled energy bands.

The electron energies in a semiconductor crystal, however, are distinctly different than for metals. Figure 3.2 (a) shows a simplified two-dimensional view of the silicon crystal that has each Si atom bonding to four neighbors. All the four valence electrons per

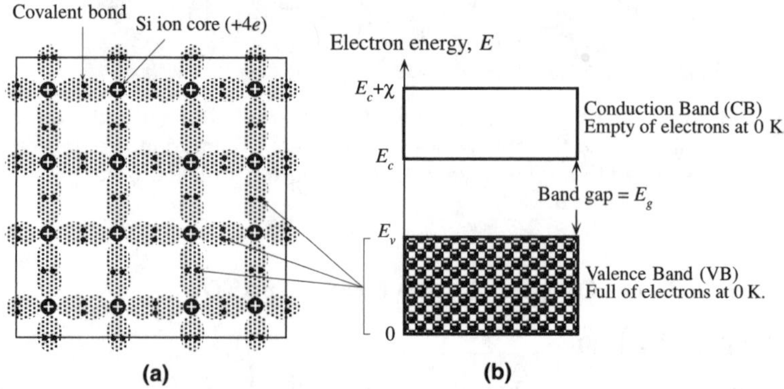

FIGURE 3.2 (a) A simplified two dimensional view of a region of the Si crystal showing covalent bonds. (b) The energy band diagram of electrons in the Si crystal at absolute zero of temperature.

atom are used in these bonds. The interactions between the Si atoms and their valence electrons result in the electron energy in the crystal falling into two distinct energy bands called the **valence band (VB)** and **conduction band (CB)** that are separated by an energy gap, **bandgap** (E_g), as shown in Figure 3.2 (b). There are no allowed electron energies in the **bandgap**; it represents the forbidden electron energies in the crystal. The valence band represents electron wavefunctions in the crystal that correspond to bonds between the atoms. Electrons that occupy these wavefunctions are the valence electrons. Since at a temperature of absolute zero all the bonds are occupied by valence electrons (there are no broken bonds), all the energy levels in the VB are normally filled with these electrons. The CB represents electron wavefunctions in the crystal that have higher energies than those in the VB and are normally empty at zero Kelvin. The top of the VB is labeled E_v, the bottom of conduction band E_c, so that $E_g = E_c - E_v$ is the bandgap. The width of the CB is called the **electron affinity** χ.

An electron placed in the CB is free to move around the crystal and also to respond to an electric field because there are plenty of neighboring empty energy levels. This electron can easily gain energy from the field and move to higher energy levels because these states are empty in the CB. In general, we can treat an electron in the CB as if it were free within the crystal by simply assigning an **effective mass** m_e^* to it. This effective mass is a quantum mechanical quantity that takes into account that the electron in the CB interacts with a periodic potential energy as it moves through the crystal, so that its inertial resistance to acceleration (definition of mass) is not the same as if it were free in vacuum.

Since the only empty states are in the CB, the excitation of an electron from the VB requires a minimum energy of E_g. Figure 3.3 illustrates what happens when an incident photon of energy $h\upsilon > E_g$ interacts with an electron in the VB. This electron absorbs the incident photon and gains sufficient energy to surmount the energy gap E_g and reach the CB. Consequently, a free electron in the CB and a "hole," corresponding to a missing electron in the VB, are created. In some semiconductors, such as Si and Ge,

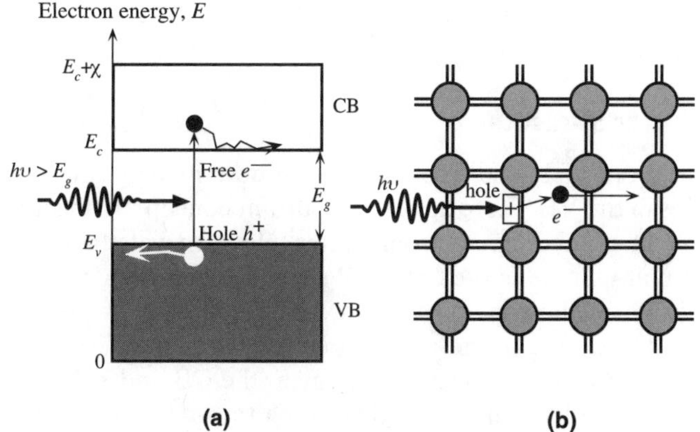

FIGURE 3.3 (a) A photon with an energy greater than E_g can excite an electron from the VB to the CB. (b) Each line between Si-Si atoms is a valence electron in a bond. When a photon breaks a Si-Si bond, a free electron and a hole in the Si-Si bond is created.

the photon absorption process also involves lattice vibrations (vibrations of the Si atoms), which we have not shown in Figure 3.3.

The empty electronic state, or the missing electron, in the bond is what we call a **hole** in the valence band. The free electron, which is in the CB, can wander around the crystal and contribute to the electrical conduction when an electric field is applied. The region remaining around the hole in the VB is positively charged because a charge of $-e$ has been removed from an otherwise neutral region of the crystal. This hole, denoted as h^+, can also wander around the crystal as if it were "free." This is because an electron in a neighboring bond can "jump", *i.e.* tunnel, into the hole to fill the vacant electronic state at this site and thereby create a hole at its original position. This is effectively equivalent to the hole being displaced in the opposite direction. Thus conduction in semiconductors occurs by both electrons and holes with charges $-e$ and $+e$ respectively and their own effective masses m_e^* and m_h^*.

Although in this specific example a photon of energy $h\upsilon > E_g$ creates an electron-hole pair, other sources of energy can also lead to an electron-hole pair creation. In fact, in the absence of radiation, there is still an electron-hole generation process going on in the sample as a result of **thermal generation.** Due to thermal energy, the atoms in the crystal are constantly vibrating, which corresponds to the bonds between the Si atoms being periodically deformed with a distribution of energies. Energetic vibrations can rupture bonds and thereby create electron and hole pairs (EHPs) by exciting electrons from the VB to the CB.

When a wandering electron in the CB meets a hole in the VB, it has found an empty electronic state of lower energy and it therefore occupies it. The electron falls from the CB to the VB to fill the hole. This is called **recombination,** which results in the annihilation of an electron from the CB and a hole in the VB. The excess energy of the electron falling from CB to VB in certain semiconductors, such as GaAs and InP, is emitted as a photon. In Si and Ge, the excess energy is lost as lattice vibrations (heat). In the steady state, the thermal generation rate is balanced by the recombination rate so that the electron concentration n in the CB and hole concentration p in the VB remain constant; both n and p depend on the temperature.

steady => n, p constant (as a function of T)

B. Semiconductor Statistics

Many important properties of semiconductors are described by considering electrons in the CB and holes in the VB. There are two important concepts. **Density of states (DOS)** $g(E)$ represents the number of electronic states (electron wavefunctions) in a band per unit energy per unit volume of the crystal. We can use quantum mechanics to calculate the DOS by considering how many electron wavefunctions there are within a given energy range per unit volume of the crystal. Figure 3.4 (a) and (b) show in a simplified way how $g(E)$ depends on the electron energy in the CB and VB. According to quantum mechanics, for an electron confined within a three-dimensional potential energy well, as a conduction electron would be in the crystal, the DOS increases with energy as $g(E) \propto ((E - E_c)^{1/2}$ in which $(E - E_c)$ is the electron energy from the bottom of the CB. DOS gives information only on available states and not on their actual occupation.

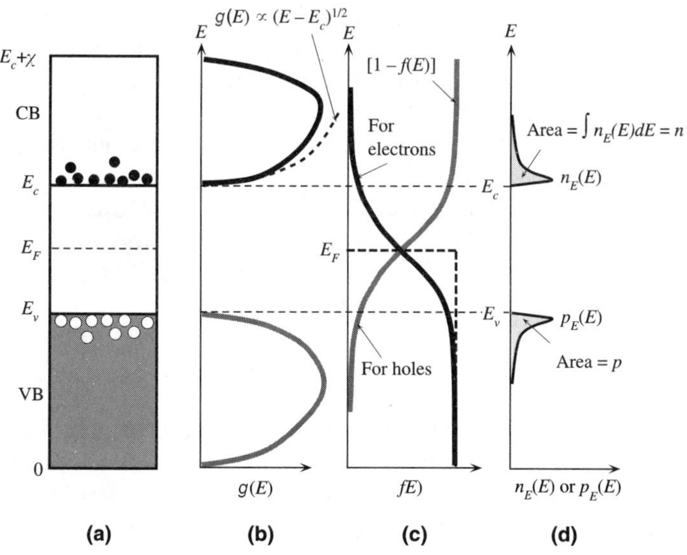

FIGURE 3.4 (a) Energy band diagram. (b) Density of states (number of states per unit energy per unit volume). (c) Fermi-Dirac probability function (probability of occupancy of a state). (d) The product of $g(E)$ and $f(E)$ is the energy density of electrons in the CB (number of electrons per unit energy per unit volume). The area under $n_E(E)$ vs. E is the electron concentration.

The **Fermi-Dirac function** $f(E)$ is the probability of finding an electron in a quantum state with energy E (state implies a wavefunction). It is a fundamental property of a collection of interacting electrons in *thermal equilibrium*. It is given by

$$f(E) = \frac{1}{1 + \exp\left(\frac{E - E_F}{k_B T}\right)} \quad \textbf{(1)}$$

Fermi-Dirac function

in which k_B is the Boltzmann constant, T is the temperature (K), and E_F is an electron energy parameter called the **Fermi energy**, which has a number of important properties. The most useful definition of E_F is in terms of a change in E_F. Any change ΔE_F across a material system represents electrical work input or output per electron.[1] If V is the potential difference between two points, then

$$\Delta E_F = eV \quad \textbf{(2)}$$

Definition of Fermi energy

For a semiconductor system in the dark, in equilibrium, and with no applied voltage or no emf generated, $\Delta E_F = 0$ and E_F must be uniform across the system. Further, as will be apparent below, E_F is related to the electron concentration n in the CB and

[1] For readers familiar with thermodynamics, its rigorous definition is that E_F is the *chemical potential* that is Gibbs free energy per electron. The definition in Equation (2) is in terms of a change in E_F.

hole concentration p in the VB. The behavior of $f(E)$ is shown in Figure 3.4 (c) assuming that the Fermi level E_F is located in the bandgap. Note that the probability of finding a hole, a missing electron, in a state with energy E is $1-f(E)$.

Although at E_F the probability of electron occupancy from Eq. (1) is 1/2, there may be no states for an electron to occupy. What is important is the product $g_{CB}(E)f(E)$, that is the actual number of electrons per unit energy per unit volume, $n_E(E)$, in the CB, which is shown in Figure 3.4 (d). Thus, $n_E\,dE = g_{CB}(E)f(E)\,dE$ is the number of electrons in the energy range E to $E + dE$. Integrating this from the bottom (E_c) to the top $(E_c + \chi)$ of the CB gives the electron concentration n in the CB. In other words,

$$n = \int_{E_c}^{E_c+\chi} g_{CB}(E)f(E)\,dE \tag{3}$$

Whenever $(E_c - E_F) \gg k_BT$, *i.e.* E_F is at least a few k_BT below E_c, then $f(E) \approx \exp[-(E - E_F)/k_BT]$. That is, the Fermi-Dirac statistics can be replaced by *Boltzmann* statistics. Such semiconductors are called **nondegenerate**. It implies that the number of electrons in the CB is far less than the number states in this band. For non-degenerate semiconductors, the above integration leads to,

Electron concentration in CB

$$n = N_c \exp\left[-\frac{(E_c - E_F)}{k_BT}\right] \tag{4}$$

in which $N_c = 2[2\pi m_e^* k_BT/h^2]^{3/2}$ is a temperature-dependent constant, called the **effective density of states at the CB edge**. The result of the integration in Eq. (4) seems to be simple but it is however an approximation as it assumes that $(E_c - E_F) \gg k_BT$. We can interpret Eq. (4) as follows. If we take all the states in the conduction band and replace them with an effective concentration N_c (number of states per unit volume) at E_c and then multiply this simply by the Boltzmann probability function, $f(E_c) = \exp[-(E_c - E_F)/k_BT]$, we obtain the concentration of electrons at E_c, *i.e.* in the conduction band. N_c is thus an effective density of states at the CB band edge.

We can carry out a similar analysis for the concentration of holes in the VB as visualized in Figure 3.4. Multiplying the density of states $g_{VB}(E)$ in the VB with the probability of occupancy by a hole, $[1 - f(E)]$, gives p_E, the hole concentration per unit energy. Integrating this over the VB gives the hole concentration by assuming that E_F is a few k_BT above E_v, we obtain,

Hole concentration in VB

$$p \approx N_v \exp\left[-\frac{(E_F - E_v)}{k_BT}\right] \tag{5}$$

in which $N_v = 2[2\pi m_h^* k_BT/h^2]^{3/2}$ is the **effective density of states at the VB edge**.

There are no specific assumptions in our derivations above, except for E_F being a few k_BT away from the band edges, which means that Equations (4) and (5) are generally valid. It is apparent from Eqs. (4) and (5) that the location of E_F determines the electron and hole concentrations. Thus, E_F is a useful material property. In an **intrinsic semiconductor** (a pure crystal), $n = p$, by using Eqs. (4) and (5) we can show that the Fermi level E_{Fi} in the intrinsic crystal is above E_v and located in the bandgap at,

Fermi-level in intrinsic crystal

$$E_{Fi} = E_v + \tfrac{1}{2}E_g - \tfrac{1}{2}kT\ln\left(\frac{N_c}{N_v}\right) \tag{6}$$

Typically, N_c and N_v values are comparable and both occur in the logarithmic term so that E_{Fi} is very approximately in the middle of the bandgap as originally sketched in Figure 3.4.

There is a useful semiconductor relation between n and p, called the **mass action law**. From Eqs. (4) and (6), the product np is

$$np = N_c N_v \exp\left(-\frac{E_g}{k_B T}\right) = n_i^2 \quad (7)$$

Mass action law

in which $E_g = E_c - E_v$ is the bandgap energy and n_i^2 has been *defined by* $N_c N_v \exp(-E_g/k_B T)$ and is a constant that depends on the temperature and the material properties, *e.g.* depends on E_g, and not on the position of the Fermi level. The intrinsic concentration n_i corresponds to the concentration of electrons or holes in an undoped (pure) crystal, *i.e.* intrinsic semiconductor. In such a semiconductor $n = p = n_i$, which is therefore called the **intrinsic concentration**. The mass action law is valid whenever we have thermal equilibrium and the sample is in the dark.

Equations (4) and (5) determine the total concentration of electrons and holes in the CB and VB respectively. The average energy of the electrons in the CB can be calculated by using $n_E(E)$, their energy distribution. The result gives the average energy as $(3/2)k_B T$ above E_c. Since the electron in the CB is "free" in the crystal with an effective mass m_e^*, it wanders around the crystal with an average kinetic energy $(3/2)k_B T$; the same as a free atom in a gas or vapor in a tank. This is not surprising as both particles are wandering freely (without interacting with each other) and obey Boltzmann statistics. If v is the electron velocity and triangular brackets represent an average, then $< (1/2) m_e^* v^2 >$ must be $(3/2) k_B T$. We can thus calculate the *root mean square veloc ity* $\sqrt{\langle v \rangle^2}$ which is called the **thermal velocity** and is typically ~10^5 m s^{-1}. The same ideas apply to holes in the VB with an effective hole mass m_h^*.

C. Extrinsic Semiconductors

By introducing small amounts of impurities into an otherwise pure crystal, it is possible to obtain a semiconductor in which the concentration of carriers of one polarity is much in excess of the other type. Such semiconductors are referred to as **extrinsic semiconductors** vis-à-vis the intrinsic case of a pure and perfect crystal. For example, by adding pentavalent impurities, such as arsenic, which have a valency one more than Si, we can obtain a semiconductor in which the electron concentration is much larger than the hole concentration. In this case we will have an ***n*-type semiconductor**. If we add trivalent impurities, such as boron, which have a valency of one less than four, we then have an excess of holes over electrons, that is a ***p*-type semiconductor**.

Arsenic has five valence electrons whereas Si has four. When the Si crystal is doped with small amounts of As, each As atom substitutes for one Si atom and is surrounded by four Si atoms. When an As atom bonds with four Si atoms, it has one electron left unbounded. This fifth electron cannot find a bond to go into, so it is left orbiting around the As atom as illustrated in Figure 3.5 (a). The As^+ ionic center with an electron e^- orbiting it resembles a hydrogen atom in a silicon environment. We can easily calculate how much energy is required to free this electron away from the As site thereby ionizing the As impurity by using our knowledge on the ionization of a hydrogen atom (removing

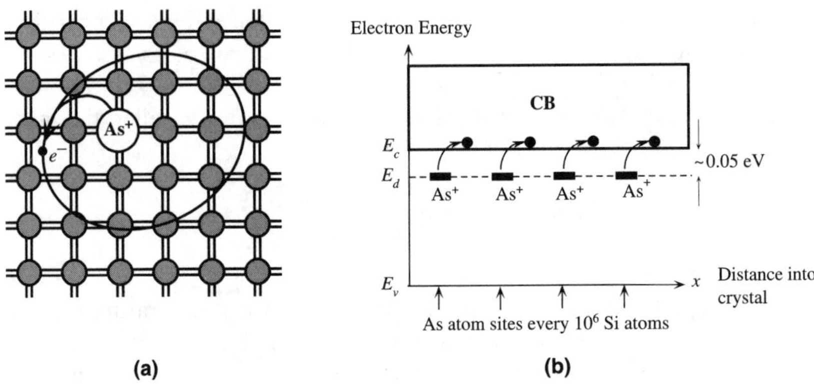

FIGURE 3.5 (a) The four valence electrons of As allow it to bond just like Si but the fifth electron is left orbiting the As site. The energy required to release to free fifth electron into the CB is very small. (b) Energy band diagram for an *n*-type Si doped with 1 ppm As. There are donor energy levels just below E_c around As^+ sites.

the electron from the H-atom). This energy turns out to be a few hundredths of an electronvolt, *i.e.* ~0.05 eV, which is comparable to the thermal energy at room temperature $(\sim k_B T = 0.025 \text{ eV})$. Thus, the fifth valence electron can be readily freed by thermal vibrations of the Si lattice. The electron will then be "free" in the semiconductor, or in other words, it will be in the CB. The energy required to excite the electron to the CB is therefore ~0.05 eV. The addition of As atoms introduces localized electronic states at the As sites because the fifth electron has a localized wavefunction, of the hydrogenic type, around As^+. The energy of these states, E_d, is ~0.05 eV below E_c because this is how much energy is required to take the electron away into the CB. Thermal excitation by lattice vibrations at room temperature is sufficient to ionize the As atom, *i.e.* excite the electron from E_d into the CB. This process creates free electrons, however the As^+ ions remain immobile as shown in the energy band diagram sketch of an *n*-type semiconductor in Figure 3.5 (b).

Because the As atom donates an electron into the CB, it is called a **donor** impurity. E_d is the electron energy around the donor atom and it is below E_c by ~0.05 eV as in Figure 3.5 (b). If N_d is the donor atom concentration in the crystal, provided that $N_d \gg n_i$, then at room temperature, the electron concentration in the CB will nearly be equal to N_d, *i.e.* $n = N_d$. The hole concentration will be $p = n_i^2/N_d$ which is less than the intrinsic concentration because a few of the large number of electrons in the CB recombine with holes in the VB to maintain $np = n_i^2$.

The conductivity σ of a semiconductor depends on both electrons and holes as both contribute to charge transport. If μ_e and μ_h are the drift mobilities of the electrons and holes respectively then

Semiconductor conductivity

$$\sigma = en\mu_e + ep\mu_h \quad (8)$$

which for an *n*-type semiconductor becomes,

n-type conductivity

$$\sigma = eN_d\mu_e + e\left(\frac{n_i^2}{N_d}\right)\mu_h \approx eN_d\mu_e \quad (9)$$

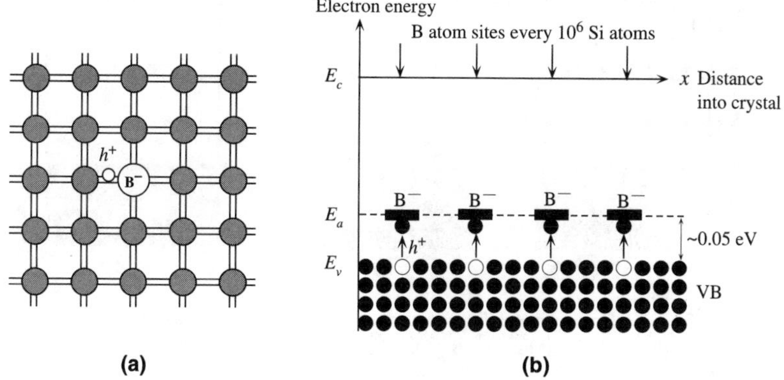

FIGURE 3.6 (a) Boron doped Si crystal. B has only three valence electrons. When it substitutes for a Si atom one of its bonds has an electron missing and therefore a hole. (b) Energy band diagram for a *p*-type Si doped with 1 ppm B. There are acceptor energy levels just above E_v around B^- sites. These acceptor levels accept electrons from the VB and therefore create holes in the VB.

We should, by similar arguments to the above, anticipate that doping a Si crystal with a trivalent atom (valency of 3) such as B (boron) will result in a *p*-type Si, which has an excess of holes in the crystal. Consider doping Si with small amounts of B as shown in Figure 3.6 (a). Because B has only three valence electrons, when it shares them with four neighboring Si atoms, one of the bonds has a missing electron, which is of course a "hole." A nearby electron can tunnel into this hole and displace it further away from the boron atom. As the hole moves away, it gets attracted by the negative charge left behind on the boron atom. The binding energy of this hole to the B^- ion can be calculated using the hydrogenic atom analogy just like in the *n*-type Si case. This binding energy also turns out to be very small, ~0.05 eV, so that at room temperature the thermal vibrations of the lattice can free the hole away from the B^- site. A free hole, we recall, exists in the VB. The escape of the hole from the B^- site involves the B atom accepting an electron from a neighboring Si-Si bond (from the VB), which effectively results in the hole being displaced away and its eventual escape to freedom in the VB. The B atom introduced into the Si crystal therefore acts as an electron **acceptor** impurity. The electron accepted by the B atom comes from a nearby bond. On the energy band diagram, an electron leaves the VB and gets accepted by a B atom that becomes negatively charged. This process leaves a hole in the VB that is free to wander away as illustrated Figure 3.6 (b).

It is apparent that doping a silicon crystal with a trivalent impurity results in a *p*-type material. We have many more holes than electrons for electrical conduction since the negatively charged B atoms are immobile and hence cannot contribute to the conductivity. If the concentration of acceptor impurities N_a in the crystal is much greater than the intrinsic concentration n_i, then at room temperature all the acceptors would have been ionized and thus $p = N_a$. The electron concentration is then determined by the mass action law, $n = n_i^2/N_a$ which is much smaller than p, and consequently the conductivity is simply given by $\sigma = eN_a\mu_h$.

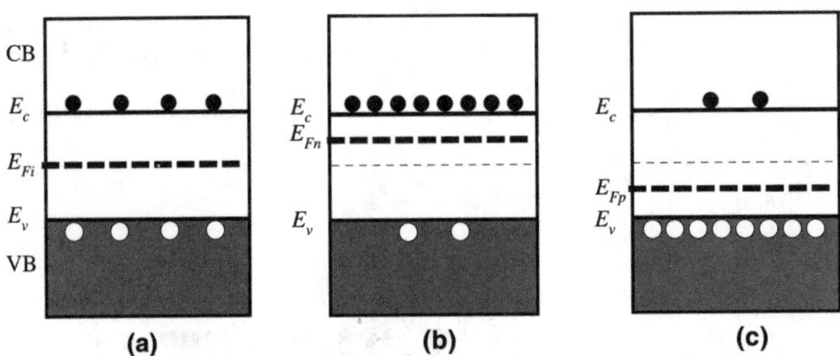

FIGURE 3.7 Energy band diagrams for (a) intrinsic (b) *n*-type and (c) *p*-type semiconductors. In all cases, $np = n_i^2$. Note that donor and acceptor energy levels are not shown.

Figure 3.7 (a) to (c) show the energy band diagrams of an intrinsic, *n*-type, and a *p*-type semiconductor. The energy distance of E_F from E_c and E_v determines the electron and hole concentrations by virtue of Eqs. (4) and (5). Note the locations of the Fermi-level in each case: E_{Fi} for intrinsic, E_{Fn} for *n*-type, and E_{Fp} for *p*-type.

The following definitions and notation are generally used to describe extrinsic semiconductors. Electrons in an *n*-type semiconductor $(n > p)$ are **majority carriers**, whereas holes are **minority carriers**. The electron, *i.e.* majority carrier, concentration in this *n*-type semiconductor in equilibrium is n_{no}, in which the subscripts refer to *n*-type semiconductor and equilibrium (which excludes photoexcitation). The hole, *i.e.* minority carrier, concentration is denoted by p_{no}. In this notation, $n_{no} = N_d$ and the mass action law is $n_{no}p_{no} = n_i^2$. Similarly, hole or majority carrier concentration in a *p*-type semiconductor $(p > n)$ is p_{po} and electron, minority carrier, concentration is n_{po}. Further, $p_{po} = N_a$ and $n_{no}p_{no} = n_i^2$.

D. Compensation Doping

Compensation doping describes the doping of a semiconductor with both donors and acceptors to control the properties. For example, a *p*-type semiconductor doped with N_a acceptors can be converted to an *n*-type semiconductor by simply adding donors until the concentration N_d exceeds N_a. The effect of donors compensates for the effect of acceptors and vice versa. The electron concentration is then given by $N_d - N_a$, provided the latter is larger than n_i. When both acceptors and donors are present what essentially happens is that electrons from donors recombine with the holes from the acceptors so that the mass action law $np = n_i^2$ is obeyed. Remember, we cannot simultaneously increase the electron and hole concentrations because that leads to an increase in the recombination rate that returns the electron and hole concentrations to satisfy $np = n_i^2$. When an acceptor atom accepts a valence band electron, a hole is created in the VB. This hole then recombines with an electron from the CB. Suppose that we have more donors than acceptors. If we take the initial electron concentration as $n = N_d$ then the recombination between the electrons from the donors and N_a holes generated by N_a acceptors results in the electron concentration reduced by N_a to $n = N_d - N_a$. By a similar argument, if we have more acceptors than donors, the hole concentration becomes $p = N_a - N_d$ with electrons from N_d donors recombining with holes from N_a acceptors.

E. Degenerate and Non-degenerate Semiconductors

In non-degenerate semiconductors, the number of states in the CB far exceeds the number of electrons, so the likelihood of two electrons trying to occupy the same state is almost nil. This means that the Pauli exclusion principle can be neglected and the electron statistics can be described by the Boltzmann statistics. N_c is a measure of the density of states in the CB. The Boltzmann expression in Eq. (4) for n is valid only when $n \ll N_c$. Those semiconductors for which $n \ll N_c$ and $p \ll N_v$ are termed **non-degenerate** semiconductors.

When the semiconductor has been excessively doped with donors, then n may be so large, typically $10^{19} - 10^{20}$ cm^{-3}, that it may be comparable to N_c. In that case, the Pauli exclusion principle becomes important in the electron statistics and we have to use the Fermi-Dirac statistics. Such a semiconductor exhibits properties that are more metal-like than semiconductor-like, *e.g.* the resistivity is approximately proportional to the absolute temperature. Semiconductors that have $n > N_c$ or $p > N_v$ are called **degenerate semiconductors**.

The large carrier concentration in a degenerate semiconductors is due to its heavy doping. For example, as the donor concentration in an n-type semiconductor is increased, at sufficiently high doping levels the donor atoms become so close to each other that their orbitals overlap to form a narrow energy band, which overlaps and becomes part of the conduction band. The valence electrons from the donors fill the band from E_c. This situation is reminiscent to the valence electrons filling overlapping energy bands in a metal. In a degenerate n-type semiconductor, the Fermi level is therefore within the CB, or above E_c just like E_F is within the band in a metal. The majority of the states between E_c and E_F are full of electrons as indicated in Figure 3.8 (a). In the case of a p-type degenerate semiconductor, the Fermi level lies in the VB below E_v as in Figure 3.8 (b). One cannot simply assume that $n = N_d$ or $p = N_a$ in a degenerate semiconductor because the dopant concentration is so large that they interact with each other. Not all dopants are able to become ionized, and the carrier concentration eventually reaches a saturation typically around ~10^{20} cm^{-3}. Furthermore, the mass action law, $np = n_i^2$, is not valid for degenerate semiconductors.

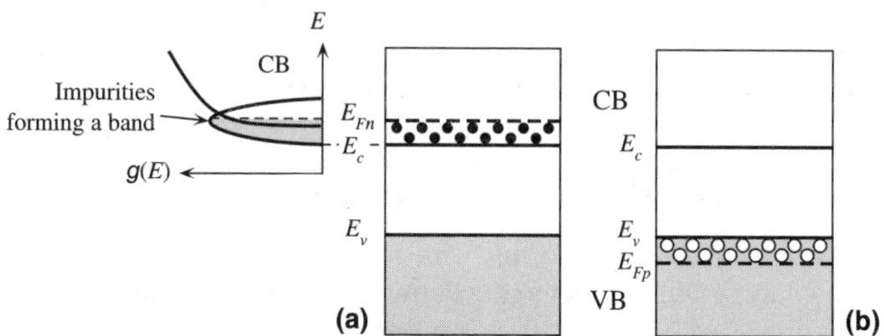

FIGURE 3.8 (a) Degenerate n-type semiconductor. Large number of donors form a band that overlaps the CB. (b) Degenerate p-type semiconductor.

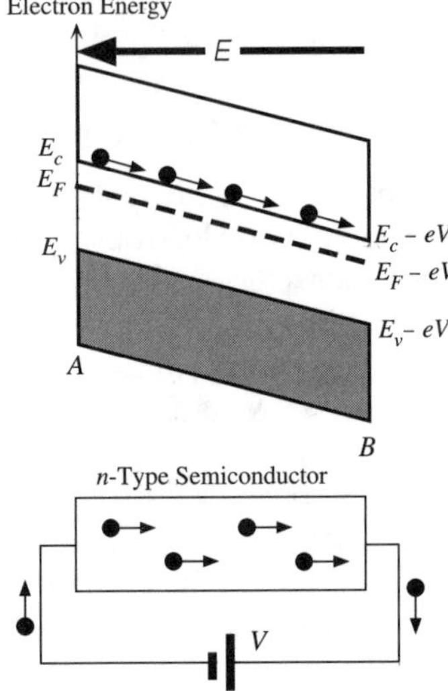

FIGURE 3.9 Energy band diagram of an n-type semiconductor connected to a voltage supply of V volts. The whole energy diagram tilts because the electron now has an electrostatic potential energy as well.

F. Energy Band Diagrams in an Applied Field

Consider the energy band diagram for an n-type semiconductor that is connected to a voltage supply of V and is carrying a current. The Fermi level E_F is above that for the intrinsic case (E_{Fi}), closer to E_c than E_v. The applied voltage drops uniformly along the semiconductor so that the electrons in the semiconductor now also have an imposed electrostatic potential energy, which decreases towards the positive terminal as depicted in Figure 3.9. The whole band structure, the CB and the VB, therefore tilts. When an electron drifts from A towards B, its PE decreases because it is approaching the positive terminal.

For a semiconductor system in the dark, in equilibrium and with no applied voltage or no emf generated, E_F must be uniform across the system since $\Delta E_F = eV = 0$. However, when electrical work is done on the system, *e.g.* when a battery is connected to a semiconductor, then E_F is not uniform throughout the whole system. A change ΔE_F in E_F within a material system is equivalent to electrical work per electron or eV. The Fermi level E_F therefore follows the electrostatic PE behavior. The change in E_F from one end to the other, $E_F(A) - E_F(B)$ is just eV, the energy needed in taking an electron through the semiconductor as shown in Figure 3.9. Electron concentration in the semiconductor is uniform so that $E_c - E_F$ must be constant from one end to the other. Thus the CB, VB, and E_F all bend by the same amount.

EXAMPLE 3.1.1 Fermi levels in semiconductors

An n-type Si wafer has been doped uniformly with 10^{16} antimony (Sb) atoms cm^{-3}. Calculate the position of the Fermi energy with respect to the Fermi energy E_{Fi} in intrinsic Si. The n-type Si sample above is further doped with 2×10^{17} boron atoms cm^{-3}. Calculate position of the Fermi energy with respect to the Fermi energy E_{Fi} in intrinsic Si at room temperature (300 K), and hence with respect to the Fermi energy in the n-type case above.

Solution Sb (Group V) gives n-type doping with $N_d = 10^{16}$ cm^{-3}, and since $N_d \gg n_i$ ($= 1.45 \times 10^{10}$ cm^{-3}), we have $n = N_d = 10^{16}$ cm^{-3}. For intrinsici Si,

$$n_i = N_c \exp\left[-(E_c - E_{Fi})/k_B T\right],$$

whereas for doped Si,

$$n = N_c \exp\left[-(E_c - E_{Fn})/k_B T\right] = N_d$$

in which E_{Fi} and E_{Fn} are the Fermi energies in the intrinsic and n-type Si. Dividing the two expressions,

$$N_d/n_i = \exp\left[(E_{Fn} - E_{Fi})/k_B T\right]$$

so that

$$E_{Fn} - E_{Fi} = k_B T \ln(N_d/n_i) = (0.0259 \text{ eV}) \ln(10^{16}/1.45 \times 10^{10}) = 0.348 \text{ eV}.$$

When the wafer is further doped with boron, the acceptor concentration, $N_a = 2 \times 10^{17}$ cm^{-3} $> N_d = 10^{16}$ cm^{-3}. The semiconductor is compensation doped and compensation converts the semiconductor to a p-type Si. Thus, $p = N_a - N_d = (2 \times 10^{17} - 10^{16}) = 1.9 \times 10^{17}$ cm^{-3}.

For intrinsic Si,

$$p = n_i = N_v \exp\left[-(E_{Fi} - E_v)/k_B T\right],$$

whereas for doped Si,

$$p = N_v \exp\left[-(E_{Fp} - E_v)/k_B T\right] = N_a - N_d$$

in which E_{Fi} and E_{Fp} are the Fermi–energies in the intrinsic and p–type Si, respectively. Dividing the two expressions,

$$p/n_i = \exp\left[-(E_{Fp} - E_{Fi})/k_B T\right]$$

so that

$$E_{Fp} - E_{Fi} = -k_B T \ln(p/n_i) = -(0.0259 \text{ eV}) \ln(1.9 \times 10^{17}/1.45 \times 10^{10}) = -0.424 \text{ eV}$$

EXAMPLE 3.1.2 Conductivity

What is the conductivity of an n-type Si crystal that has been doped uniformly with 10^{16} cm^{-3} phosphorus (P) atoms (donors) if the drift mobility of electrons is about 1350 cm^2 V^{-1} s^{-1}?

Solution Since $N_d = 10^{16}$ cm^{-3} $> n_i = 1.45 \times 10^{10}$ cm^{-3}, the electron concentration $n = N_d$ and we can neglect the hole concentration $p = n_i^2/N_d \ll n$. Thus,

$$\sigma = eN_d\mu_e = (1.6 \times 10^{-19}\text{ C})(1 \times 10^{16}\text{ cm}^{-3})(1350\text{ cm}^2\text{ V}^{-1}\text{ s}^{-1}) = 2.16\ \Omega^{-1}\text{ cm}^{-1}.$$

3.2 DIRECT AND INDIRECT BANDGAP SEMICONDUCTORS: *E-k* DIAGRAMS

We know from quantum mechanics that when the electron is within an infinite potential energy well of spatial width L, its energy is quantized and given by

$$E_n = \frac{(\hbar k_n)^2}{2m_e}$$

in which m_e is the mass of the electron and the wavevector k_n is essentially a quantum number determined by

$$k_n = \frac{n\pi}{L}$$

in which $n = 1, 2, 3 \ldots$. The energy increases parabolically with the wavevector k_n. We also know that the electron momentum is given by $\hbar k_n$. This description can be used to represent the behavior of electrons in a metal within which their average potential energy can be taken to be very roughly zero. In other words, we take $V(x) = 0$ within the metal crystal and $V(x)$ to be large, for example $V(x) = V_o$ (several electron volts) outside, so that the electron is contained within the metal. This is the *nearly free electron* model of a metal that has been quite successful in interpreting many of the metallic properties. Indeed, we calculate the density of states $g(E)$ based on the three-dimensional potential well problem. However, it is quite obvious that this model is too simple since it does not take into account the actual variation of the electron potential energy in the crystal.

The potential energy of the electron depends on its location within the crystal and is periodic due to the regular arrangement of the atoms. How does a periodic potential energy affect the relationship between E and k? It will no longer be simply $E_n = (\hbar k_n)^2/2m_e$.

To find the energy of the electron in a crystal we need to solve the Schrödinger equation for a periodic potential energy function in three dimensions. We first consider the hypothetical one-dimensional crystal shown in Figure 3.10. The electron potential energy functions for each atom add to give an overall potential energy function $V(x)$, which is clearly periodic in x with the periodicity of the crystal, a. Thus, $V(x) = V(x + a) = V(x + 2a) = \ldots$ and so on. Our task is therefore to solve the Schrödinger equation,

$$\frac{d^2\psi}{dx^2} + \frac{2m_e}{\hbar^2}[E - V(x)]\psi = 0 \qquad \textbf{(1)}$$

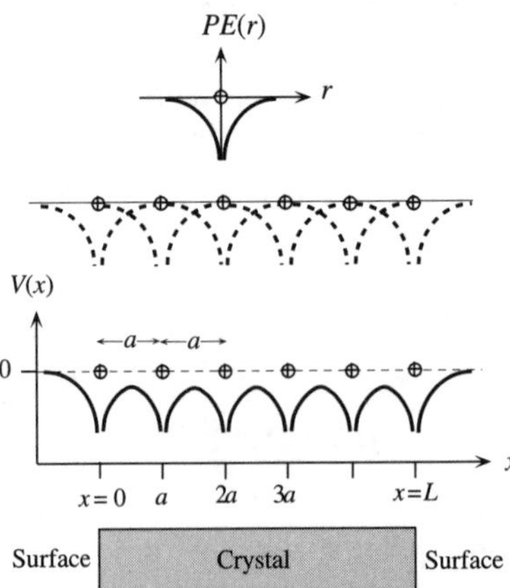
PE(r)
r
V(x)
a
a
0
x
x = 0
a
2a
3a
x = L
Surface
Crystal
Surface
PE of the electron around an isolated atom
When *N* atoms are arranged to form the crystal then there is an overlap of individual electron *PE* functions.
PE of the electron, *V*(*x*), inside the crystal is periodic with a period *a*.

FIGURE 3.10 The electron potential energy *(PE), V(x),* inside the crystal is periodic with the same periodicity as that of the crystal, *a*. Far away outside the crystal, by choice, $V = 0$ (the electron is free and $PE = 0$).

subject to the condition that the potential energy, $V(x)$, is periodic in a, *i.e.*

$$V(x) = V(x + ma); \qquad m = 1, 2, 3 \ldots \tag{2}$$

Periodic PE in the crystal

The solution of Eq. (1) will give the electron wavefunction in the crystal and hence the electron energy. Since $V(x)$ is periodic, we should expect, by intuition at least, the solution $\psi(x)$ to be periodic. It turns out that the solutions to Eq. (1), which are called **Bloch wavefunctions**, are of the form

$$\psi_k(x) = U_k(x) \exp(jkx) \tag{3}$$

Bloch wave in the crystal

in which $U_k(x)$ is a periodic function that depends on $V(x)$ and has the same periodicity a as $V(x)$. The term $\exp(jkx)$, of course, represents a traveling wave whose wavevector is k. We should remember that we have to multiply this by $\exp(-jE/\hbar)$, in which E is the energy, to get the overall wavefunction $\Psi(x, t)$. Thus, the electron wavefunction in the crystal is a traveling wave that is modulated by $U_k(x)$. Further, both $\exp(jkx)$ and $\exp(-jkx)$ are possible and represent left and right traveling waves.

There are many such Bloch wavefunction solutions to the one-dimensional crystal each identified with a particular k value, say k_n which acts as a kind of quantum number. Each $\psi_k(x)$ solution corresponds to a particular k_n and represents a state with an energy E_k. The dependence of the energy E_k on the wavevector k is illustrated in an *E-k* diagram. Figure 3.11 shows a typical *E-k* diagram for the hypothetical one-dimensional solid for k values in the range $-\pi/a$ to $+\pi/a$. Just as $\hbar k$ is the momentum of a free electron, $\hbar k$ for the Bloch electron is the momentum involved in its interaction with external fields, such as those involved in the photon absorption processes. Indeed the rate of change of $\hbar k$ is the externally applied force F_{ext} on the electron such as that due to an electric field $\mathbf{E}$(*i.e.* $F_{\text{ext}} = e\mathbf{E}$). Thus, for the electron within the crystal, $d(\hbar k)/dt = F_{\text{ext}}$ and consequently we call $\hbar k$ the **crystal momentum** of the electron.[2]

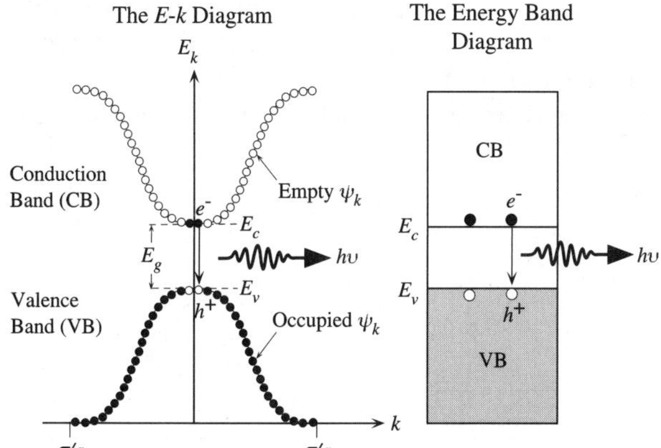

FIGURE 3.11 The *E-k* diagram of a direct bandgap semiconductor such as GaAs. The *E-k* curve consists of many discrete points with each point corresponding to a possible state, wavefunction $\psi_k(x)$, that is allowed to exist in the crystal. The points are so close that we normally draw the *E-k* relationship as a continuous curve. In the energy range E_v to E_c there are no points ($\psi_k(x)$ solutions).

[2]The actual momentum of the electron, however, is not $\hbar k$ because $d(\hbar k)/dt \neq F_{\text{external}} + F_{\text{internal}}$. The true momentum p_e satisfies $dp_e/dt = F_{\text{external}} + F_{\text{internal}}$ (all forces on the electron). However, as we are interested in interactions with external forces such as an applied field, we treat $\hbar k$ as if it were the momentum of the electron in the crystal and use the name *crystal momentum.*

Inasmuch as the momentum of the electron in the x–direction in the crystal is given by $\hbar k$, the E-k diagram is an energy vs. crystal momentum plot. The states $\psi_k(x)$ in the lower E-k curve constitute the wavefunctions for the valence electrons and thus correspond to the states in the **valence band** (VB). Those in the upper E-k curve, on the other hand, correspond to the states in the **conduction band** (CB), since they have higher energies. All the valence electrons at 0 K therefore fill the states (particular k_n values) in the lower E-k diagram.

It should be emphasized that an E-k curve in the diagram consists of many discrete points each point corresponding to a possible state, wavefunction $\psi_k(x)$, that is allowed to exist in the crystal. The points are so close that we draw the E-k relationship as a continuous curve. It is clear from the E-k diagram that there is a range of energies, from E_v to E_c, for which there are no solutions to the Schrödinger equation and hence there are no $\psi_k(x)$ with energies in E_v to E_c. Furthermore, we also note that the E-k behavior is not a simple parabolic relationship except near the bottom of the CB and the top of the VB.

However, above a temperature of absolute zero, due to thermal excitation, some of the electrons from the top of the valence band will be excited to the bottom of the conduction band. According to the E-k diagram in Figure 3.11, when an electron and hole recombine, the electron simply drops from the bottom of the CB to the top of the VB without any change in its k value so that this transition is quite acceptable in terms of momentum conservation. We should recall that the momentum of the emitted photon is negligible compared with the momentum of the electron. The E-k diagram in Figure 3.11 is therefore for a **direct bandgap semiconductor**. The minimum of the CB is directly above the maximum of the VB.

The simple E-k diagram sketched in Figure 3.11 is for the hypothetical one–dimensional crystal in which each atom simply bonds with two neighbors. In real crystals, we have a three-dimensional arrangement of atoms with $V(x, y, z)$ showing periodicity in more than one direction. The E-k curves are then not as simple as that in Figure 3.11 and often show unusual features. The E-k diagram for GaAs, which is shown in Figure 3.12 (a), as it turns out, has general features that are quite similar to that sketched in Figure 3.11. GaAs is therefore a direct bandgap semiconductor in which electron–hole

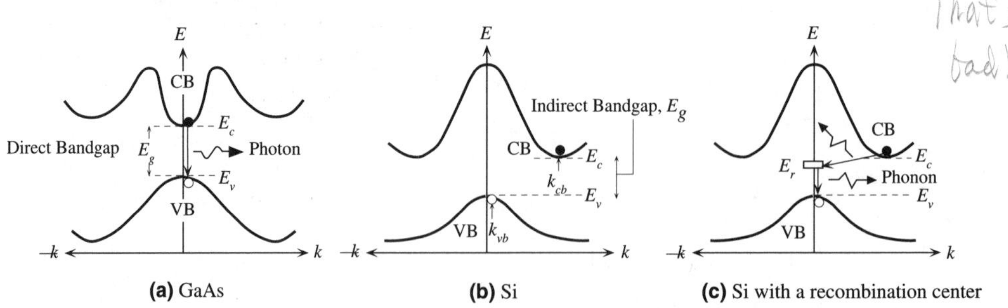

FIGURE 3.12 (a) In GaAs the minimum of the CB is directly above the maximum of the VB. GaAs is therefore a direct bandgap semiconductor. (b) In Si, the minimum of the CB is displaced from the maximum of the VB and Si is an indirect bandgap semiconductor. (c) Recombination of an electron and a hole in Si involves a recombination center.

pairs can recombine directly and emit a photon. The majority of light-emitting devices use direct bandgap semiconductors to make use of direct recombination.

In the case of Si, the diamond crystal structure leads to an *E-k* diagram that has the essential features depicted in Figure 3.12 (b). We notice that the minimum of the CB is *not* directly above the maximum of the VB, but it is displaced on the *k*-axis. Such crystals are called **indirect bandgap semiconductor.** An electron at the bottom of the CB cannot therefore recombine directly with a hole at the top of the VB because for the electron to fall down to the top of the VB, its momentum must change from k_{cb} to k_{vb}, which is not allowed by the law of conservation of momentum. Thus, direct electron-hole recombination does not take place in Si and Ge. The recombination process in these elemental semiconductors occurs via a **recombination center** at an energy level E_r within the bandgap as illustrated in Figure 3.12 (c). These recombination centers may be crystal defects or impurities. The electron is first captured by the defect at E_r. The change in the energy and momentum of the electron by this capture process is transferred to lattice vibrations, that is, to **phonons**. As much as an electromagnetic radiation is quantized in terms of photons, lattice vibrations in the crystal are quantized in terms of phonons. Lattice vibrations travel in the crystal just like a wave and these waves are called phonons. The captured electron at E_r can readily fall down into an empty state at the top of the VB and thereby recombine with a hole as in Figure 3.12 (c). Typically, the electron transition from E_c to E_v involves the emission of further lattice vibrations.

In some indirect bandgap semiconductors, such as GaP, however, the recombination of the electron with a hole at certain recombination centers results in photon emission. The *E-k* diagram is similar to that shown in Figure 3.12 (c) except that the recombination centers at E_r are generated by the purposeful addition of nitrogen impurities to GaP. The electron transition from E_r to E_v involves photon emission.

3.3 *pn* JUNCTION PRINCIPLES

A. Open Circuit

Consider what happens when one side of a sample of Si is doped *n*-type and the other *p*-type as shown in Figure 3.13 (a). Assume that there is an *abrupt* discontinuity between the *p* and *n* regions, which we call the **metallurgical junction**, *M* in Figure 3.13 (a), in which the fixed (immobile) ionized donors and the free electrons (in the conduction band, CB) in the *n*-region and fixed ionized acceptors and holes (in the valence band, VB) in the *p*-region are also shown.

Due to the hole concentration gradient from the *p*-side, in which $p = p_{po}$, to the *n*-side in which $p = p_{no}$, holes *diffuse* towards the right and enter the *n*-region and recombine with the electrons (majority carriers) in this region. The *n*-side near the junction therefore becomes depleted of majority carriers and therefore has exposed positive donor ions (As^+) of concentration N_d. Similarly, the electron concentration gradient drives the electrons by diffusion towards the left. Electrons diffusing into the *p*-side recombine with the holes (majority carriers), which exposes negative acceptor ions (B^-) of concentration N_a in this region. The regions on both sides of the junction *M* consequently becomes *depleted* of free carriers in comparison with the bulk *p* and *n* regions

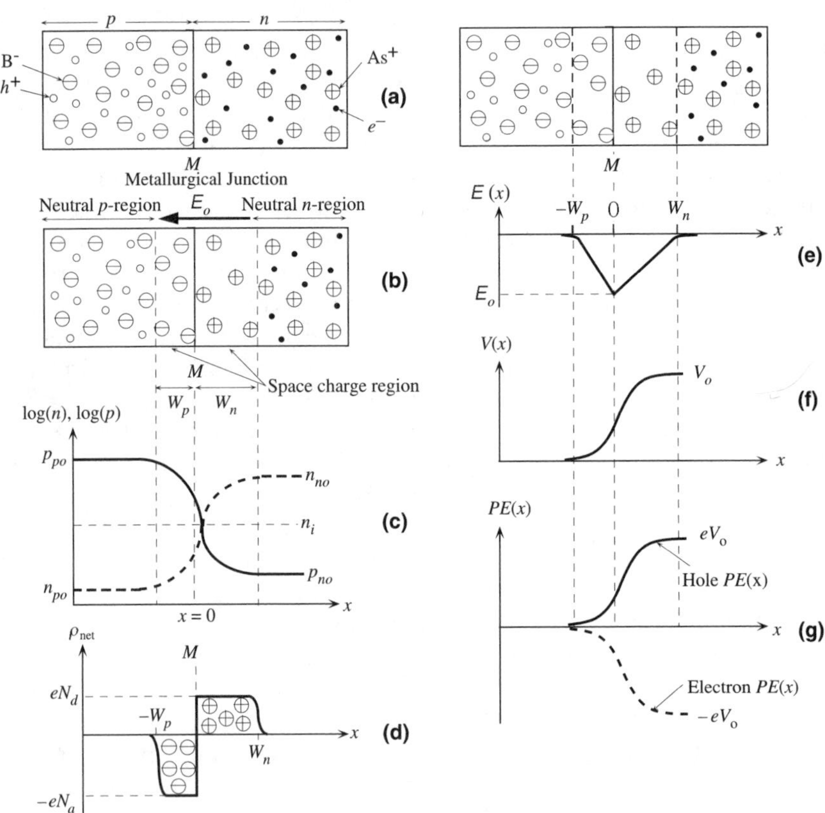

FIGURE 3.13 Properties of the pn junction.

far away from the junction. There is therefore a **space charge layer** (SCL) around M. Figure 3.13 (b) shows the SCL, also known as the **depletion region** around M. Figure 3.13 (c) illustrates the hole and electron concentration profiles in which the vertical concentration scale is logarithmic. Note that we must, under equilibrium conditions (*e.g.*, no applied bias, or photoexcitation), have $pn = n_i^2$ everywhere.

It is clear that there is an internal electric field $\boldsymbol{E}_o$ from positive ions to negative ions, *i.e.* $-x$ direction, which tries to drift the holes back into the p-region and electrons back into the n-region. This field drives the holes in the opposite direction to their diffusion. As shown in Figure 3.13 (b), $\boldsymbol{E}_o$ imposes a drift force on holes in the $-x$ direction whereas the hole diffusion flux is in the $+x$ direction. A similar situation also applies for electrons with the electric field attempting to drift the electrons against diffusion from n to the p-region. It is apparent that as more and more holes diffuse towards the right, and electrons towards the left, the internal field around M will increase until eventually an "equilibrium" is reached when the rate of holes diffusing towards the right is just balanced by holes drifting back to the left, driven by the field $\boldsymbol{E}_o$. The electron diffusion and drift fluxes will also be balanced in equilibrium. For uniformly doped p and n regions, the net space charge density $\rho_{\text{net}}(x)$ across the semiconductor will be as shown in Figure 3.13 (d). The net space charge density ρ_{net} is negative and equal to $-eN_a$ in the SCL from $x = -W_p$ to $x = 0$ (M is at $x = 0$) and then positive and equal to $+eN_d$ from $x = 0$ to W_n. The total charge on the left-hand side must equal to that on the right-hand side for overall charge neutrality, so that

$$N_a W_p = N_d W_n \tag{1}$$

Depletion widths

Figure 3.13 arbitrarily assumes that the donor concentration is less than the acceptor concentration, $N_d < N_a$. From Eq. (1) this implies that $W_n > W_p$, *i.e.*, the depletion region penetrates the *n*-side, lightly doped side, more than the *p*-side, heavily doped side. Indeed, if $N_a \gg N_d$, then the depletion region is almost entirely on the *n*-side. We generally indicate heavily doped regions with the superscript plus sign as p^+. The electric field, $\mathbf{E}(x)$, and the net space charge density $\rho_{net}(x)$ at a point are related in electrostatics[3] by $d\mathbf{E}/dx = \rho_{net}(x)/\varepsilon$ in which $\varepsilon = \varepsilon_o \varepsilon_r$ is the permittivity of the medium and ε_o and ε_r are the absolute permittivity and relative permittivity of the semiconductor material. We can thus integrate $\rho_{net}(x)$ across the diode and thus determine the electric field. The variation of the electric field across the *pn* junction is shown in Figure 3.13 (e). The negative field means that it is in the $-x$ direction. Note that $\mathbf{E}(x)$ reaches a maximum value $\mathbf{E}_o$ at M.

The potential $V(x)$ at any point x can be found by integrating the electric field since by definition $\mathbf{E} = -dV/dx$. Taking the potential on the *p*-side far away from M as zero (we have no applied voltage), which is an arbitrary reference level, then $V(x)$ increases in the depletion region towards the *n*-side as indicated in Figure 3.13 (f). Notice that on the *n*-side the potential reaches V_o, which is called the **built-in potential.**

In an abrupt *pn* junction, $\rho_{net}(x)$ can simply and approximately be described by step functions as displayed in Figure 3.13 (d). Using the step form of $\rho_{net}(x)$ in Figure 3.13 (d) and integrating it gives the electric field and the built-in potential,

$$\mathbf{E}_o = -\frac{eN_d W_n}{\varepsilon} = -\frac{eN_a W_p}{\varepsilon} \tag{2}$$

Built-in field

and

$$V_o = -\frac{1}{2}\mathbf{E}_o W_o = \frac{eN_a N_d W_o^2}{2\varepsilon(N_a + N_d)} \tag{3}$$

Built-in potential

in which $\varepsilon = \varepsilon_o \varepsilon_r$ and $W_o = W_n + W_p$ is the total width of the depletion region under a zero-applied voltage. If we know W_o, then W_n or W_p follow readily from Eq. (1). Equation (3) is a relationship between the built-in voltage V_o and the depletion region width W_o. If we know V_o, we can calculate W_o.

The simplest way to relate V_o to the doping parameters is to make use of Boltzmann statistics. For the system consisting of *p*- and *n*-type semiconductors together, in equilibrium, the Boltzmann statistics[4] demands that the concentrations n_1 and n_2 of carriers at potential energies E_1 and E_2 are related by

[3] This is called *Gauss's law in point* form and comes from Gauss's law in electrostatics. The integration of the electric field $\mathbf{E}$ over a closed surface S is related to the total enclosed charge $Q_{enclosed}$, $\int \mathbf{E}dS = Q_{enclosed}/\varepsilon$, in which ε is the permittivity of the medium.

[4] We use Boltzmann statistics, *i.e.* $n(E) \propto \exp(-E/k_BT)$, because the concentration of electrons in the conduction band whether on the *n*-side or *p*-side is never so large that the Pauli exclusion principle becomes important. As long as the carrier concentration in the conduction band is much smaller than N_c, we can use Boltzmann statistics.

$$\frac{n_2}{n_1} = \exp\left[\frac{-(E_2 - E_1)}{k_B T}\right]$$

in which E is the potential energy, that is qV in which q is charge and V voltage. Considering electrons ($q = -e$), we see from Figure 3.13 (g) that $E = 0$ on the p-side far away from M where $n = n_{po}$, and $E = -eV_o$ on the n-side away from M where $n = n_{no}$. Thus

$$n_{po}/n_{no} = \exp(-eV_o/k_B T) \tag{4}$$

This shows that V_o depends on n_{no} and n_{po} and hence on N_d and N_a. The corresponding equation for hole concentrations is similarly,

$$p_{no}/p_{po} = \exp(-eV_o/k_B T) \tag{5}$$

Thus, rearranging Eqs (4) and (5) we obtain,

$$V_o = \frac{k_B T}{e}\ln\left(\frac{n_{no}}{n_{po}}\right) \quad \text{and} \quad V_o = \frac{k_B T}{e}\ln\left(\frac{p_{po}}{p_{no}}\right)$$

We can now write p_{po} and p_{no} in terms of the dopant concentrations inasmuch as $p_{po} = N_a$, $p_{no} = n_i^2/n_{no} = n_i^2/N_d$, so that V_o becomes

Built-in potential

$$V_o = \frac{k_B T}{e}\ln\left(\frac{N_a N_d}{n_i^2}\right) \tag{6}$$

Clearly, V_o has been conveniently related to the dopant and host material properties via N_a, N_d, and n_i^2, which is given by $(N_c N_v)\exp(-E_g/k_B T)$. The built-in voltage (V_o) is the potential across a pn junction, going from p to n-type semiconductor, in an open circuit. It is *not* the voltage across the diode that is made up of V_o as well as the contact potentials at the metal to semiconductor junctions at the electrodes. If we add V_o and the contact potentials at the electroded ends, we will find zero. Once we know the built-in potential V_o from Eq.(6), we can then calculate the width W_o of the depletion region from Eq. (3).

B. Forward Bias

Consider what happens when a battery with a voltage V is connected across a pn junction so that the positive terminal of the battery is attached to the p-side and the negative terminal to the n-side (forward bias). The negative polarity of the supply will reduce the potential barrier V_o by V as shown in Figure 3.14 (a) and (b). The reason is that the bulk regions outside the SCL have high conductivities, due to plenty of majority carriers in the bulk in comparison with the depletion region in which there are mainly immobile ions. Thus, the applied voltage drops mostly across the depletion width W. Consequently, V directly opposes V_o and the potential barrier against diffusion is reduced to $(V_o - V)$ as depicted in Figure 3.14 (b). This has drastic consequences because the probability that a hole in the p-side will surmount this potential barrier and diffuse to the n-side now becomes proportional to $\exp[-e(V_o - V)/k_B T]$. In other words, the applied voltage effectively reduces the built-in potential and hence the built-in field that acts against diffusion. Consequently, many holes can now diffuse across the depletion region and enter the n-side. This results in the **injection of excess minority carriers**, holes

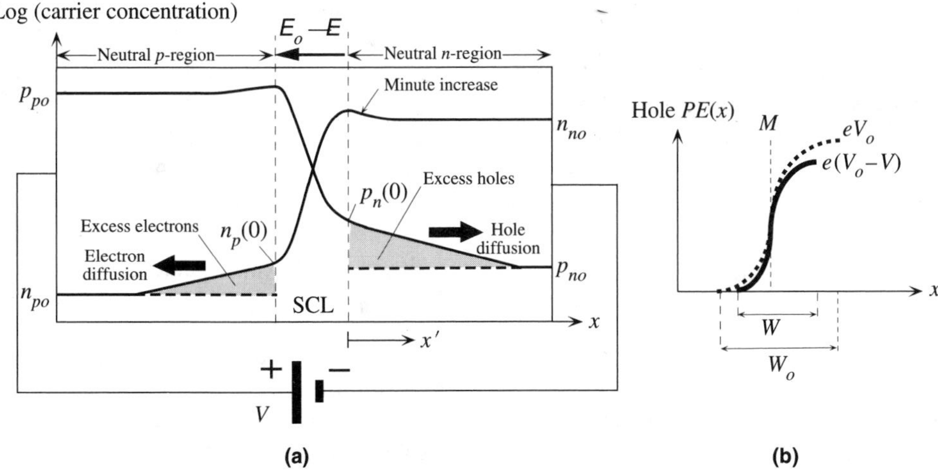

(a) (b)

FIGURE 3.14 Forward biased *pn* junction and the injection of minority carriers (a) Carrier concentration profiles across the device under forward bias. (b) The hole potential energy with and without an applied bias. W is the width of the SCL with forward bias

into n-region. Similarly, excess electrons can now diffuse towards the p-side and enter this region and thereby become injected minority carriers.

When holes are injected into the neutral n-side, they draw some electrons from the bulk of n-side (and hence from the battery) so there is a small increase in the electron concentration. This small increase in the majority carriers is necessary to balance the hole charges and maintain neutrality in the n-side.

The hole concentration, $p_n(0) = p_n(x' = 0)$, just outside the depletion region at $x' = 0$ (x' is measured from W_n) is due to excess holes diffusing as a result of the reduction in the built-in potential barrier. This concentration, $p_n(0)$, is determined by the probability of surmounting the new potential energy barrier $e(V_o - V)$,

$$p_n(0) = p_{po} \exp\left[\frac{-e(V_o - V)}{k_B T}\right] \tag{7}$$

This follows directly from the Boltzmann equation, by virtue of the hole potential energy rising by $e(V_o - V)$ from $x = -W_p$ to $x = W_n$, as indicated in Figure 3.14 (b), and at the same time the hole concentration falling from p_{po} to $p_n(0)$. By dividing Eq. (7) by Eq. (5) we get the effect of the applied voltage out directly, which shows how the voltage V determines the amount of *excess* holes diffusing and arriving at the n-region,

$$p_n(0) = p_{no} \exp\left(\frac{eV}{k_B T}\right) \tag{8}$$

Law of the junction

which is called the **law of the junction**. Equation (8) describes the effect of the applied voltage V on the injected minority carrier concentration just outside the depletion region, $p_n(0)$. Obviously, with no applied voltage, $V = 0$ and $p_n(0) = p_{no}$, which is exactly what we expect.

Injected holes diffuse in the n-region and eventually recombine with electrons in this region; there are many electrons in the n-side. Those electrons lost by recombination are readily replenished by the negative terminal of the battery connected to this side. The current due to holes diffusing in the n-region can be sustained because more holes can be supplied by the p-region, which can be replenished by the positive terminal of the battery.

Electrons are similarly injected from the n-side to the p-side. The electron concentration $n_p(0)$ just outside the depletion region at $x = -W_p$ is given by the equivalent of Eq.(8) for electrons, *i.e.*

Law of the Junction

$$n_p(0) = n_{po} \exp\left(\frac{eV}{k_B T}\right) \tag{9}$$

In the p-region, the injected electrons diffuse toward the positive terminal where their concentration is n_{po}. As they diffuse they recombine with some of the many holes in this region. Those holes lost by recombination can be readily replenished by the positive terminal of the battery connected to this side. The current due to the diffusion of electrons in the p-side can be maintained by the supply of electrons from the n-side, which can be replenished by the negative terminal of the battery. It is apparent that an electric current can be maintained through a pn junction under forward bias, and that the current flow seems to be due to the *diffusion of minority carriers*. There is, however, some drift of majority carriers as well.

If the lengths of the p- and n-regions are longer than the minority carrier diffusion lengths, then we will be justified to expect the hole concentration $p_n(x')$ profile on the n-side to fall exponentially towards the thermal equilibrium value, p_{no}, as depicted in Figure 3.14 (a). If $\Delta p_n(x') = p_n(x') - p_{no}$ is the **excess minority carrier concentration**, then

$$\Delta p_n(x') = \Delta p_n(0) \exp(-x'/L_h) \tag{10}$$

in which L_h is the **hole diffusion length** defined by $L_h = \sqrt{(D_h \tau_h)}$, in which D_h is the diffusion coefficient of holes and τ_h is the mean hole recombination lifetime (minority carrier lifetime) in the n-region. The diffusion length is the average distance diffused by a minority carrier before it disappears by recombination. Equation (10) can be derived by rigorous means but it is stated here as a reasonable result. The rate of recombination of injected holes at any point x' in the neutral n-region is proportional to the excess hole concentration at that point x'. In the steady state, this recombination rate at x' is just balanced by the rate of holes brought to x' by diffusion. This is the physical argument that leads to Eq. (10).

The hole diffusion current density $J_{D,\text{hole}}$ is the *hole diffusion flux* multiplied by the hole charge[5]

$$J_{D,\text{hole}} = -eD_h \frac{dp_n(x')}{dx'} = -eD_h \frac{d\Delta p_n(x')}{dx'}$$

i.e

$$J_{D,\text{hole}} = \left(\frac{eD_h}{L_h}\right) \Delta p_n(0) \exp\left(-\frac{x'}{L_h}\right) \tag{11}$$

[5] The hole diffusion flux is $-D_h(dp/dx)$ and the diffusing charge is $+e$.

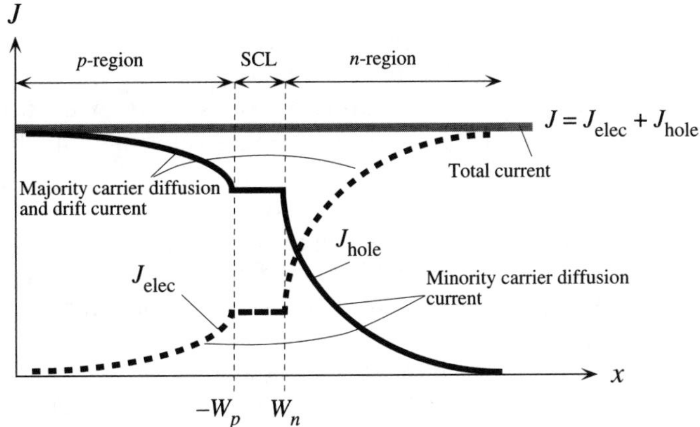

FIGURE 3.15 The total current anywhere in the device is constant. Just outside the depletion region it is primarily due to the diffusion of minority carriers and near the contacts it is primarily due to majority carrier drift.

Although the equation above shows that hole diffusion current depends on location, the total current at any location is, however, the sum of hole and electron contributions, which is independent of x as indicated in Figure 3.15. The decrease in the minority carrier diffusion current with x' is made up by the increase in the current due to the drift of the majority carriers as schematically shown in Figure 3.15. The field in the neutral region is not totally zero but a small value, just sufficient to drift the huge number of majority carriers there and maintain a constant current.

We can now use the law of the junction to substitute for $\Delta p_n(0)$ in Eq. (11) in terms of the applied voltage V in Eq. (8). Further, we can eliminate p_{no} by $p_{no} = n_i^2/n_{no} = n_i^2/N_d$. Thus, at $x' = 0$, just outside the depletion region, from Eq. (11) the hole diffusion current is

$$J_{D,\text{hole}} = \left(\frac{eD_h n_i^2}{L_h N_d}\right)\left[\exp\left(\frac{eV}{k_B T}\right) - 1\right]$$

Hole diffusion current

There is a similar expression for the electron diffusion current density $J_{D,\text{elec}}$ in the p-region. We will assume that the electron and hole currents do not change across the depletion region because, in general, the width of this region is narrow (and, for the time being, we neglect the recombination in the SCL). The electron current at $x = -W_p$ is the same as that at $x = W_n$. The total current density is then simply given by $J_{D,\text{hole}} + J_{D,\text{elec}}$, *i.e.*

$$J = \left(\frac{eD_h}{L_h N_d} + \frac{eD_e}{L_e N_a}\right) n_i^2 \left[\exp\left(\frac{eV}{k_B T}\right) - 1\right]$$

or

$$J = J_{so}\left[\exp\left(\frac{eV}{k_B T}\right) - 1\right] \quad \textbf{(12)}$$

Shockley diode equation

This is the familiar diode equation with $J_{so} = [(eD_h/L_hN_d) + (eD_e/L_eN_a)]n_i^2$. It is frequently called the **Shockley equation.** It represents the *diffusion of minority carriers* in the neutral regions. The constant J_{so} depends not only on the doping, N_d, N_a, but also on the material via n_i, D_h, D_e, L_h, and L_e. It is known as the **reverse saturation current density**, because if we apply a reverse bias $V = -V_r$ greater than the thermal voltage k_BT/e (= 25 mV), Eq. (12) becomes $J = -J_{so}$.

So far we have assumed that, under a forward bias, the minority carriers diffusing and recombining in the neutral regions are supplied by the external current. However, some of the minority carriers will recombine in the depletion region. The external current must therefore also supply the carriers lost in the recombination process in the SCL. Consider for simplicity a symmetrical *pn* junction as in Figure 3.16 under forward bias. At the metallurgical junction at the center C, the hole and electron concentrations are p_M and n_M and are equal. We can find the SCL recombination current by considering electrons recombining in the *p*-side in W_p and holes recombining in the *n*-side in W_n as shown by the shaded areas ABC and BCD, respectively, in Figure 3.16. Suppose that the **mean hole recombination time** in W_n is τ_h and **mean electron recombination time** in W_p is τ_e. The rate at which the electrons in ABC are recombining is the area ABC (nearly all injected electrons) divided by τ_e. The electrons are replenished by the diode current. Similarly, the rate at which holes in BCD are recombining is the area BCD divided by τ_h. Thus, the recombination current density is

$$J_{\text{recom}} = \frac{eABC}{\tau_e} + \frac{eBCD}{\tau_h}$$

We can evaluate the areas ABC and BCD by taking them as triangles, $ABC \approx \left(\frac{1}{2}\right) W_pn_M$, *etc.* so that

$$J_{\text{recom}} \approx \frac{e\frac{1}{2}W_pn_M}{\tau_e} + \frac{e\frac{1}{2}W_np_M}{\tau_h} \tag{13}$$

Under steady state and equilibrium conditions, assuming a non-degenerate semiconductor, we can use Boltzmann statistics to relate these concentrations to the potential energy. At A, the potential is zero and at M it is $\frac{1}{2}e(V_o - V)$ so that

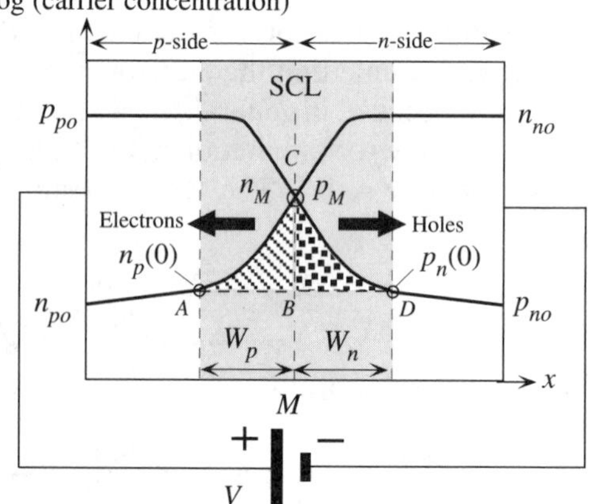

FIGURE 3.16 Forward biased *pn* junction and the injection of carriers and their recombination in the SCL.

$$\frac{p_M}{p_{po}} = \exp\left[-\frac{e(V_o - V)}{2k_BT}\right]$$

Since V_o depends on dopant concentrations and n_i as in Eq. (6) and further $p_{po} = N_a$, we can simplify the above to

$$p_M = n_i \exp\left(\frac{eV}{2k_BT}\right)$$

This means that the recombination current for $V > k_BT/e$ is given by,

$$J_{\text{recom}} = \frac{en_i}{2}\left[\frac{W_p}{\tau_e} + \frac{W_n}{\tau_h}\right]\exp\left(\frac{eV}{2k_BT}\right) \tag{14}$$

Recombination current

From a better quantitative analysis, the expression for the recombination current can be shown to be[6]

$$J_{\text{recom}} = J_{ro}[\exp(eV/2k_BT) - 1]$$

in which J_{ro} is the pre-exponential constant in Eq. (14).

Equation (14) is the current that supplies the carriers that recombine in the depletion region. The total current into the diode will supply carriers for minority carrier diffusion in the neutral regions and recombination in the space charge layer so that it will be the sum of Eqs. (12) and (14). In general, the diode current is written as

$$I = I_o\left[\exp\left(\frac{eV}{\eta k_BT}\right) - 1\right] \tag{15}$$

Diode equation

in which I_o is a constant and η, called the **diode ideality factor**, is 1 for diffusion controlled and 2 for SCL recombination controlled characteristics. Figure 3.17 shows both the forward and the reverse $I - V$ characteristics of a typical *pn* junction.

C. Reverse Bias

When a *pn* junction is reverse biased, the reverse current is typically very small, as shown in Figure 3.17. The reverse bias across a *pn* junction is illustrated in Figure 3.18 (a). The applied voltage drops mainly across the resistive depletion region, which becomes wider. The negative terminal will cause holes in the *p*-side to move away from the SCL, which results in more exposed negative acceptor ions and thus a wider SCL. Similarly, the positive terminal will attract electrons away from the SCL, which exposes more positively charged donors. The depletion width on the *n*-side therefore also widens. The movement of electrons in the *n*-region towards the positive battery terminal cannot be sustained because there is no electron supply to this *n*-side. The *p*-side cannot supply electrons to the *n*-side because it has almost none. However, there is a small reverse current due to two causes.

The applied voltage increases the built-in potential barrier, as depicted in Figure 3.18 (b). The electric field in the SCL is larger than the built-in internal field E_o. The small number of holes on the *n*-side near the depletion region become extracted and

[6] This is generally proved in advanced texts.

FIGURE 3.17 Forward and reverse I-V characteristics of a pn junction (the positive and negative current axes have different scales and hence the discontinuity at the origin)

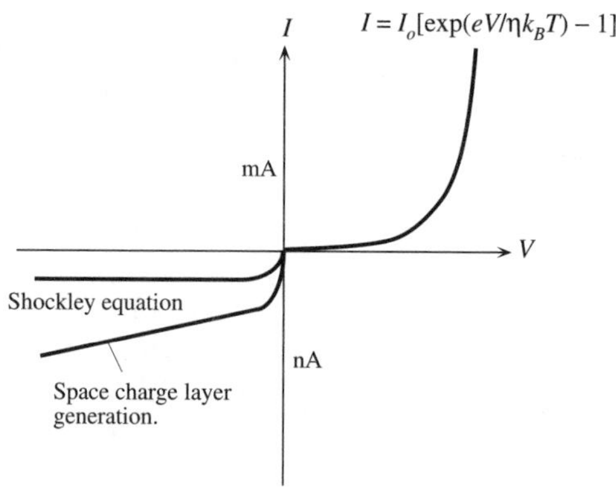

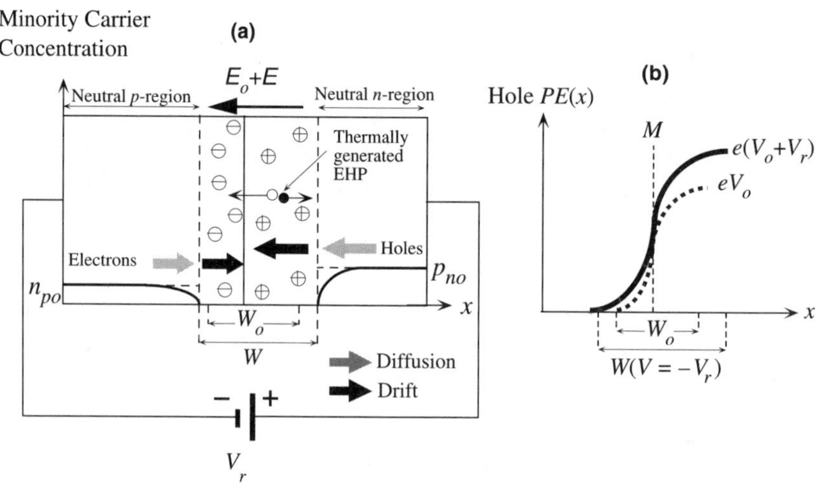

FIGURE 3.18 Reverse biased pn junction. (a) Minority carrier profiles and the origin of the reverse current. (b) Hole PE across the junction under reverse bias

swept by the field across the SCL over to the p-side. This small current can be maintained by the diffusion of holes from the n-side bulk to the SCL boundary.

Assume that the reverse bias $V_r > 25\text{ mV} = k_BT/e$. The hole concentration $p_n(0)$ just outside the SCL is nearly zero by the law of the junction, Eq. (8), whereas the hole concentration in the bulk (or near the negative terminal) is the equilibrium concentration p_{no}, which is small. There is therefore a small concentration gradient and hence a small hole diffusion current towards the SCL as shown in Figure 3.18 (a). Similarly, there is a small electron diffusion current from bulk p-side to the SCL. Within the SCL, these carriers are drifted by the field. This minority carrier diffusion current is the Shockley model. The reverse current is given by Eq. (12) with a negative voltage, which leads to a diode current density of $-J_{so}$ called the **reverse saturation current density**. The value of J_{so} depends only on the material via n_i, μ_h, μ_e, the dopant concentrations, *etc.*, but not on the voltage $(V_r > k_B T/e)$. Furthermore, as J_{so} depends on n_i^2, it is strongly temperature dependent. In some books it is stated that thermal generation of minority carriers in the neutral region within a diffusion length to the SCL, the diffusion of these carriers to the SCL, and their subsequent drift through the SCL is the cause of the reverse current. This description, in essence, is identical to the Shockley model.

The thermal generation of electron hole pairs (EHPs) in the space charge region, as shown in Figure 3.18 (a), can also contribute to the observed reverse current since the internal field in this layer will separate the electron and hole and drift them toward the neutral regions. This drift will result in an external current in addition to the reverse current due to the diffusion of minority carriers. The theoretical evaluation of SCL generation current involves an in-depth knowledge of the charge carrier generation processes via recombination centers, which is discussed in advanced texts. Suppose that τ_g is the **mean time to generate an electron-hole pair** by virtue of the thermal vibrations of the lattice; τ_g is also called the **mean thermal generation time**. Given τ_g, the rate of thermal generation per unit volume must be n_i/τ_g because it takes on average τ_g seconds to create n_i number of EHPs per unit volume. Furthermore, since WA, in which A is the cross-sectional area, is the volume of the depletion region, the rate of EHP, or charge carrier, generation is $(AWn_i)/\tau_g$. Both holes and electrons drift in the depletion region and each contributes equally to the current. The observed current density must be $e(Wn_i)/\tau_g$. Therefore, the reverse current density component due to thermal generation of electron-hole pairs within the SCL should be given by

$$J_{\text{gen}} = \frac{eWn_i}{\tau_g} \tag{16}$$

EHP thermal generation in SCL

The reverse bias widens the width W of the depletion layer and hence increases J_{gen}. The total reverse current density J_{rev} is the sum of the diffusion and generation components, *i.e.*

$$J_{\text{rev}} = \left(\frac{eD_h}{L_h N_d} + \frac{eD_e}{L_e N_a}\right) n_i^2 + \frac{eWn_i}{\tau_g} \tag{17}$$

Total reverse current

which is shown schematically in Figure 3.17. The thermal generation component J_{gen} in Eq. (16) increases with reverse bias V_r because the SCL width W increases with V_r.

The terms in the reverse current in Eq. (17) are predominantly controlled by n_i^2 and n_i. Their relative importance depends not only on the semiconductor properties but also on the temperature since $n_i \sim \exp(-E_g/2k_B T)$. Figure 3.19 shows the reverse current I_{rev} in dark in a Ge *pn* junction (a photodiode) plotted as $\ln(I_{\text{rev}})$ vs. $1/T$ to highlight the two different processes in Eq. (17). The measurements in Figure 3.19 show that above 238 K, I_{rev} is controlled by n_i^2 because the slope of $\ln(I_{\text{rev}})$ vs. $1/T$ yields an E_g of approximately 0.63 eV, close to the expected E_g of about 0.66 eV in Ge. Below 238 K, I_{rev} is controlled by n_i because the slope of $\ln(I_{\text{rev}})$ vs. $1/T$ is equivalent to an $E_g/2$ of approximately 0.33 eV. In this range, the reverse current is due to EHP generation in the SCL via defects and impurities (recombination centers).

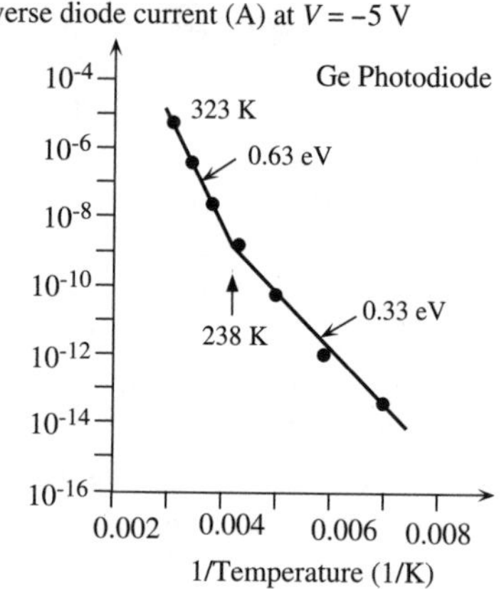

FIGURE 3.19 Reverse diode current in a Ge *pn* junction as a function of temperature in a $\ln(I_{rev})$ vs $1/T$ plot. Above 238 K, I_{rev} is controlled by n_i^2 and below 238 K it is controlled by n_i. The vertical axis is a logarithmic scale with actual current values.
(From D. Scansen and S.O. Kasap, *Cnd. J. Physics.* **70**, 1070–1075, 1992.)

D. Depletion Layer Capacitance

It is apparent that the depletion region of a *pn* junction has positive and negative charges separated over a distance W similar to a parallel plate capacitor as indicated in Figure 3.13 (d). If A is the cross-sectional area, the stored charge in the depletion region is $+Q = eN_dW_nA$ on the *n*-side and $-Q = -eN_aW_pA$ on the *p*-side. Unlike in the case of a parallel plate capacitor, Q does not depend linearly on the voltage V across the device. It is useful to define an incremental capacitance that relates the incremental charge stored to an incremental voltage change across the *pn* junction. When the voltage V across a *pn* junction changes by dV to $V + dV$, then W also changes and, as a result, the amount of charge in the depletion region becomes $Q + dQ$. The **depletion layer capacitance** C_{dep} is defined by

Depletion layer capacitance

$$C_{dep} = \left|\frac{dQ}{dV}\right| \tag{18}$$

If the applied voltage is V, then the voltage across the depletion layer W is $V_o - V$ and Eq. (3) in this case becomes,

SCL width and voltage

$$W = \left[\frac{2\varepsilon(N_a + N_d)(V_o - V)}{eN_aN_d}\right]^{1/2} \tag{19}$$

The amount of charge (on any one side of the depletion layer) is $|Q| = eN_dW_nA = eN_aW_pA$, and $W = W_n + W_p$. We can therefore substitute for W in Eq. (19) in terms of Q and then differentiate it to obtain dQ/dV. The final result for the depletion capacitance is

Depletion layer capacitance

$$C_{dep} = \frac{\varepsilon A}{W} = \frac{A}{(V_o - V)^{1/2}}\left[\frac{e\varepsilon(N_aN_d)}{2(N_a + N_d)}\right]^{1/2} \tag{20}$$

We should note that C_{dep} is given by the same expression as that for the parallel plate capacitor, $\varepsilon A/W$, but with W being voltage-dependent by virtue of Eq. (19). Putting a reverse bias $V = -V_r$ in Eq. (20) shows that C_{dep} decreases with increasing V_r. Typically, C_{dep} under reverse bias is of the order of a few picofarads.

E. Recombination Lifetime

Consider recombination in a direct bandgap semiconductor, for example, doped GaAs. Recombination involves a direct meeting of an electron and a hole pair. Suppose that excess electrons and holes have been injected, as would be in a *pn*-junction under forward bias, and that Δn_p is the excess electron concentration and Δp_p is the excess hole concentration in the *neutral* *p*-side of a GaAs *pn* junction. Injected electron and hole concentrations would be the same to maintain charge neutrality, that is, $\Delta n_p = \Delta p_p$. Thus, at any instant,

$$n_p = n_{po} + \Delta n_p = \text{instantaneous } \textit{minority} \text{ carrier concentration,}$$

and

$$p_p = p_{po} + \Delta n_p = \text{instantaneous } \textit{majority} \text{ carrier concentration.}$$

The instantaneous recombination rate will be proportional to both the electron and hole concentrations at that instant, that is, $n_p p_p$. Suppose that the thermal generation rate of EHPs is G_{thermal}. The net rate of *change* of Δn_p is

$$\partial \Delta n_p / \partial t = -B n_p p_p + G_{\text{thermal}} \tag{21}$$

in which B is called the **direct recombination capture coefficient**. In equilibrium $\partial \Delta n_p / \partial t = 0$ so that setting Eq. (21) to zero and using $n_p = n_{po}$ and $p_p = p_{po}$, in which the subscript o refers to thermal equilibrium concentrations, we find $G_{\text{thermal}} = B n_{po} p_{po}$. Thus, the rate of change in Δn_p in Eq. (21) is

$$\frac{\partial \Delta n_p}{\partial t} = -B\left(n_p p_p - n_{po} p_{po}\right) \tag{22}$$

Rate of change due to recombination

In many instances the rate of change $\partial \Delta n_p / \partial t$ is proportional to Δn_p, and an **excess minority carrier recombination time (lifetime)** τ_e is defined by

$$\frac{\partial \Delta n_p}{\partial t} = -\frac{\Delta n_p}{\tau_e} \tag{23}$$

Recombination time definition

Consider practical cases in which injected excess minority carrier concentration Δn_p is much greater than the actual equilibrium minority carrier concentration n_{po}. There are two conditions on Δn_p corresponding to weak and strong injection based on Δn_p compared with the majority carrier concentration p_{po}.

In **weak injection**, $\Delta n_p \ll p_{po}$. Then $n_p \approx \Delta n_p$ and $p_p \approx p_{po} + \Delta p_p \approx p_{po} \approx N_a =$ acceptor concentration. Therefore, with these approximation in Eq. (22) we obtain,

$$\partial \Delta n_p / \partial t = -B N_a \Delta n_p \tag{24}$$

Thus, comparing with Eq. (23),

$$\tau_e = 1 / B N_a \tag{25}$$

Weak injection recombination lifetime

and is constant under weak injection conditions as here.

In **strong injection**, $\Delta n_p \gg p_{po}$. Then it is easy to show that with this condition, Eq. (22) becomes

$$\partial \Delta n_p / \partial t = B \Delta p_p \Delta n_p = B\left(\Delta n_p\right)^2 \tag{26}$$

so that under high level injection conditions the lifetime τ_e is inversely proportional to the injected carrier concentration. When a light-emitting diode (LED) is modulated, under high injection levels for example, the lifetime of the minority carriers is therefore not constant, which in turn leads to distortion of the modulated light output.

EXAMPLE 3.3.1 A direct bandgap *pn* junction

A symmetrical GaAs *pn* junction with a cross sectional area $A = 1\ \text{mm}^2$ has the following properties: N_a (*p*-side doping) $= N_d$ (*n*-side doping) $= 10^{23}\ \text{m}^{-3}$; $B = 7.21 \times 10^{-16}\ \text{m}^3\ \text{s}^{-1}$; $n_i = 1.8 \times 10^{12}\ \text{m}^{-3}$; $\varepsilon_r = 13.2$; μ_h (in the *n*-side) $= 250\ \text{cm}^2\ \text{V}^{-1}\ \text{s}^{-1}$; μ_e (in the *p*-side) $= 5000\ \text{cm}^2\ \text{V}^{-1}\ \text{s}^{-1}$. Diffusion coefficients are related to drift mobilities via the **Einstein relation**: $D_h = \mu_h k_B T/e$, $D_e = \mu_e k_B T/e$. The forward voltage across the diode is 1 V. What is the diode current due to the minority carrier diffusion at 300 K assuming direct recombination? If the *mean* minority carrier recombination time in the depletion region is of the order of ~10 ns, estimate the recombination component of the current?

Solution Assuming weak injection, we can readily calculate the recombination times τ_e and τ_h for electrons and holes recombining in the neutral *p* and *n*-regions respectively. Using S.I. units and $k_B T/e = 0.02585$ V, for a symmetric device,

$$\tau_h = \tau_e = \frac{1}{BN_a} = \frac{1}{(7.21 \times 10^{-16}\ \text{m}^3\text{s}^{-1})(1 \times 10^{23}\ \text{m}^{-3})} = 1.39 \times 10^{-8}\ \text{s}$$

Diffusion coefficients are

$$D_h = \mu_h k_B T/e = (0.2585)(250 \times 10^{-4}) = 6.46 \times 10^{-4}\ \text{m}^2\ \text{s}^{-1}$$

and

$$D_e = \mu_e k_B T/e = (0.2585)(5000 \times 10^{-4}) = 1.29 \times 10^{-2}\ \text{m}^2\ \text{s}^{-1}$$

The diffusion lengths are

$$L_h = (D_h\tau_h)^{1/2} = [(6.46 \times 10^{-4}\ \text{m}^2\ \text{s}^{-1})(1.39 \times 10^{-8}\ \text{s})]^{1/2} = 3.00 \times 10^{-6}\ \text{m}.$$

and

$$L_e = (D_e\tau_e)^{1/2} = [(1.29 \times 10^{-2}\ \text{m}^2\ \text{s}^{-1})(1.39 \times 10^{-8}\ \text{s})]^{1/2} = 1.34 \times 10^{-5}\ \text{m}.$$

The reverse saturation current due to diffusion in the neutral regions is

$$I_{so} = A\left(\frac{D_h}{L_h N_d} + \frac{D_e}{L_e N_a}\right) e\, n_i^2$$

$$= (10^{-6})\left[\frac{6.46 \times 10^{-4}}{(3.00 \times 10^{-6})(10^{23})} + \frac{1.29 \times 10^{-2}}{(1.34 \times 10^{-5})(10^{23})}\right](1.6 \times 10^{-19})(1.8 \times 10^{12})^2 = 6.13 \times 10^{-21}\,\text{A}$$

Thus, the forward diffusion current is

$$I_{\text{diff}} = I_{so}\exp\left(\frac{eV}{k_B T}\right) = (6.13 \times 10^{-21}\ \text{A})\exp\left[\frac{1.0\ \text{V}}{0.02585\ \text{V}}\right] = 3.9 \times 10^{-4}\ \text{A}$$

The built-in voltage V_o is given by

$$V_o = \frac{k_B T}{e}\ln\left(\frac{N_a N_d}{n_i^2}\right) = (0.02585)\ln\left[\frac{10^{23}10^{23}}{(1.8 \times 10^{12})^2}\right] = 1.28\ \text{V}$$

The depletion layer width W is

$$W = \left[\frac{2\varepsilon(N_a + N_d)(V_o - V)}{e\, N_a N_d}\right]^{1/2}$$

$$= \left[\frac{2(13.2)(8.85 \times 10^{-12}\ \text{F m}^{-1})(10^{23} + 10^{23}\ \text{m}^{-3})(1.28 - 1\ \text{V})}{(1.6 \times 10^{-19}\ \text{C})(10^{23}\ \text{m}^{-3})(10^{23}\ \text{m}^{-3})} \right]^{1/2}$$

$$= 9.0 \times 10^{-8}\ \text{m}, \quad \text{or} \quad 0.090\ \mu\text{m}.$$

For a symmetric diode, $W_p = W_n = \frac{1}{2}W$, and taking $\tau_e = \tau_h = \tau_r \approx 10$ ns,

$$I_{ro} = \frac{A e n_i}{2}\left[\frac{W_p}{\tau_e} + \frac{W_n}{\tau_h}\right] = \frac{A e n_i}{2}\left(\frac{W}{\tau_r}\right)$$

$$\approx \frac{(10^{-6})(1.6 \times 10^{-19})(1.8 \times 10^{12})}{2}\left(\frac{9.0 \times 10^{-8}}{10 \times 10^{-9}}\right) = 1.3 \times 10^{-12}\ \text{A}$$

so that

$$I_{\text{recom}} \approx I_{ro}\exp\left[\frac{eV}{2k_BT}\right] = (1.3 \times 10^{-12}\ \text{A})\exp\left[\frac{1.0\ \text{V}}{2(0.02585\ \text{V})}\right] = 3.3 \times 10^{-4}\ \text{A}$$

In this example, the diffusion and recombination components are about the same order of magnitude.

3.4 THE *pn* JUNCTION BAND DIAGRAM

A. Open Circuit

Figure 3.20 (a) shows the energy band diagram of a *pn* junction under open circuit. If E_{Fp} and E_{Fn} are the Fermi levels in the *p*- and *n*-sides, then in equilibrium and in the dark, the Fermi level must be uniform through the two materials as depicted in Figure 3.20 (a). Far away from the metallurgical junction M, in the bulk of the *n*-type semiconductor we should still have an *n*-type semiconductor and $E_c - E_{Fn}$ should be the same as in the isolated *n*-type material. Similarly, $E_{Fp} - E_v$ far away from M inside the *p*-type material should also be the same as in the isolated *p*-type material. These features are sketched in Figure 3.20 (a) keeping E_{Fp} and E_{Fn} the same through the whole system and, of course, keeping the bandgap, $E_c - E_v$, the same. Clearly, to draw the energy band diagram we have to bend the bands, E_c and E_v, near the junction at M because E_c on the *n*-side is close to E_{Fn} whereas on the *p*-side it is far away from E_{Fp}.

The instant the two semiconductors are brought together to form the junction, electrons diffuse from the *n*-side to the *p*-side and as they do so they deplete the *n*-side near the junction. Thus, E_c must move away from E_{Fn} as we move toward M, which is exactly what is sketched in Figure 3.20 (a). Holes diffuse from the *p*-side to the *n*-side and the loss of holes in the *p*-type material near the junction means that E_v moves away from E_{Fp} as we move towards M, which is also shown in the figure. Furthermore, as electrons and holes diffuse toward each other, most of them recombine and disappear around M, which leads to the formation of a **space charge layer** (SCL) as we saw in Figure 3.13 (b). The SCL zone around the metallurgical junction has therefore been *depleted* of carriers compared with the bulk.

The electrostatic potential energy (PE) of the electron decreases from 0 inside the *p*-region to $-eV_o$ inside the *n*-region as shown in Figure 3.13 (g). The total energy of the electron must therefore decrease going from the *p*- to the *n*-region by an amount eV_o. In other words, the electron in the *n*-side at E_c must overcome a PE barrier to go over to E_c in the *p*-side. This PE barrier is eV_o in which V_o is the built-in potential that we

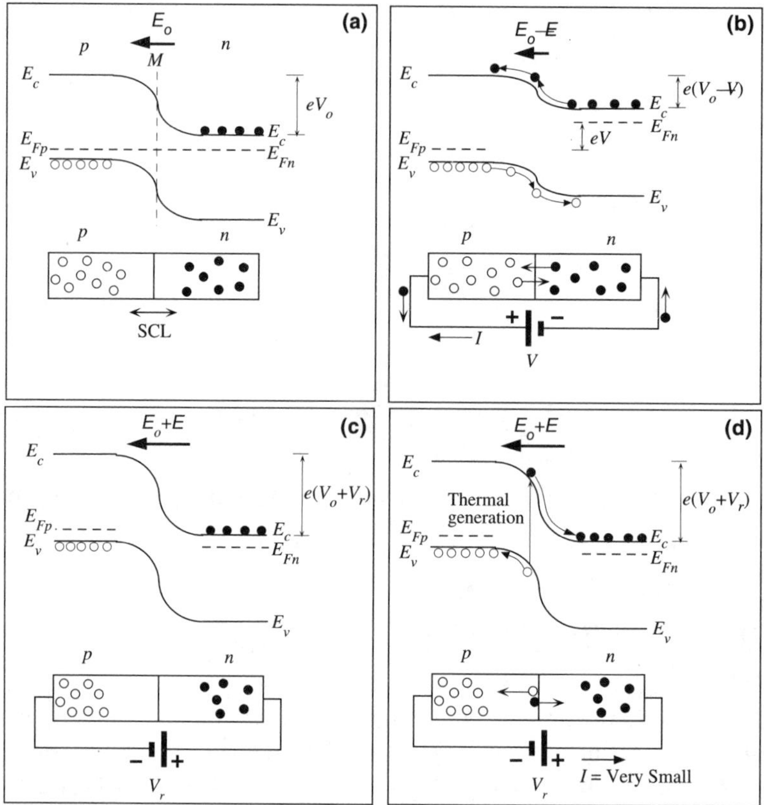

FIGURE 3.20 Energy band diagrams for a *pn* junction under (a) open circuit, (b) forward bias and (c) reverse bias conditions. (d) Thermal generation of electron hole pairs in the depletion region results in a small reverse current.

evaluated previously. Band bending around M therefore accounts not only for the variation of electron and hole concentrations in this region but also for the effect of the built-in potential (and hence the built-in field, as the two are related). The diffusion of electrons from n-side to p-side is prevented by the built-in PE barrier eV_o. This barrier also prevents holes from diffusing from the p- to the n-side.

It should be noted that in the SCL region, the Fermi level is neither close to E_c nor E_v compared with the bulk or neutral semiconductor regions. This means that both n and p in this zone are much less than their bulk values n_{no} and p_{po}. The metallurgical junction region has been depleted of carriers compared with the bulk. Any applied voltage must therefore drop across the SCL.

B. Forward and Reverse Bias

When the pn junction is forward biased, majority of the applied voltage drops across the depletion region so that the applied voltage is in opposition to the built-in potential, V_o. Figure 3.20 (b) shows the effect of forward bias, which is to reduce the PE barrier from eV_o to $e(V_o - V)$. The electrons at E_c in the n-side can now readily overcome the PE barrier and diffuse to the p-side. The diffusing electrons from the n-side can be replenished easily by the negative terminal of the battery connected to this side. Similarly, holes can now diffuse from p- to n-side. The positive terminal of the battery can replenish those holes diffusing away from the p-side. There is therefore a current flow through the junction and around the circuit.

The probability that an electron at E_c in the n-side overcomes the new PE barrier and diffuses to E_c in the p-side is now proportional to the Boltzmann factor $\exp[-e(V_o - V)/k_BT]$. The latter increases enormously even for small forward voltages. Thus, an extensive diffusion of electrons from n- to p-side takes place. Similar ideas also apply to holes at E_v in the p-side that also overcome the barrier $e(V_o - V)$ to diffuse into the n-side. Since the forward current is due to the number of electrons and holes overcoming the barrier it is also proportional to $\exp[-e(V_o - V)/k_BT]$ or $\exp(eV/k_BT)$.

When a reverse bias, $V = -V_r$, is applied to the pn junction the voltage again drops across the SCL. In this case, however, V_r adds to the built-in potential V_o so that the PE barrier becomes $e(V_o + V_r)$, as shown in Figure 3.20 (c). The field in the SCL at M increases to $\mathbf{E}_o + \mathbf{E}$ in which $\mathbf{E}$ is the applied field (it is not simply V/W). There is hardly any reverse current because if an electron were to leave the n-side to travel to the positive terminal, it cannot be replenished from the p-side (virtually no electrons on the p-side). There is, however, a small reverse current arising from thermal generation of electron hole pairs (EHPs) in the SCL and thermal generation of minority carriers within a diffusion length to the SCL. When an EHP is thermally generated in the SCL, as shown in Figure 3.20 (d), the field here separates the pair. The electron falls down the PE hill, down E_c, to the n-side to be collected by the battery. Similarly, the hole falls down its own PE hill (energy increases downwards for holes) to make it to the p-side. The process of falling down a PE hill is the same process as being driven by a field, in this case by $\mathbf{E}_o + \mathbf{E}$. A thermally generated hole in the n-side within a diffusion length can diffuse to the SCL and then drift across the SCL, which would result in a reverse current. Similarly, a thermally generated electron in the p-side within a diffusion length to the SCL can also contribute to a reverse current. All these reverse current components are very small compared with the forward current as they depend on the rate of thermal generation.

3.5 LIGHT EMITTING DIODES

A. Principles

A light emitting diode (LED) is essentially a pn junction diode typically made from a direct bandgap semiconductor, for example GaAs, in which the electron hole pair (EHP) recombination results in the emission of a photon. The emitted photon energy is therefore approximately equal to the bandgap energy, $h\upsilon \approx E_g$. Figure 3.21 (a) shows the energy band diagram of an unbiased pn^+ junction device in which the n- side is more heavily doped than the p-side. The band diagram is drawn to keep the Fermi level uniform through the device, which is a requirement of equilibrium with no applied bias. The depletion region in a pn^+ device extends mainly into the p-side. There is a potential

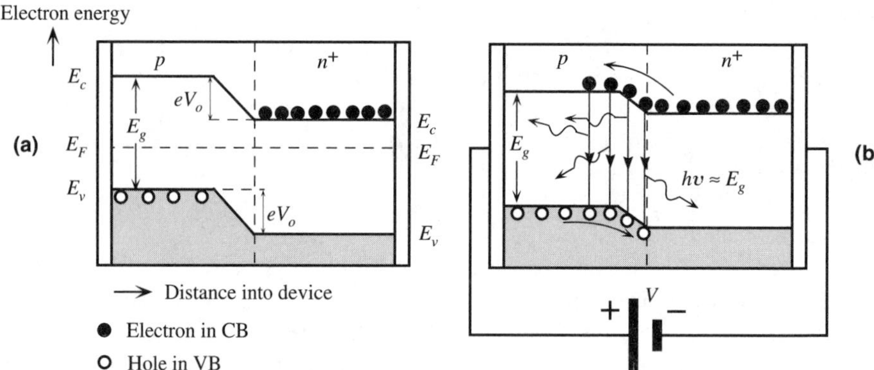

FIGURE 3.21 (a) The energy band diagram of a $p - n^+$ (heavily n-type doped) junction without any bias. Builtin potential V_o prevents electrons from diffusing from n^+ to p side. (b) The applied bias reduces V_o and thereby allows electrons to diffuse, be injected, into the p-side. Recombination around the junction and within the diffusion length of the electrons in the p-side leads to photon emission.

energy (PE) barrier eV_o from E_c on the n-side to E_c on the p-side, that is, $\Delta E_c = eV_o$, in which V_o is the *built-in voltage*. The higher concentration of conduction (free) electrons in the n-side encourages the diffusion of conduction electrons from the n- to the p-side. However, this net electron diffusion is prevented by the electron PE barrier eV_o.

As soon as a forward bias V is applied, this voltage drops across the depletion region since this is the most resistive part of the device. Consequently, the built-in potential V_o is reduced to $V_o - V$, which then allows the electrons from the n^+ side to diffuse, or become injected, into the p-side as depicted in Figure 3.21 (b). The hole injection component from p into the n^+ side is much smaller than the electron injection component from the n^+ to p-side. The recombination of injected electrons in the depletion region as well as in the neutral p-side results in the spontaneous emission of photons. Recombination primarily occurs within the depletion region and within a volume extending over the diffusion length L_e of the electrons in the p-side. This recombination zone is frequently called the **active region**.[7] The phenomenon of light emission from EHP recombination as a result of minority carrier injection as in this case is called **injection electroluminescence**. Because of the statistical nature of the recombination process between electrons and holes, the emitted photons are in random directions; they result from spontaneous emission processes in contrast to stimulated emission. The LED structure has to be such that the emitted photons can escape the device without being reabsorbed by the semiconductor material. This means the p-side has to be sufficiently narrow or we have to use *heterostructure* devices as discussed later in § 3.7.

[7] The "active region" term is probably more appropriate for laser diodes in which there is a photon amplification in this region.

B. Device Structures

In its simplest technological form, LEDs are typically fabricated by *epitaxially* growing doped semiconductor layers on a suitable substrate (*e.g.* GaAs or GaP) as depicted in Figure 3.22 (a). This type of planar *pn* junction is formed by the epitaxial growth of first the *n*-layer and then the *p*-layer. The substrate is essentially a mechanical support for the *pn* junction device (the layers) and can be of different material. The *p*-side is on the surface from which light is emitted and is therefore made narrow (a few microns) to allow the photons to escape without being reabsorbed. To ensure that most of the recombination takes place in the *p*-side, the *n*-side is heavily doped (n^+). Those photons that are emitted towards the *n*-side become either absorbed or reflected back at the substrate interface depending on the substrate thickness and the exact structure of the LED. The use of a segmented back electrode as in Figure 3.22 (a) will encourage reflections from the semiconductor-air interface. It is also possible to form the *p*-side by diffusing dopants into the epitaxial n^+-layer, which is a diffused junction planar LED as illustrated in Figure 3.22 (b).

If the epitaxial layer and the substrate crystals have different crystal lattice parameters, then there is a lattice mismatch between the two crystal structures. This causes lattice strain in the LED layer and hence leads to crystal defects. Such crystal defects encourage radiationless EHP recombinations. That is, a defect acts as a recombination center. Such defects are reduced by lattice-matching the LED epitaxial layer to the substrate crystal. It is therefore important to lattice-match the LED layer to the substrate crystal. For example, one of the AlGaAs alloys is a direct bandgap semiconductor that has a bandgap in the red-emission region. It can be grown on GaAs substrates with excellent lattice match, which results in high efficiency LED devices.

Figure 3.22 (a) and (b) both show the planar *pn* junction-based simple LED structures. However, not all light rays reaching the semiconductor-air interface can escape because of total internal reflection (TIR). Those rays with angles of incidence greater than the critical angle θ_c become reflected as depicted in Figure 3.23 (a). For the GaAs-air interface, for example, θ_c is only 16°, which means that much of the light suffers TIR. It is

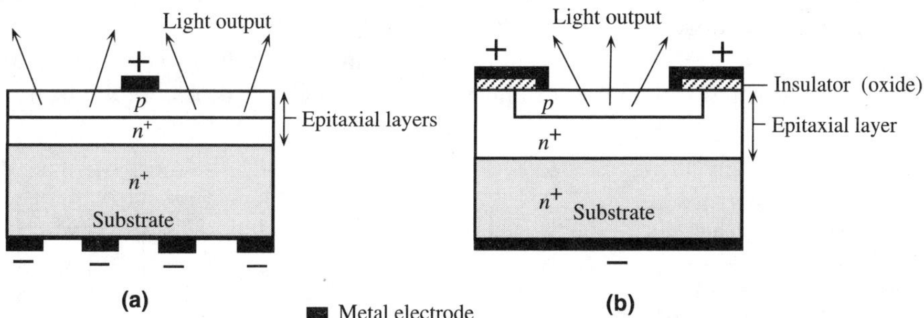

FIGURE 3.22 A schematic illustration of typical planar surface emitting LED devices. (a) *p*-layer grown epitaxially on an n^+ substrate. (b) First n^+ is epitaxially grown and then *p* region is formed by dopant diffusion into the epitaxial layer.

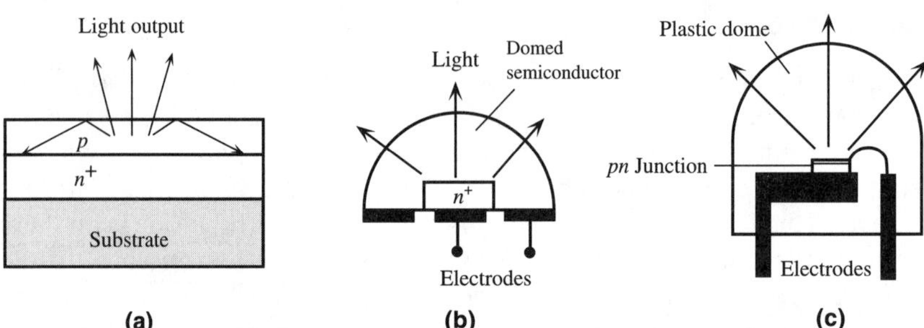

FIGURE 3.23 (a) Some light suffers total internal reflection and cannot escape. (b) Internal reflections can be reduced and hence more light can be collected by shaping the semiconductor into a dome so that the angles of incidence at the semiconductor-air surface are smaller than the critical angle. (b) An economic method of allowing more light to escape from the LED is to encapsulate it in a transparent plastic dome.

possible to shape the surface of the semiconductor into a dome, or hemisphere, so that light rays strike the surface at angles less than θ_c and therefore do not experience TIR as illustrated in Figure 3.23 (b). The main drawback, however, is the additional difficult process in fabricating such domed LEDs and the associated increase in expense. An inexpensive and common procedure that reduces TIR is the encapsulation of the semiconductor junction within a transparent plastic medium (an epoxy) that has a higher refractive index than air and, further, also has a domed surface on one side of the *pn* junction as shown in Figure 3.23 (c). Many individual LEDs are sold in similar types of plastic bodies.

3.6 LED MATERIALS

There are various direct bandgap semiconductor materials that can be readily doped to make commercial *pn* junction LEDs that emit radiation in the red and infrared range of wavelengths. An important class of commercial semiconductor materials that cover the visible spectrum is the **III-V ternary alloys** based on alloying GaAs and GaP, which are denoted as $GaAs_{1-y}P_y$. In this compound, As and P atoms from group V are distributed randomly at normal As sites in the GaAs crystal structure. When $y < 0.45$, the alloy $GaAs_{1-y}P_y$ is a direct bandgap semiconductor and hence the EHP recombination process is direct, which is depicted in Figure 3.24 (a). The rate of recombination is di-

FIGURE 3.24 (a) Photon emission in a direct bandgap semiconductor. (b) GaP is an indirect bandgap semiconductor. When doped with nitrogen there is an electron trap at E_N. Direct recombination between a trapped electron at E_N and a hole emits a photon. (c) In Al doped SiC, EHP recombination is through an acceptor level like E_a.

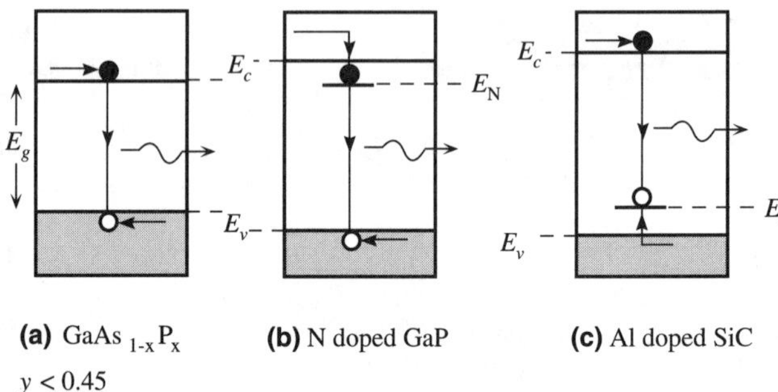

rectly proportional to the product of electron and hole concentrations. The emitted wavelengths range from about 630 nm, red, for $y = 0.45$ ($GaAs_{0.55}P_{0.45}$) to 870 nm for $y = 0$, GaAs.

$GaAs_{1-y}P_y$ alloys (which include GaP) with $y > 0.45$ are indirect bandgap semiconductors. The EHP recombination processes occur through recombination centers and involve lattice vibrations rather than photon emission. However, if we add **isoelectronic impurities** such as nitrogen (in the same group V as P) into the semiconductor crystal then some of these N atoms substitute for P atoms. Since N and P have the same valency, N atoms substituting for P atoms form the same number of bonds and do not act as donors or acceptors. The electronic cores of N and P are, however, different. The positive nucleus of N is less shielded by electrons compared with that of the P atom. This means that a conduction electron in the neighborhood of an N atom will be attracted and may become trapped at this site. N atoms therefore introduce localized energy levels, or electron traps, E_N, near the conduction band edge as depicted in Figure 3.24 (b). When a conduction electron is captured at E_N, it can attract a hole (in the valence band) in its vicinity by Coulombic attraction and eventually recombine with it directly and emit a photon. The emitted photon energy is only slightly less than E_g as E_N is typically close to E_c. As the recombination process depends on N doping, it is not as efficient as direct recombination. Thus, the efficiency of LEDs from N doped indirect bandgap $GaAs_{1-y}P_y$ semiconductors is less than those from direct bandgap semiconductors. Nitrogen doped indirect bandgap $GaAs_{1-y}P_y$ alloys are widely used in inexpensive green, yellow, and orange LEDs.

There are two types of blue LED materials. GaN is a direct bandgap semiconductor with an E_g of 3.4 eV. The blue GaN LEDs actually use the GaN alloy; InGaN has a bandgap of about 2.7 eV which corresponds to blue emission. The less efficient type is the Al doped *silicon carbide* (SiC), which is an indirect bandgap semiconductor. The acceptor type localized energy level captures a hole from the valence band and a conduction electron then recombines with this hole to emit a photon, as schematically shown in Figure 3.24 (c). As the recombination process is not direct and therefore not as efficient, the brightness of blue SiC LEDs is limited. Recently, there has been considerable progress made toward more efficient blue LEDs using direct bandgap compound semiconductors such as II-VI semiconductors, for example ZnSe (Zn and Se are in groups II and VI in the Periodic Table). The main problem in using II-VI compounds is the current technological difficulty in appropriately doping these semiconductors to fabricate efficient *pn* junctions.

There are various commercially important direct bandgap semiconductor materials that emit in the red and infrared wavelengths, which are typically **ternary** (containing three elements) and **quarternary** (four elements) alloys based on III and V elements, so called **III-V alloys**. For example, GaAs with a bandgap of about 1.43 eV emits radiation at around 870 nm in the infrared. But ternary alloys based on $Al_{1-x}Ga_xAs$ in which $x < 0.43$ are direct bandgap semiconductors. The composition can be varied to adjust the bandgap and hence the emitted radiation from about 640 – 870 nm, from deep red light to infrared.

In-Ga-Al-P is a quarternary III-V alloy (In, Ga, and Al from III and P from V) that has a direct bandgap variation with composition over the visible range. It can be lattice-matched to GaAs substrates when in the composition range $In_{0.49}Al_{0.17}Ga_{0.34}P$ to

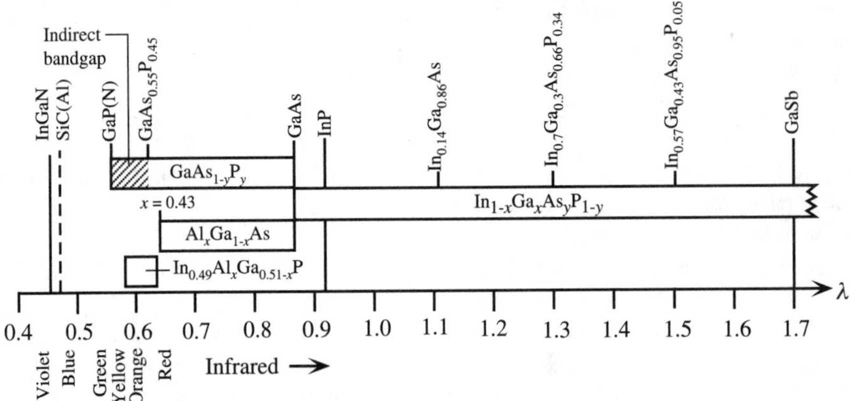

FIGURE 3.25 Free space wavelength coverage by different LED materials from the visible spectrum to the infrared including wavelengths used in optical communications. Hatched region and dashed lines are indirect E_g materials.

$In_{0.49}Al_{0.058}Ga_{0.452}P$. Recent high-intensity LEDs have been based on this material, which is likely to eVentually dominate the high-intensity visible LED range.

The bandgap of quarternary alloys $In_{1-x}Ga_xAs_{1-y}P_y$ can be varied with composition (x and y) to span wavelengths from 870 nm (GaAs) to 3.5 μm (InAs), which includes the optical communication wavelengths of 1.3 and 1.55 μm. Figure 3.25 summarizes some typical wavelengths that can be emitted for a few selected semiconductor materials over the range from 0.4 to 1.7 μm.

External efficiency η_{external} of an LED quantifies the efficiency of conversion of electrical energy into an emitted external optical energy. It incorporates the "internal" efficiency of the radiative recombination process and the subsequent efficiency of photon extraction from the device. The input of electrical power into an LED is simply the diode current and diode voltage product (IV). If P_{out} is the optical power emitted by the device, then

External quantum efficiency

$$\eta_{\text{external}} = \frac{P_{\text{out}}(\text{Optical})}{IV} \times 100\% \quad \textbf{(1)}$$

and some typical values are listed in Table 3.1. For indirect bandgap semiconductors η_{external} are generally less than 1% whereas for direct bandgap semiconductors with the right device structure, η_{external} can be substantial.

3.7 HETEROJUNCTION HIGH INTENSITY LEDS

A junction (such as a *pn* junction) between two differently doped semiconductors that are of the same material (same bandgap E_g) is called a **homojunction**. A junction between two different bandgap semiconductors is called a **heterojunction**. A semiconductor device structure that has junctions between different bandgap materials is called a **heterostructure device** (HD). The refractive index of a semiconductor material depends

TABLE 3.1 Selected LED semiconductor materials. Optical communication channels are at 850 nm (local network) and at 1.3 and 1.55 μm (long distance). D = Direct, I = Indirect bandgap, DH = Double heterostructure. $\eta_{external}$ is typical and may vary substantially depending on the device structure.

Semiconductor	Substrate	D or I	λ (nm)	$\eta_{external}$ (%)	Comment
GaAs	GaAs	D	870–900	10	Infrared LEDs
$Al_xGa_{1-x}As$ $(0 < x < 0.4)$	GaAs	D	640–870	5–20	Red to IR LEDs. DH
$In_{1-x}Ga_xAs_yP_{1-y}$ $(y \approx 2.20x, 0 < x < 0.47)$	InP	D	1–1.6 μm	>10	LEDs in communications
InGaN alloys	GaN or SiC	D	430–460	2	Blue LED
	Saphire		500–530	3	Green LED
SiC	Si; SiC	I	460–470	0.02	Blue LED. Low efficiency
$In_{0.49}Al_xGa_{0.51-x}P$	GaAs	D	590–630	1–10	Amber, green, red LEDs
$GaAs_{1-y}P_y$ $(y < 0.45)$	GaAs	D	630–870	<1	Red–IR
$GaAs_{1-y}P_y$ $(y > 0.45)$ (N or Zn, O doping)	GaP	I	560–700	<1	Red, orange, yellow LEDs
GaP (Zn-O)	GaP	I	700	2–3	Red LED
GaP (N)	GaP	I	565	<1	Green LED

on its bandgap. A wider bandgap semiconductor has a lower refractive index. This means that by constructing LEDs from heterostructures, we can engineer a dielectric waveguide within the device and thereby channel photons out from the recombination region.

The homojunction LED shown in Figure 3.23 (a) has two drawbacks. The *p*-region must be narrow to allow the photons to escape without much reabsorption. When the *p*-side is narrow, some of the injected electrons in the *p*-side reach the surface by diffusion and recombine through crystal defects near the surface. This radiationless recombination process decreases the light output. In addition, if the recombination occurs over a relatively large volume (or distance), due to long electron diffusion lengths, then the chances of reabsorption of emitted photons becomes higher; the amount of reabsorption increases with the material volume.

LED constructions for increasing the intensity of the output light make use of the double heterostructure (DH) structure. Figure 3.26 (a) shows a **double-heterostructure** (DH) device based on two junctions between different semiconductor materials with different bandgaps. In this case, the semiconductors are AlGaAs with $E_g \approx 2$ eV and GaAs with $E_g \approx 1.4$ eV. The double heterostructure in Figure 3.26 (a) has an n^+p heterojunction between n^+-AlGaAs and *p*-GaAs. There is another heterojunction between *p*-GaAs and *p*-AlGaAs. The *p*-GaAs region is a thin layer, typically a fraction of a micron and it is lightly doped.

The simplified energy band diagram for the whole device in the absence of an applied voltage is shown in Figure 3.26 (b). The Fermi level E_F is continuous through the whole structure. There is a potential energy barrier eV_o for electrons in the CB of n^+-AlGaAs against diffusion into *p*-GaAs. There is a bandgap change at the junction between *p*-GaAs and *p*-AlGaAs that results in a step change, ΔE_c in E_c, between the two bands of *p*-GaAs and *p*-AlGaAs. This, ΔE_c, is effectively a potential energy barrier that prevents any electrons in the CB in *p*-GaAs passing to the CB of *p*-AlGaAs.

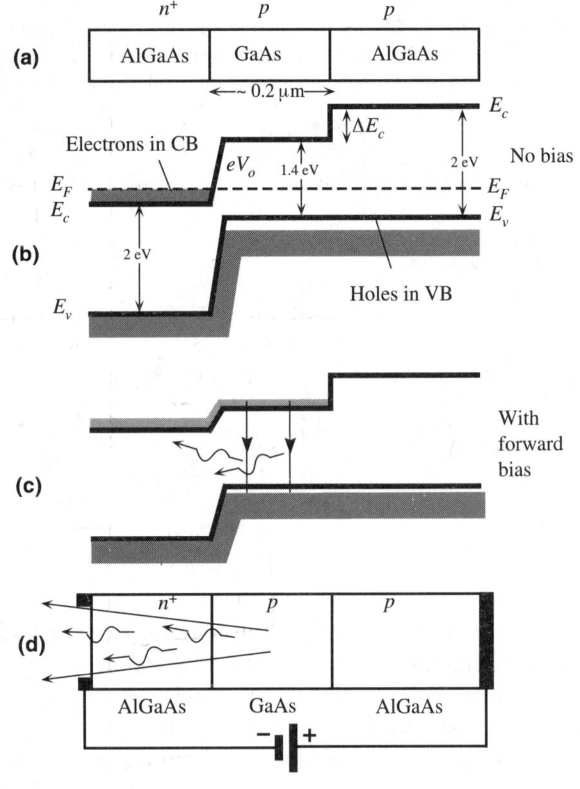

FIGURE 3.26 (a) A double heterostructure diode has two junctions which are between two different bandgap semiconductors (GaAs and AlGaAs). (b) A simplified energy band diagram with exaggerated features. E_F must be uniform. (c) Forward biased simplified energy band diagram. (d) Forward biased LED. Schematic illustration of photons escaping reabsorption in the AlGaAs layer and being emitted from the device.

When a forward bias is applied, majority of this voltage drops between the n^+-AlGaAs and p-GaAs and reduces the potential energy barrier eV_o, just as in the normal pn junction diode. This allows electrons in the CB of n^+-AlGaAs to be injected (by diffusion) into p-GaAs as shown in Figure 3.26 (c). These electrons, however, are *confined* to the CB of p-GaAs since there is a barrier ΔE_c between p-GaAs and p-AlGaAs. The wide bandgap AlGaAs layers therefore act as **confining layers** that restrict injected electrons to the p-GaAs layer. The recombination of injected electrons and the holes already present in this p-GaAs layer results in spontaneous photon emission. Since the bandgap E_g of AlGaAs is greater than GaAs, the emitted photons do not get reabsorbed as they escape the active region and can reach the surface of the device as depicted in Figure 3.26 (d). Since light is also not absorbed in p-AlGaAs, it can be reflected to increase the light output. Another advantage of the AlGaAs/GaAs heterojunction is that there is only a small lattice mismatch between the two crystal structures. Hence, negligible strain induced interfacial defects (*e.g.* dislocations) in the device compared with the defects at the surface of the semiconductor in conventional homojunction LED structure shown in Figure 3.23 (a). The DH LED is much more efficient than the homojunction LED.

3.8 LED CHARACTERISTICS

The energy of an emitted photon from an LED is not simply equal to the bandgap energy E_g because electrons in the conduction band are distributed in energy and so are the holes in the valence band. Figure 3.27 (a) and (b) illustrate the energy band diagram and the energy distributions of electrons and holes in the conduction band (CB) and valence band (VB) respectively. The electron concentration as a function of energy in the CB is given by $g(E)f(E)$ in which $g(E)$ is the density of states and $f(E)$ is the Fermi-Dirac function (probability of finding an electron in a state with energy E). The product $g(E)f(E)$ represents the electron concentration per unit energy or the concentration in energy, and is plotted along the horizontal axis in Figure 3.27 (b). There is a similar energy distribution for holes in the VB.

The electron concentration in the CB as a function of energy is asymmetrical and has a peak at $(1/2)k_BT$ above E_c. The energy spread of these electrons is typically about $\sim 2k_BT$ from E_c as shown in Figure 3.27 (b). The hole concentration is similarly spread from E_v in the valence band. We should recall that the rate of direct recombination is proportional to both the electron and hole concentrations at the energies involved. The transition, which is identified as 1 in Figure 3.27 (a), involves the direct recombination of an electron at E_c and a hole at E_v. But the carrier concentrations near the band edges are very small and hence this type of recombination does not occur frequently. The relative intensity of light at this photon energy $h\upsilon_1$ is small, as shown in Figure 3.27 (c). The transitions that involve the largest electron and hole concentrations occur most frequently. For example, the transition 2 in Figure 3.27 (a) has the maximum probability as both electron and hole concentrations are largest at these energies, as shown in Figure 3.27 (b). The relative intensity of light corresponding to this transition energy, $h\upsilon_2$, is then maximum, or close to maximum, as indicated in Figure 3.27 (c).[8] The transitions

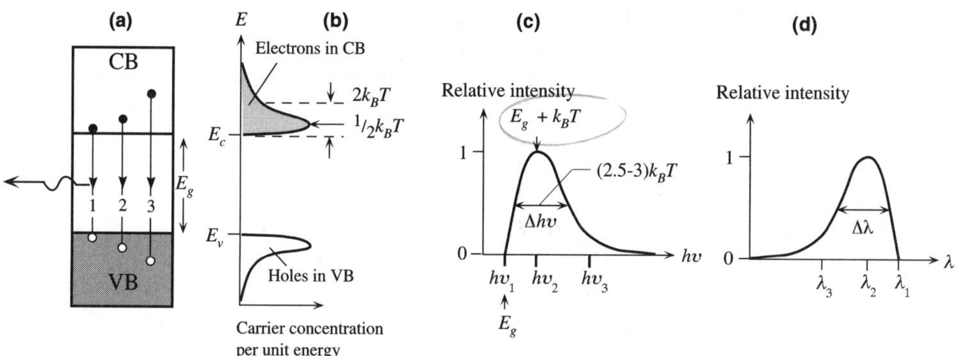

FIGURE 3.27 (a) Energy band diagram with possible recombination paths. (b) Energy distribution of electrons in the CB and holes in the VB. The highest electron concentration is $(1/2)k_BT$ above E_c. (c) The relative light intensity as a function of photon energy based on (b). (d) Relative intensity as a function of wavelength in the output spectrum based on (b) and (c).

[8] The intensity is not necessarily maximum when both the electron and hole concentrations are maximum, but it will be close.

marked as 3 in Figure 3.27 (a) that emit relatively high energy photons, $h\upsilon_3$, involve energetic electrons and holes whose concentrations are small, as apparent in Figure 3.27 (b). Thus, the light intensity at these relatively high photon energies is small. The fall in light intensity with photon energy is shown in Figure 3.27 (c). The relative light intensity vs. photon energy characteristic of the output spectrum is shown in Figure 3.27 (c) and represents an important LED characteristic. Given the spectrum in Figure 3.27 (c), we can also obtain the relative light intensity vs. wavelength characteristic as shown in Figure 3.27 (d) since $\lambda = c/\upsilon$. The **linewidth** of the output spectrum, $\Delta\upsilon$ or $\Delta\lambda$, is defined as width between half-intensity points as defined in Figure 3.27 (c) and (d).

The wavelength for the peak intensity and the linewidth $\Delta\lambda$ of the spectrum are obviously related to the energy distributions of the electrons and holes in the conduction and valence bands and therefore to the density of states in these bands (and hence to individual semiconductor properties). The photon energy for the peak emission is roughly $E_g + k_BT$ inasmuch as it corresponds to peak-to-peak transitions in the energy distributions of the electrons and holes in Figure 3.27 (b). The linewidth $\Delta(h\upsilon)$ is typically between $2.5k_BT$ to $3k_BT$ as shown in Figure 3.27 (c).

The output spectrum, or the relative intensity vs. wavelength characteristics, from an LED depends not only on the semiconductor material but also on the structure of the *pn* junction diode, including the dopant concentration levels. The spectrum in Figure 3.27 (d) represents an idealized spectrum without including the effects of heavy doping on the energy bands. For a heavily doped *n*-type semiconductor there are so many donors that the electron wavefunctions at these donors overlap to generate a narrow impurity band centered at E_d but extending into the conduction band. Thus, the donor impurity band overlaps the conduction band and hence effectively lowers E_c. The minimum emitted photon energy from heavily doped semiconductors is therefore less than E_g and depends on the amount of doping.

Typical characteristics of a red LED (655 nm) as an example are shown in Figure 3.28 (a) to (c). The output spectrum in Figure 3.28 (a) exhibits less asymmetry than the idealized spectrum in Figure 3.27 (d). The width of the spectrum is about 24 nm,

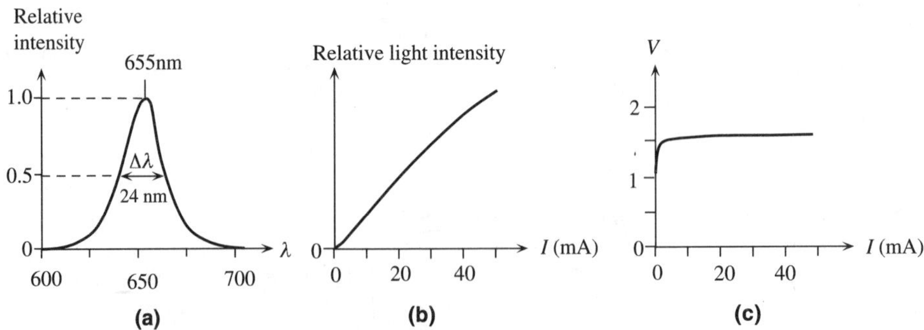

FIGURE 3.28 (a) A typical output spectrum (relative intensity vs. wavelength) from a red GaAsP LED. (b) Typical output light power vs. forward current. (c) Typical *I-V* characteristics of a red LED. The turn-on voltage is around 1.5 V.

which corresponds to a width of about $2.7k_BT$ in the energy distribution of the emitted photons. As the LED current increases, so does the injected minority carrier concentration, thus the rate of recombination and hence the output light intensity. However, the increase in the output light power is not linear with the LED current as apparent in Figure 3.28 (b). At high current levels, strong injection of minority carriers leads to the recombination time depending on the injected carrier concentration and hence on the current itself; this leads to a nonlinear recombination rate with current. Typical current-voltage characteristics are shown in Figure 3.28 (c), where it can be seen that the **turn-on** or the **cut-in voltage** is about 1.5 V from which point the current increases sharply with voltage. The turn-on voltage depends on the semiconductor and generally increases with the energy bandgap E_g. For example, typically, for a blue LED it is about 3.5–4.5 V, for a yellow LED it is about 2 V, and for a GaAs infrared LED it is around 1 V.

EXAMPLE 3.8.1 LED Output Spectrum

Given that the width of the relative light intensity vs. photon energy spectrum of an LED is typically around $\sim 3k_BT$, what is the linewidth $\Delta\lambda_{1/2}$ in the output spectrum in terms of wavelength?

Solution We note that the emitted wavelength λ is related to the photon energy E_{ph} by

$$\lambda = c/\upsilon = hc/E_{ph}.$$

If we differentiate λ with respect to the photon energy E_{ph} we get

$$\frac{d\lambda}{dE_{ph}} = -\frac{hc}{E_{ph}^2}$$

We can represent small changes or intervals (or Δ) by differentials, *e.g.* $\Delta\lambda/\Delta E_{ph} \approx |d\lambda/dE_{ph}|$, then

$$\Delta\lambda \approx \frac{hc}{E_{ph}^2}\Delta E_{ph}$$

We are given the energy width of the output spectrum, $\Delta E_{ph} = \Delta(h\upsilon) \approx 3k_BT$. Then, using the latter and substituting for E_{ph} in terms of λ we find,

$$\Delta\lambda \approx \lambda^2 \frac{3k_BT}{hc}$$

Thus, at

$$\lambda = 870\text{ nm}, \qquad \Delta\lambda = 47\text{ nm},$$
$$\lambda = 1300\text{ nm}, \qquad \Delta\lambda = 105\text{ nm},$$
$$\lambda = 1550\text{ nm}, \qquad \Delta\lambda = 149\text{ nm}.$$

These linewidths are typical values and the exact values depend on the LED structure.

EXAMPLE 3.8.2 LED output wavelength variations

Consider a GaAs LED. The bandgap of GaAs at 300 K is 1.42 eV, which changes (decreases) with temperature as $dE_g/dT = -4.5 \times 10^{-4}$ eV K^{-1}. What is the change in the emitted wavelength if the temperature change is 10°C?

Solution Neglecting the k_BT term and taking $\lambda = hc/E_g$ we have,

$$\frac{d\lambda}{dT} = -\frac{hc}{E_g^2}\left(\frac{dE_g}{dT}\right) = -\frac{(6.626 \times 10^{-34})(3 \times 10^8)}{(1.42 \times 1.6 \times 10^{-19})^2}(-4.5 \times 10^{-4} \times 1.6 \times 10^{-19})$$

so that,

$$\frac{d\lambda}{dT} = 2.77 \times 10^{-10}\ \text{m K}^{-1}, \text{ or } 0.277\ \text{nm K}^{-1}.$$

The change in the wavelength $\Delta\lambda$ for $\Delta T = 10°\text{C}$ is

$$\Delta\lambda = (d\lambda/dT)\Delta T = (0.277\ \text{nm K}^{-1})(10\ \text{K}) \approx 2.8\ \text{nm}.$$

Since E_g decreases with temperature, the wavelength *increases* with temperature. This calculated change is within 10% of typical values for GaAs LEDs quoted in data books.

EXAMPLE 3.8.3 InGaAsP on InP substrate

The ternary alloy $In_{1-x}Ga_xAs_yP_{1-y}$ grown on an InP crystal substrate is a suitable commercial semiconductor material for infrared wavelength LED and laser diode applications. The device requires that the InGaAsP layer is lattice matched to the InP crystal substrate to avoid crystal defects in the InGaAsP layer. This in turn requires that $y \approx 2.2x$. The bandgap E_g of the ternary alloy in eV is then given by the empirical relationship,

$$E_g \approx 1.35 - 0.72y + 0.12y^2; \quad 0 \leq x \leq 0.47$$

Calculate the compositions of InGaAsP ternary alloys for peak emission at a wavelength of 1.3 μm.

Solution We first note that we need the required bandgap E_g at the wavelength of interest. The photon energy at peak emission is $hc/\lambda = E_g + k_BT$. Then in electron volts,

$$E_g = \frac{ch}{e\lambda} - \frac{k_BT}{e}$$

and at $\lambda = 1.3 \times 10^{-6}$ m, taking $T = 300$ K,

$$E_g = \frac{(3 \times 10^8)(6.626 \times 10^{-34})}{(1.6 \times 10^{-19})(1.3 \times 10^{-6})} - 0.0259\ \text{eV} = 0.928\ \text{eV}.$$

The InGaAsP then must have y satisfying,

$$0.928 = 1.35 - 0.72y + 0.12y^2$$

Solving this quadratic equation on a calculator gives $y = 0.66$. Then $x = 0.66/2.2 = 0.3$. The quarternary alloy is $In_{0.7}Ga_{0.3}As_{0.66}P_{0.34}$.

3.9 LEDS FOR OPTICAL FIBER COMMUNICATIONS

The type of light source suitable for optical communications depends not only on the communication distance but also on the bandwidth requirement. For short haul applications, *e.g.* local networks, LEDs are preferred as they are simpler to drive, more economic, have a longer lifetime, and provide the necessary output power even though their output spectrum is much wider than that of a laser diode. LEDs are frequently used

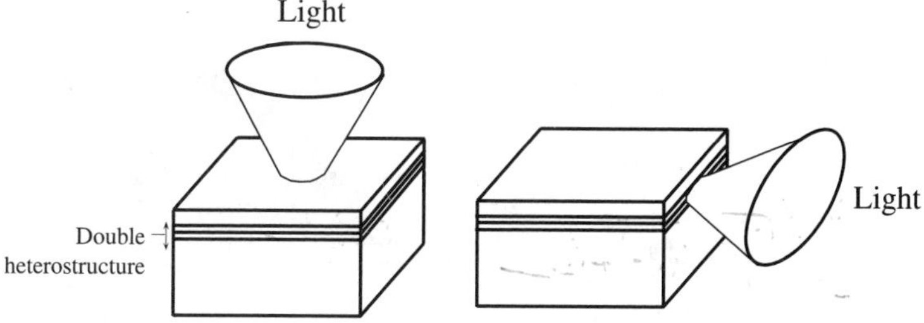

FIGURE 3.29 (a) Surface emitting LED. (b) Edge emitting LED.

with graded index fibers inasmuch as dispersion in a graded index fiber is primarily intermodal rather than intramodal. For long-haul and wide bandwidth communications, laser diodes are invariably used because of their narrow linewidth, high output power, and higher signal bandwidth capability.

There are essentially two types of LED devices that are illustrated in Figure 3.29. If the emitted radiation emerges from an area in the plane of the recombination layer as in (a) then the device is a **surface emitting LED** (SLED). If the emitted radiation emerges from an area on an edge of the crystal as in (b), *i.e.* from an area on a crystal face perpendicular to the active layer, then the LED is an **edge emitting LED** (ELED).

The simplest method of coupling the radiation from a surface-emitting LED into an optical fiber is to etch a well in the planar LED structure and lower the fiber into the well as close as possible to the **active region** where emission occurs. This type of structure, as shown in Figure 3.30 (a) is called a **Burrus** type device (after its originator). An epoxy resin is used to bond the fiber and provide refractive index matching between the glass fiber and the LED material to capture as much of the light rays as possible. Note that in the double heterostructure LED used in this way, the photons emitted from the active region (*e.g. p*-GaAs) do not get absorbed by the neighboring layer (AlGaAs), which has a wider bandgap. Another method is to use a truncated **spherical lens** (a microlens) with a high refractive index ($n = 1.9 - 2$) to focus the light into the fiber as

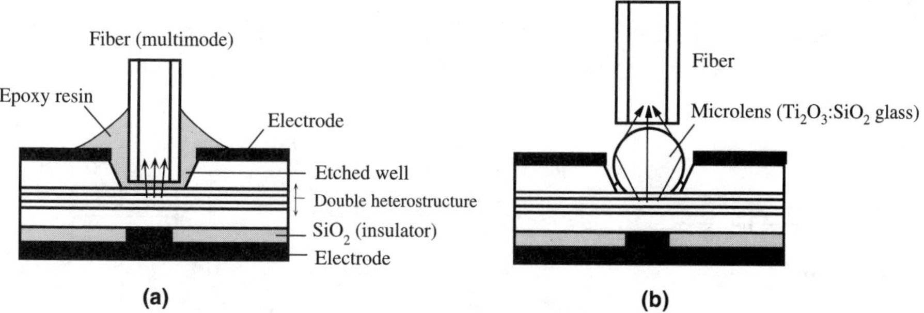

FIGURE 3.30 (a) Light is coupled from a surface emitting LED into a multimode fiber using an index matching epoxy. The fiber is bonded to the LED structure. (b) A microlens focuses diverging light from a surface emitting LED into a multimode optical fiber.

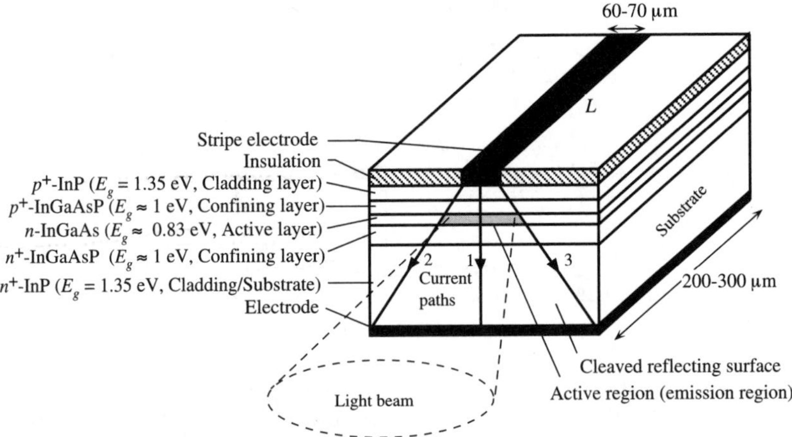

FIGURE 3.31 Schematic illustration of the the structure of a double heterojunction stripe contact edge emitting LED.

shown in Figure 3.30 (b). The lens is bonded to the LED with a refractive index matching cement and, in addition, the fiber can be bonded to the lens with a similar cement.

Edge emitting LEDs provide a greater intensity light and also a beam that is more collimated than the surface emitting LEDs. Figure 3.31 shows the structure of a typical edge emitting LED for operation at ~1.5 μm. The light is guided to the edge of the crystal by a **dielectric waveguide** formed by wider bandgap semiconductors surrounding a double heterostructure. The recombination of injected carriers occurs in the InGaAs active region, which has a bandgap $E_g \approx 0.83$ eV. Recombination is confined to this layer because the surrounding InGaAsP layers, **confining layers**, have a wider bandgap $(E_g \approx 1 \text{ eV})$ and the InGaAsP/InGaAs/InGaAsP layers form a double heterostructure. The light emitted in the active region (InGaAs) spreads into the neighboring layers (InGaAsP), which act to contain the light and guide it along the crystal to the edge. InP has a wider bandgap $(E_g \approx 1.35 \text{ eV})$ and thus a lower refractive index than InGaAsP. The two InP layers adjoining the InGaAsP layers therefore act as **cladding layers** and thereby confine the light to the DH structure.

In general, some kind of lens system is used to conveniently couple the emitted radiation from an ELED into a fiber. For example, in Figure 3.32 (a), a hemispherical lens

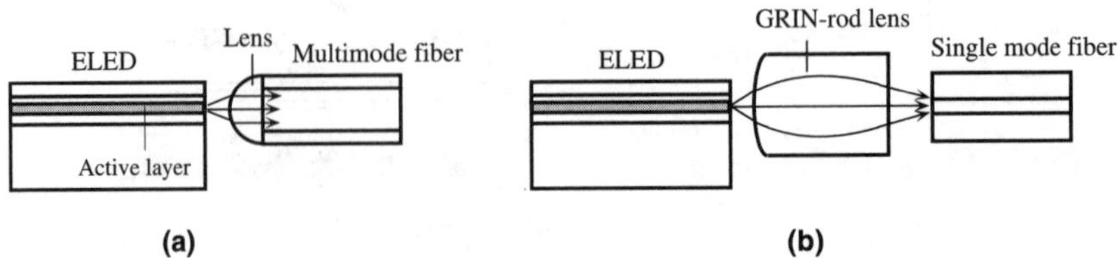

FIGURE 3.32 Light from an edge emitting LED is coupled into a fiber typically by using a lens or a GRIN rod lens.

attached to the fiber end is used for collimating the beam into the fiber. **A graded index (GRIN) rod lens** is a glass rod that has a parabolic refractive index profile across its cross-section with the maximum index on the rod axis. It is like a large diameter short length graded index "fiber" (typical diameters are 0.5-2 mm). A GRIN rod lens can be used to focus the light from an ELED into a fiber as depicted in Figure 3.31 (b). This coupling is particularly useful for single mode fibers inasmuch as their core diameters are typically ~10 μm.

The output spectra from surface and edge emitting LEDs using the same semiconductor material is not necessarily the same. The first reason is that the active layers have different doping levels. Second, there is the self-absorption of some of the photons guided along the active layer as in ELED. Typically, the linewidth of the output spectrum from an ELED is less than that from a SLED. In one set of experiments, for example, an InGaAsP ELED operating near 1300 nm was observed to have a linewidth of 75 nm, whereas the corresponding SLED had a linewidth of 125 nm.

QUESTIONS AND PROBLEMS

3.1 Electrons in the CB of a semiconductor

(a) Consider the energy distribution of electrons $n_E(E)$ in the conduction band. Assuming that the density of state $g_{CB}(E) \propto (E - E_c)^{1/2}$ and using Boltzmann statistics $f(E) \approx \exp[-(E - E_F)/k_B T]$, show that the energy distribution of the electrons in the CB can be written as,

$$y(x) = Cx^{1/2}\exp(-x)$$

in which $x = E/k_B T$ is electron energy in terms of $k_B T$ measured from E_c and C is a temperature dependent constant (independent of E).

(b) Setting arbitrarily $C = 1$, plot $y(x)$ vs. x. Where is the maximum and what is the FWHM (full width at half maximum, *i.e.* between half maximum points)?

(c) Show that the average electron energy in the CB is $(3/2)k_B T$, by using the definition of average,

$$x_{\text{average}} = \left(\int xy\,dx\right)/\left(\int y\,dx\right)$$

in which the integration is from $x = 0\ (E_c)$ to say $x = 10$ (far away from E_c where $y \to 0$). You need to use a numerical integration.

(d) Show that the maximum in the energy distribution is at $x = 1/2$ or at $E_{\max} = (1/2)k_B T$.

(e) Given that the electron effective mass m_e^* for GaAs is $0.067m_e$, calculate the thermal velocity of the CB electrons. If μ_e is the drift mobility and τ_e is the mean free time between electron scattering events (between electrons and lattice vibrations) and if $\mu_e = e\tau_e/m_e^*$, calculate τ_e, given $\mu_e = 8500$ cm^2 V^{-1} s^{-1}. Calculate the **drift velocity** $v_d = \mu_e E$ of the CB electrons in an applied field E of 10^5 V m^{-1}. What is your conclusion?

3.2 GaAs GaAs has an effective density of states at the CB edge N_c of 4.7×10^{17} cm^{-3} and an effective density of states at the VB edge N_v of 7×10^{18} cm^{-3}. Given its bandgap, E_g of 1.42 eV, calculate the intrinsic concentration and the intrinsic resistivity at room temperature (take as 300 K). Where is the Fermi level? Assuming that N_c and N_v scale as $T^{3/2}$, what would be the intrinsic concentration at 100 °C? If this GaAs crystal is doped with 10^{18}

TABLE 3.2 Dopant impurities scatter charge carriers and reduce the drift mobility (μ_e for electrons and μ_h for holes).

Dopant concentration (cm^{-3})	0	10^{15}	10^{16}	10^{17}	10^{18}
$\mu_e\ (\text{cm}^2\ \text{V}^{-1}\ \text{s}^{-1})$	8500	8000	7000	4000	2400
$\mu_h\ (\text{cm}^2\ \text{V}^{-1}\ \text{s}^{-1})$	400	380	310	220	160

donors cm^{-3} (such as Te), where is the new Fermi level and what is the resistivity of the sample? The drift mobilities in GaAs are shown in Table 3.2.

3.3 Direct bandgap *pn* junction Consider a GaAs *pn* junction that has the following properties: $N_a = 10^{16}$ cm^{-3} (*p*-side), $N_d = 10^{16}$ cm^{-3} (*n*-side), $B = 7.21 \times 10^{-16}$ m^3 s^{-1}, cross-sectional area $A = 0.1$ mm^2. What is the diode current due to diffusion in the neutral regions at 300 K when the forward voltage across the diode is 1 V? See Question 3.2 and Table 3.2 for GaAs properties.

3.4 Si *pn* junction Consider a long *pn* junction diode with an acceptor doping, N_a, of 10^{18} cm^{-3} on the *p*-side and donor concentration of N_d on the *n*-side. The diode is forward biased and has a voltage of 0.6 V across it. The diode cross-sectional area is 1 mm^2. The minority carrier recombination time, τ, depends on the dopant concentration, N_{dopant} (cm^{-3}), through the following approximate relation

$$\tau = \frac{5 \times 10^{-7}}{(1 + 2 \times 10^{-17} N_{\text{dopant}})}$$

(a) Suppose that $N_d = 10^{15}$ cm^{-3}. Then the depletion layer extends essentially into the *n*-side and we have to consider minority carrier recombination time, τ_h, in this region. Calculate the diffusion and recombination contributions to the total diode current given that when $N_a = 10^{18}$ cm^{-3}, $\mu_e \approx 250$ cm^2 V^{-1} s^{-1}, and when $N_d = 10^{15}$ cm^{-3}, $\mu_h \approx 450$ cm^2 V^{-1} s^{-1}. What is your conclusion?

(b) Suppose that $N_d = N_a$. Then W extends equally to both sides and, further, $\tau_e = \tau_h$. Calculate the diffusion and recombination contributions to the diode current given that when $N_a = 10^{18}$ cm^{-3}, $\mu_e \approx 250$ cm^2 V^{-1} s^{-1}, and when $N_d = 10^{18}$ cm^{-3}, $\mu_h \approx 130$ cm^2 s^{-1}. What is your conclusion?

3.5 AlGaAs LED emitter An AlGaAs LED emitter for use in a local optical fiber network has the output spectrum shown in Figure 3.33. It is designed for peak emission at 820 nm at 25°C.

(a) What is the linewidth $\Delta\lambda$ between half power points at temperatures −40°C, 25°C, and 85°C? What is the empirical relationship between $\Delta\lambda$ and T given three temperatures and how does this compare with $\Delta(h\nu) \approx 2.5k_BT - 3k_BT$?

(b) Why does the peak emission wavelength increase with temperature?

(c) Why does the peak intensity decrease with temperature?

(d) What is the bandgap of AlGaAs in this LED?

(e) The bandgap, E_g, of the ternary alloys $\text{Al}_x\text{Ga}_{1-x}\text{As}$ follows the empirical expression,

$$E_g(eV) = 1.424 + 1.266x + 0.266x^2.$$

What is the composition of the AlGaAs in this LED?

(f) When the forward current is 40 mA, the voltage across the LED is 1.5 V, and the optical power that is coupled into a multimode fiber through a lens is 25 μW. What is the overall efficiency?

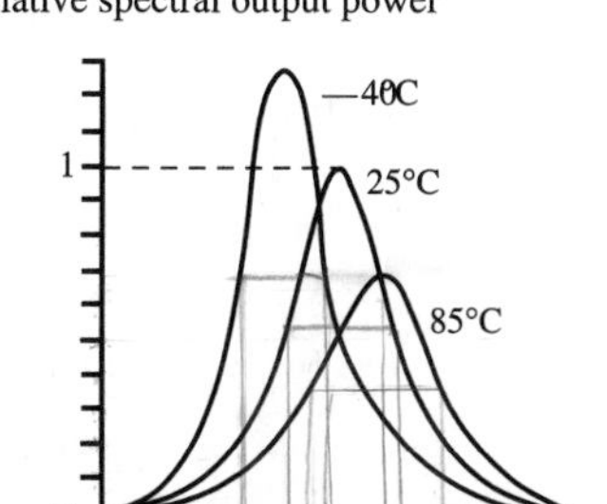

FIGURE 3.33 The output spectrum from AlGaAs LED. Values normalized to peak emission at 25°C.

3.6 III-V compound semiconductors in optoelectronics Figure 3.34 represents the bandgap E_g and the lattice parameter a in the quarternary III-V alloy system. A line joining two points represents the changes in E_g and a with composition in a ternary alloy composed of the compounds at the ends of that line. For example, starting at GaAs point, $E_g = 1.42$ eV and $a = 0.565$ nm, and E_g decreases and a increases as GaAs is alloyed with InAs and we move along the line joining GaAs to InAs. Eventually, at InAs, $E_g = 0.35$ eV and $a = 0.606$ nm. Point X in Figure 3.34 is composed of InAs and GaAs and it is the ternary alloy $In_xGa_{1-x}As$. It has $E_g = 0.7$ eV and $a = 0.587$ nm, which is the same a as that for InP. $In_xGa_{1-x}As$ at X is therefore lattice matched to InP and hence can be grown on an InP substrate without creating defects at the interface.

Further, $In_xGa_{1-x}As$ at X can be alloyed with InP to obtain a quarternary alloy, $In_xGa_{1-x}As_yP_{1-y}$, whose properties lie on the line joining X and InP and therefore all have the same lattice parameter as InP but different bandgap. Layers of $In_xGa_{1-x}As_yP_{1-y}$ with composition between X and InP can be grown epitaxially on an InP substrate by various techniques such as liquid phase epitaxy (LPE) or molecular beam expitaxy (MBE).

The hatched area between the solid lines represents the possible values of E_g and a for the quarternary III-V alloy system in which the bandgap is direct and hence suitable for direct electron and hole recombination.

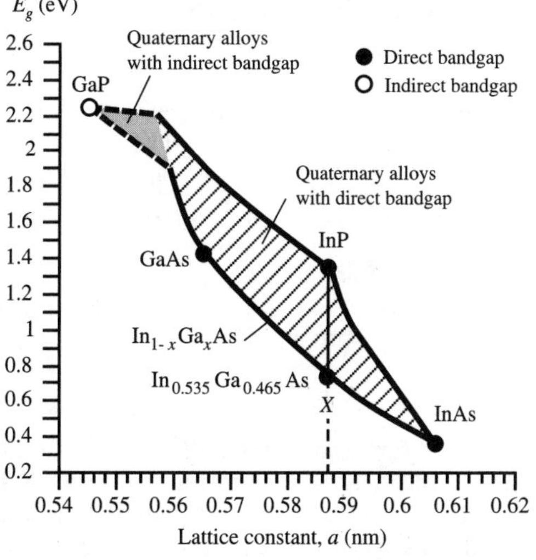

FIGURE 3.34 Bandgap energy E_g and lattice constant a for various III-V alloys of GaP, GaAs, InP and InAs. A line represents a ternary alloy formed with compounds from the end points of the line. Solid lines are for direct bandgap alloys whereas dashed lines for indirect bandgap alloys. Regions between lines represent quaternary alloys. The line from X to InP represents quaternary alloys $In_{1-x}Ga_xAs_{1-y}P_y$ made from $In_{0.535}Ga_{0.465}As$ and InP which are lattice matched to InP.

The compositions of the quarternary alloy lattice matched to InP follow the line from X to InP.

(a) Given that the $In_xGa_{1-x}As$ at X is $In_{0.535}Ga_{0.465}As$, show that quarternary alloys $In_xGa_{1-x}As_yP_{1-y}$ are lattice matched to InP when $y = 2.15x$.

(b) The bandgap energy E_g, in eV for $In_xGa_{1-x}As_yP_{1-y}$ lattice matched to InP is given by the empirical relation,

$$E_g(\text{eV}) = 1.35 - 0.72y + 0.12y^2$$

Find the composition of the quarternary alloy suitable for an emitter operating at 1.55 μm.

3.7 External conversion efficiency The *external power or conversion efficiency* η_{ext} is defined as

$$\eta_{\text{ext}} = \frac{\text{Oprical power output}}{\text{Electrical power output}} = \frac{P_o}{IV}$$

One of the major factors reducing the external power efficiency is the loss of photons in extracting the emitted photons that suffer reabsorption in the *pn* junction materials, absorption outside the semiconductors, and various reflections at interfaces.

The total light output power from a particular AlGaAs red LED is 2.5 mW when the current is 50 mA and the voltage is 1.6 V. Calculate its external conversion efficiency.

3.8 Linewidth of LEDs Experiments carried out on various direct bandgap semiconductor LEDs give the output spectral linewidth (between half intensity points as in Figure 3.33) listed in Table 3.3. From Example 3.8.1, we know that a spread in the wavelength is related to a spread in the photon energy,

$$\Delta\lambda \approx \frac{hc}{E_{ph}^2}\Delta E_{ph} \qquad \textbf{(1)}$$

Suppose that we write $E_{ph} = hc/\lambda$ and $\Delta E_{ph} = \Delta(h\upsilon) \approx mk_BT$ in which m is a numerical constant. Show that,

$$\Delta\lambda \approx \lambda^2\frac{mk_BT}{hc} \qquad \textbf{(2)}$$

and by appropriately plotting the data in Table 3.3, and assuming $T = 300$ K, find m.

TABLE 3.3 Linewidth $\Delta\lambda_{1/2}$ between half points in the output spectrum (Intensity vs. wavelength) of various LEDs.

Peak wavelength of emission (λ) nm	650	810	820	890	950	1150	1270	1500
$\Delta\lambda_{1/2}$ nm	22	36	40	50	55	90	110	150
Material (Direct E_g)	AlGaAs	AlGaAs	AlGaAs	GaAs	GaAs	InGaAsP	InGaAsP	InGaAsP

Table 3.4 shows the linewidth $\Delta\lambda_{1/2}$ for various visible LEDs. Radiative recombination is obtained by appropriately doping the material. Using $m \approx 3$, $T = 300$ K, in Eq. (2), calculate the expected spectral width for each and compare with the experimental value. What is your conclusion? Do you think E_N in Figure 3.24 (b) is a discrete level?

TABLE 3.4 Linewidth $\Delta\lambda_{1/2}$ between half points in the output spectrum (intensity vs. wavelength) of various visible LEDs using SiC and GaAsP materials.

Peak wavelength of emission (λ) nm	468	565	583	600	635
$\Delta\lambda_{1/2}$ nm	66	28	36	40	40
Color	Blue	Green	Yellow	Orange	Red
Material	SiC (Al)	GaP (N)	GaAsP (N)	GaAsP (N)	GaAsP

3.9 SLEDs and ELEDs Experiments carried out on an AlGaAs SLED (surface emitting LED) and an ELED (edge emitting LED) give the light output power vs. current data in Table 3.5.

(a) Show that the output light power vs. current characteristics are not linear.

(b) By plotting the optical power output (P_o) vs. current (I) data on a log-log plot show that $P_o \propto I^n$. Find n for each LED.

TABLE 3.5 Light output power vs. DC current for surface and edge emitting LEDs.

SLED I (mA)	25	50	75	100	150	200	250	300
SLED Light output power (mW) P_o	1.04	2.07	3.1	4.06	5.8	7.6	9.0	10.2
ELED I (mA)	25	50	75	100	150	200	250	300
ELED Light output power (mW) P_o	0.46	0.88	1.28	1.66	2.32	2.87	3.39	3.84

3.10 LED-Fiber coupling Efficiency

(a) It is found that approximately 200 μW is coupled into a multimode step index fiber from a surface emitting LED when the current is 75 mA and the voltage across the LED is about 1.5 V. What is the overall efficiency of operation?

(b) Experiments are carried out on coupling light from a 1310 nm ELED (edge emitting LED) into multimode and single mode fibers.

(i) At room temperature, when the ELED current is 120 mA, the voltage is 1.3 V, and light power coupled into a 50 μm multimode fiber with NA (numerical aperture) = 0.2 is 48 μW. What is the overall efficiency?

(ii) At room temperature, when the ELED current is 120 mA, the voltage is 1.3 V, and light power coupled into a 9 μm single mode fiber is 7 μW. What is the overall efficiency?

3.11 Internal quantum efficiency The internal efficiency η_{int} gauges what fraction of electron hole recombinations in the forward biased pn junction are radiative and therefore lead to photon emission. Nonradiative transitions are those in which an electron and a hole recombine through a recombination center such as a crystal defect or an impurity and emit phonons (lattice vibrations). By definition,

$$\eta_{\text{int}} = \frac{\text{Rate of radiative recombination}}{\text{Total rate of recombination (radiative and nonradiative)}} \quad \textbf{(1)}$$

or

$$\eta_{\text{int}} = \frac{\dfrac{1}{\tau_r}}{\dfrac{1}{\tau_r} + \dfrac{1}{\tau_{nr}}} \tag{2}$$

in which τ_r is the mean lifetime of a minority carrier before it recombines radiatively and τ_{nr} is the mean lifetime before it recombines via a recombination center without emitting a photon. The total current I is determined by the total rate of recombinations whereas the number of photons emitted per second (Φ_{ph}) is determined by the rate of radiative recombinations.

$$\eta_{\text{int}} = \frac{\text{Photons emitted per second}}{\text{Total carriers lost per second}} = \frac{\Phi_{ph}}{I/e} = \frac{P_{o(\text{int})}/h\upsilon}{I/e} \tag{3}$$

in which $P_{o(\text{int})}$ is the optical power generated internally (not yet extracted).

For a particular AlGaAs LED emitting at 850 nm it is found that $\tau_r = 50$ ns and $\tau_{nr} = 100$ ns. What is the internal optical power generated at a current of 100 mA?

"We consider alloyed or point contact junctions on n-type GaP. Then the light emitted with forward bias has a spectrum which is a comparatively narrow band, the position of the band depending on the impurities present in the GaP."

—J.W. Allen and P.E. Gibbons (1959)[9]

[9] Allen and Gibbons (talking about their invention of the LED) in "Breakdown and light emission in gallium phosphide diodes," *Journal of Electronics*, Vol. 7, No. 6, p. 518, December 1959.

CHAPTER 4

Stimulated Emission Devices LASERS

"We thought it [the laser] might have some communications and scientific uses, but we had no application in mind. If we had, it might have hampered us and not worked out as well" —Arthur Schawlow[1]

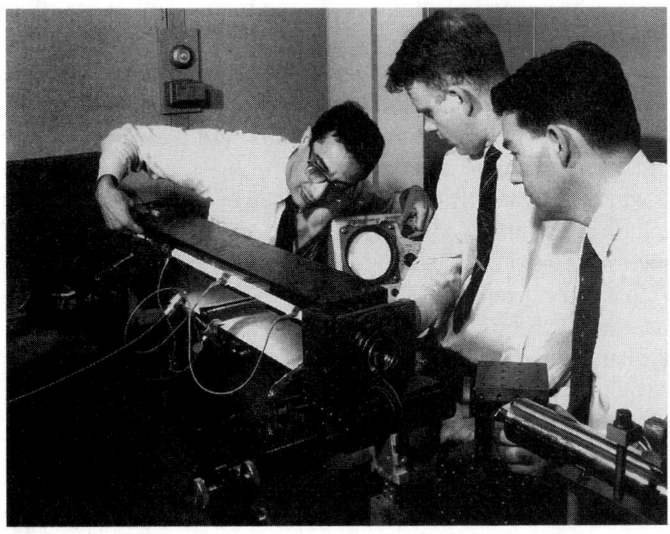

Ali Javan and his associates William Bennett Jr. and Donald Herriott at Bell Labs were first to successfully demonstrate a continuous wave (cw) helium-neon laser operation (1960). *(Courtesy of Bell Labs, Lucent Technologies.)*

4.1 STIMULATED EMISSION AND PHOTON AMPLIFICATION

An electron in an atom can be excited from an energy level E_1 to a higher energy level E_2 by the absorption of a photon of energy $h\upsilon = E_2 - E_1$ as shown in Figure 4.1 (a). When an electron at a higher energy level transits down in energy to an unoccupied energy level, it emits a photon. There are essentially two possibilities for the emission process. The electron can undergo the downward transition by itself quite *spontaneously*, or it can be *induced* to do so by another photon.

[1] Arthur Schawlow (1921–1999; Nobel Laureate, 1981) talking about the invention of the laser.

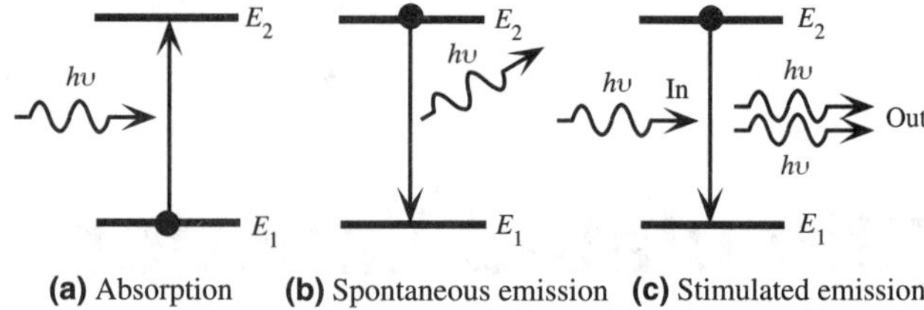

FIGURE 4.1 Absorption, spontaneous (random photon) emission and stimulated emission.

In **spontaneous emission**, the electron falls down in energy from level E_2 to E_1 and emits a photon of energy $h\upsilon = E_2 - E_1$ in a random direction as indicated in Figure 4.1 (b). Thus, a random photon is emitted. The transition is spontaneous provided that the state with energy E_1 is not already occupied by another electron. In classical physics when a charge accelerates and decelerates as in an oscillatory motion with a frequency υ it emits an electromagnetic radiation also of frequency υ. The emission process during the transition of the electron from E_2 to E_1 can be thought of as if the electron is oscillating with a frequency υ.

In **stimulated emission**, an incoming photon of energy $h\upsilon = E_2 - E_1$ stimulates the whole emission process by inducing the electron at E_2 to transit down to E_1. The emitted photon is *in phase* with the incoming photon, it is in the *same direction*, it has the *same polarization* and it has the *same energy* since $h\upsilon = E_2 - E_1$ as shown in Figure 4.1 (c). To get a feel of what is happening during stimulated emission one can think of the electric field of the incoming photon coupling to the electron and thereby driving it with the same frequency as the photon. The forced oscillation of the electron at a frequency $\upsilon = (E_2 - E_1)/h$ causes it to emit electromagnetic radiation whose electric field is in total phase with that of the stimulating photon. When the incoming photon leaves the site, the electron can return to E_1 because it has emitted a photon of energy $h\upsilon = E_2 - E_1$. Although we considered the transitions of an electron in an atom, we could have just well described photon absorption, spontaneous and stimulated emission in Figure 4.1 in terms of energy transitions of the atom itself in which case E_1 and E_2 represent the energy levels of the atom.

Stimulated emission is the basis for obtaining photon amplification since one incoming photon results in two outgoing photons which are in phase. How does one achieve a practical light amplifying device based on this phenomenon? It should be quite apparent from Figure 4.1 (c) that to obtain stimulated emission, the incoming photon should *not* be absorbed by another atom at E_1. When we are considering a collection of atoms to amplify light, we must therefore have the majority of the atoms at the energy level E_2. If this were not the case, the incoming photons would be absorbed by the atoms at E_1. When there are more atoms at E_2 than at E_1 we then have what is called a **population inversion**. It should be apparent that with *only two* energy levels we can never achieve population at E_2 greater than that at E_1 because, in the steady state, the incoming photon flux will cause as many upward excitations as downward stimulated emissions.

Let us consider the three energy level system shown in Figure 4.2. Suppose that an external excitation causes the atoms in this system to become excited to the energy level E_3. This is called the pump energy level and the process of exciting the atoms to E_3 is called **pumping**. In the present case, **optical pumping** is used though this is not the only means

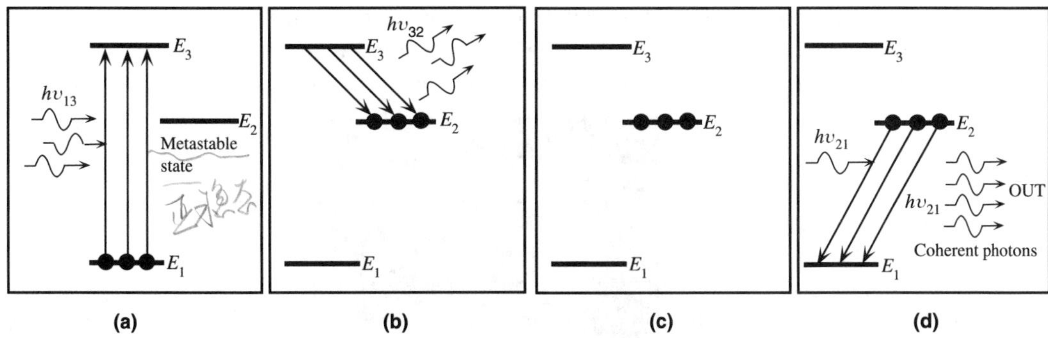

FIGURE 4.2 The principle of the LASER. (a) Atoms in the ground state are pumped up to the energy level E_3 by incoming photons of energy $h\upsilon_{13} = E_3 - E_1$. (b) Atoms at E_3 rapidly decay to the long-lived state at energy level E_2 by emitting photons or emitting lattice vibrations; $h\upsilon_{32} = E_3 - E_2$. (c) As the states at E_2 are long-lived, they quickly become populated and there is a population inversion between E_2 and E_1. (d) A random photon (from a spontaneous decay) of energy $h\upsilon_{21} = E_2 - E_1$ can initiate stimulated emission. Photons from this stimulated emission can themselves further stimulate emissions leading to an avalanche of stimulated emissions and coherent photons being emitted.

of taking the atoms to E_3. Suppose further that from E_3 the atoms decay rapidly to an energy level E_2 which happens to correspond to a state that does not rapidly and spontaneously decay to lower-energy state. In other words, the state at E_2 is a **long-lived state**.[2] Since the atoms cannot decay rapidly from E_2 to E_1 they accumulate at this energy level causing a population inversion between E_2 and E_1 as pumping takes more and more atoms to E_3 and hence E_2. When one atom at E_2 decays spontaneously, it emits a photon (a "random photon") which can go on to a neighboring atom and cause that to execute stimulated emission. The photons from the latter can go on to the next atom at E_2 and cause that to emit by stimulated emission and so on. The result is an avalanche effect of stimulated emission processes with all the photons in phase so that the light output is a large collection of coherent photons. This is the principle of the ruby laser in which the energy levels E_1, E_2 and E_3 are those of the Cr^{+3} ion in the Al_2O_3 crystal. At the end of the avalanche of stimulated emission processes, the atoms at E_2 would have dropped to E_1 and can be pumped again to repeat the stimulated emission cycle again. The emission from E_2 to E_1 is called the **lasing emission.**[3] The system we have just described for photon amplification is a LASER, an acronym for **Light Amplification by Stimulated Emission of Radiation**. In the ruby laser, pumping is achieved by using a xenon flash light. The lasing atoms are chromium ions (Cr^{3+}) in a crystal of alumina Al_2O_3 (saphire). The ends of the ruby crystal are silvered to reflect back and forward the stimulated radiation so that its intensity builds up in much the same way we build up voltage oscillations in an electrical oscillator circuit. One of the mirrors is partially silvered to allow some of this radiation to be tapped out. What comes out is a highly coherent radiation which has a high

[2] We will not examine what causes certain states to be long-lived but simply accept that these states do not decay rapidly and spontaneously decay to lower-energy states.

[3] Arthur Schawlow, one of the co-inventors of the laser, was well-known for his humor and has, apparently, said that "Anything will lase if you hit it hard enough". In 1971, Schawlow and Ted Hänsch were able to develop the first edible laser made from Jell-O (IEEE Journal of Quantum Electronics)

intensity. The coherency and the well defined wavelength of this radiation is what makes it distinctly different than a random stream of different wavelength photons emitted from a tungsten bulb, or randomly phased photons from an LED.

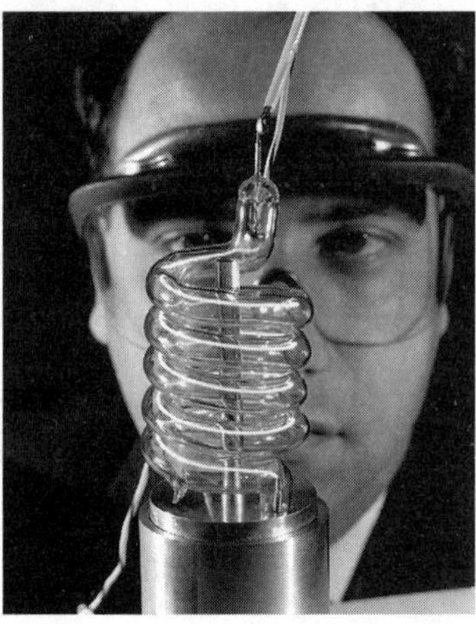

Theodore Harold Maiman was born in 1927 in Los Angeles, son of an electrical engineer. He studied engineering physics at Colorado University, while repairing electrical appliances to pay for college, and then obtained a Ph.D. from Stanford. Theodore Maiman constructed this first laser in 1960 while working at Hughes Research Laboratories (T.H. Maiman, "Stimulated optical radiation in ruby lasers", *Nature*, **187**, 493, 1960). There is a vertical chromium ion doped ruby rod in the center of a helical xenon flash tube. The ruby rod has mirrored ends. The xenon flash provides optical pumping of the chromium ions in the ruby rod. The output is a pulse of red laser light. *(Courtesy of HRL Laboratories, LLC, Malibu, California.)*

4.2 STIMULATED EMISSION RATE AND EINSTEIN COEFFICIENTS

A useful LASER medium must have a higher efficiency of stimulated emission compared with the efficiencies of spontaneous emission and absorption. We need to determine the controlling factors for the rates of stimulated emission, spontaneous emission and absorption. Consider a medium as in Figure 4.1 that has N_1 atoms per unit volume with energy E_1 and N_2 atoms per unit volume with energy E_2. Then the rate of upward transitions from E_1 to E_2 by photon absorption will be proportional to the number of atoms N_1 and also to the number of photons per unit volume with energy $h\upsilon = E_2 - E_1$. Put differently, this rate will depend on the energy density in the radiation. Thus, the upward transition rate is,

$$R_{12} = B_{12}N_1\rho(h\upsilon) \qquad \textbf{(1)}$$

where B_{12} is a proportionality constant termed the **Einstein B_{12} coefficient**, and $\rho(h\upsilon)$ is the photon energy density per unit frequency[4] which represents the number of photons per unit volume with an energy $h\upsilon\,(=E_2 - E_1)$. The rate of downward transitions

[4] Using $\rho(h\upsilon)$ defined in this way simplifies the evaluation of the proportionality constants. $\rho(h\upsilon)$ is the energy in the radiation per unit volume per unit frequency due to photons with energy $h\upsilon = E_2 - E_1$.

from E_2 to E_1 involves spontaneous and stimulated emission. First depends on the concentration N_2 of atoms at E_2 and the second depends on both N_2 and the photon concentration $\rho(h\upsilon)$ with energy $h\upsilon(=E_2 - E_1)$. Thus, the total downward transition rate is

$$R_{21} = A_{21}N_2 + B_{21}N_2\rho(h\upsilon) \tag{2}$$

where the first term is due to spontaneous emission (does not depend on the photon energy density $\rho(h\upsilon)$ to drive it) and the second term is due to stimulated emission which requires photons to drive it. A_{21} and B_{21} are the proportionality constants termed the **Einstein coefficients** for spontaneous and stimulated emissions respectively.

To find the coefficients A_{21}, B_{12} and B_{21}, we consider the events in equilibrium, that is the medium in thermal equilibrium (no external excitation). There is no net change with time in the populations at E_1 and E_2 which means

$$R_{12} = R_{21} \tag{3}$$

and furthermore in thermal equilibrium Boltzmann statistics demands that

$$\frac{N_2}{N_1} = \exp\left[-\frac{(E_2 - E_1)}{k_BT}\right] \tag{4}$$

where k_B is the Boltzmann constant and T is the absolute temperature.

Now, *in thermal equilibrium*, in the collection of atoms we are considering, radiation from the atoms must give rise to an equilibrium photon energy density, $\rho_{eq}(h\upsilon)$, that is given by *Planck's black body radiation distribution law*,[5]

$$\rho_{eq}(h\upsilon) = \frac{8\pi h\upsilon^3}{c^3\left[\exp\left(\frac{h\upsilon}{k_BT}\right) - 1\right]} \tag{5}$$

It is important to emphasize that the Planck's law in Eq. (5) applies only in thermal equilibrium; we are using this condition to determine the Einstein coefficients. During the laser operation, of course, $\rho(h\upsilon)$ is not described by Eq. (5); in fact it is much larger. From Eqs. (1) to (5) we can readily show that

$$B_{12} = B_{21} \tag{6}$$

and

$$A_{21}/B_{21} = 8\pi h\upsilon^3/c^3 \tag{7}$$

Now consider the ratio of stimulated to spontaneous emission,

$$\frac{R_{21}(\text{stim})}{R_{21}(\text{spon})} = \frac{B_{21}N_2\rho(h\upsilon)}{A_{21}N_2} = \frac{B_{21}\rho(h\upsilon)}{A_{21}} \tag{8}$$

which, by Eq. (7), can be written as

$$\frac{R_{21}(\text{stim})}{R_{21}(\text{spon})} = \frac{c^3}{8\pi h\upsilon^3}\rho(h\upsilon) \tag{9}$$

---

[5] See, for example, any modern physics textbook.

In addition, the ratio of stimulated emission to absorption is

$$\frac{R_{21}(\text{stim})}{R_{12}(\text{absorp})} = \frac{N_2}{N_1} \quad \textbf{(10)}$$

There are two important conclusions. For stimulated photon emission to exceed photon absorption, by Eq. (10), we need to achieve **population inversion**, that is $N_2 > N_1$. For stimulated emission to far exceed spontaneous emission, by Eq. (9), we must have a large photon concentration which is achieved by building an **optical cavity** to contain the photons.

It is important to point that the population inversion requirement $N_2 > N_1$ means that we depart from thermal equilibrium. According to Boltzmann statistics in Eq. (4), $N_2 > N_1$ implies a *negative absolute temperature*! The laser principle is based on **non-thermal equilibrium**.[6]

4.3 OPTICAL FIBER AMPLIFIERS

A light signal that is traveling along an optical fiber over a long distance suffers marked attenuation. It becomes necessary to regenerate the light signal at certain intervals for long haul communications over several thousand kilometers. Instead of regenerating the optical signal by photodetection, conversion to an electrical signal, amplification and then conversion back from electrical to light energy by a laser diode, it becomes practical to amplify the signal directly by using an optical amplifier.

One practical **optical amplifier** is based on the **erbium (Er^{3+} ion) doped fiber amplifier (EDFA)**.[7] The core region of an optical fiber is doped with Er^{3+} ions. Other rare earth ion dopants can also be used such as a neodymium ion (Nd^{3+}). The host fiber core material is a glass based on SiO_3-GeO_2 and perhaps some other glass forming oxides such as Al_2O_3. It is easily fused to a single mode long distance optical fiber by a technique called splicing.

When the Er^{3+} ion is implanted in the host glass material it has the energy levels indicated in Figure 4.3 where E_1 corresponds to the lowest energy possible for the Er^{3+} ion.[8] There are two convenient energy levels for optically pumping the Er^{3+} ion which are at approximately 1.27 eV and 1.54 eV above the ground energy level. These are labeled respectively as E_3 and E_3'. The Er^{3+} ions are optically pumped, usually from a laser diode, to excite them to E_3. The wavelength for this pumping is about 980 nm. The Er^{3+} ions decay rapidly from E_3 to a **long-lived** energy level at E_2 which has a long lifetime of about ~10 ms (very long on the atomic scale). The decays from E_3' to E_3 and from E_3

[6] "But I thought, now wait a minute! The second law of thermodynamics assumes thermal equilibrium. We don't have that!" Charles D Townes (born 1915; Nobel Laureate, 1964). The laser idea occurred to Charles Townes, apparently, while he was taking a walk one early morning in Franklin Park in Washington DC while attending a scientific committee meeting. Non-thermal equilibrium (population inversion) is critical to the principle of the laser.

[7] EDFA was first reported in 1987 by E. Desurvire, J.R. Simpson and P.C. Becker and in 1994 AT&T began deploying EDFA repeaters in long-haul fiber communications.

[8] The valence electrons of the Er^{3+} ion are arranged to satisfy the Pauli exclusion principle and Hund's rule and this arrangement has the energy E_1.

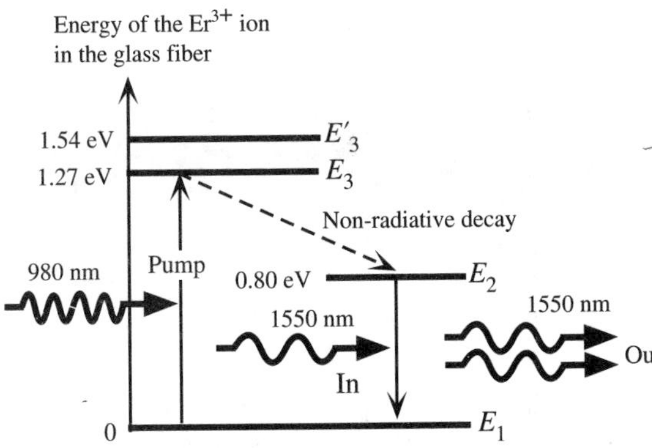

FIGURE 4.3 Energy diagram for the Er^{3+} ion in the glass fiber medium and light amplification by stimulated emission from E_2 to E_1. Dashed arrows indicate radiationless transitions (energy emission by lattice vibrations).

to E_2 involve energy losses by radiationless transitions (phonon[9] emissions) and are very rapid. Thus more and more Er^{3+} ions accumulate at E_2 which is 0.80 eV above the ground energy. The accumulation of Er^{3+} ions at E_2 leads to a population inversion between E_2 and E_1. Signal photons at 1550 nm have an energy of 0.80 eV, or $E_2 - E_1$, and give rise to *stimulated transitions* of Er^{3+} ions from E_2 to E_1. Any Er^{3+} ions left at E_1, however, will *absorb* the incoming 1550 nm photons to reach E_2. To achieve light amplification we must therefore have stimulated emission exceeding absorption. This is only possible if there are more Er^{3+} ions at the E_2 level than at the E_1 level; if we have population inversion. If N_2 and N_1 are the number of Er^{3+} ions at E_2 and E_1 then it is clear that the difference between stimulated emission (from E_2 to E_1) and absorption (E_1 to E_2) rate controls the net optical gain G_{op},

$$G_{op} = K(N_2 - N_1)$$

where K is a constant that, amongst other factors, depends on the pumping intensity.

In practice the erbium doped fiber is inserted into the fiber communications line by splicing as shown in the simplified schematic diagram in Figure 4.4 and it is pumped from a laser diode through a coupling fiber arrangement which allows only the pumping wavelength to be coupled. Some of the Er^{3+} ions at E_2 will decay spontaneously from E_2 to E_1 which will give rise to unwanted noise in the amplified light signal. Further, if the EDFA is not pumped at any time it will act as an attenuator as the 1550 nm photons will be absorbed by Er^{3+} ions which will become excited from E_1 to E_2. In returning back to E_1 by spontaneous emission they will emit 1550 nm photons randomly and not along the fiber axis. Although the Er^{3+} ions can also be pumped to the E'_3 level using a pumping wavelength of 810 nm, this process is much less efficient than the 980 nm pumping to the E_3 level. Optical isolators inserted at the entry and exit end of the amplifier allow only the optical signals at 1550 nm to pass in one direction and prevent the 980 pump light from propagating back or forward into the communication system. There may be another pump diode coupled at the right end of the EDFA similar to that on the

[9] A phonon is a quantum of lattice vibrational energy just as a photon is a quantum of electromagnetic energy.

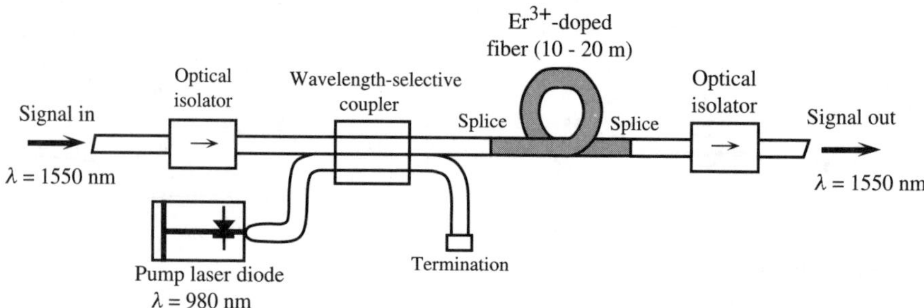

FIGURE 4.4 A simplified schematic illustration of an EDFA (optical amplifier). The erbium-ion doped fiber is pumped by feeding the light from a laser pump diode, through a coupler, into the erbium ion doped fiber.

left side in Figure 4.4. In addition, there is usually a photodetector system coupled to monitor the pump power or the EDFA output power. These are not shown in Figure 4.4.

There are a few important facts about the EDFA that are not shown in Figure 4.3. First is that the energy levels E_1, E_2, E_3 *etc.* are not single unique levels but rather each consists of a closely spaced collection of several levels. Consequently there is a range of stimulated transitions from E_2 to E_1 which corresponds to a wavelength range of about 1525–1565 nm that can be amplified; an optical bandwidth of about 40 nm. Thus, the EDFA can be used as an optical amplifier in wavelength division multiplexed (WDM) systems if the wavelength range is within this optical bandwidth. However, the gain is not uniform over the whole bandwidth and special techniques must be used to "flatten" the gain. In addition, it is possible to excite the Er^{3+} ion from the "bottom" of E_1 levels to the "top" of the E_2 levels with a 1480 nm excitation which can also be used as a possible pump wavelength, though the 980 nm pumping is more efficient.

The **gain efficiency** of an EDFA is the maximum optical gain achievable per unit optical pumping power and are quoted in dB/mW. Typical gain efficiencies are around 8–10 dB/mW at 980 nm pumping. A 30 dB or 10^3 gain is easily attainable with a few milliwatts of pumping at 980 nm.

4.4 GAS LASERS: THE He-Ne LASER

With the HeNe laser one has to confess that the actual explanation is by no means simple since we have to know such things as the energy states of the whole atom. We will consider only the lasing emission at 632.8 nm which gives the well-known red color to the HeNe laser light. The actual stimulated emission occurs from the Ne atoms. He atoms are used to excite the Ne atoms by atomic collisions.

Ne is an inert gas with a ground state $(1s^2 2s^2 2p^6)$ which will be represented as $(2p^6)$ by ignoring the inner closed $1s$ and $2s$ subshells. If one of the electrons from the $2p$ orbital is excited to a $5s$-orbital then the excited configuration $(2p^5 5s^1)$ is a state of the Ne atom that has higher energy. Similarly He is also an inert gas which has the ground state configuration of $(1s^2)$. The state of He when one electron is excited to a $2s$-orbital can be represented as $(1s^1 2s^1)$ and has higher energy.

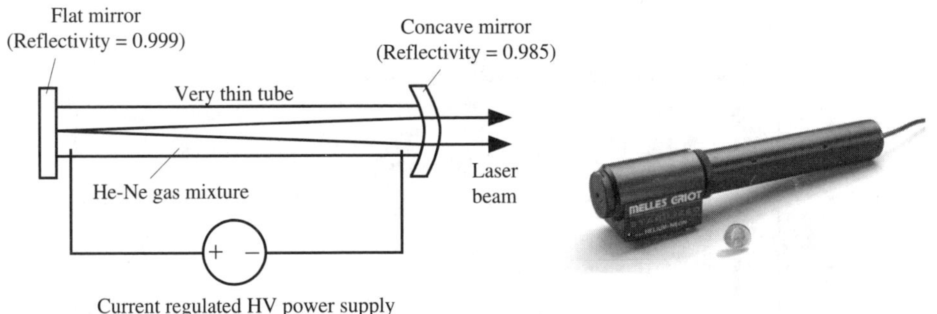

FIGURE 4.5 A schematic illustration of the principle of the He-Ne laser. Right: A modern stabilized compact He-Ne laser. *(Courtesy of Melles Griot.)*

The HeNe laser consists of a gaseous mixture of He and Ne atoms in a gas discharge tube as sketched schematically in Figure 4.5. The ends of the tube are mirrored to reflect the stimulated radiation and buildup intensity within the cavity. In other words, an *optical* cavity is formed by the end-mirrors so that reflection of photons back into the lasing medium builds up the photon concentration in the cavity; a requirement of an efficient stimulated emission process as discussed above. By using dc or RF high voltage, electrical discharge is obtained within the tube which causes the He atoms to become excited by collisions with the drifting electrons. Thus,

$$\text{He} + e^- \rightarrow \text{He}^* + e^-$$

where He* is an excited He atom.

The excitation of the He atom by an electron collision puts the second electron in He into a $2s$ state and changes its spin so that the excited He atom, He*, has the configuration $(1s^1 2s^1)$ with parallel spins which is *metastable* (long lasting) with respect to the $(1s^2)$ state as shown schematically in Figure 4.6 He* cannot spontaneously emit a photon and decay down to the $(1s^2)$ ground state because the *orbital quantum number* l of the electron must change by ±1, *i.e.* Δl must be ±1 for any photon emission or absorption process. (Question 4.1 and Figure 4.35 provide a further discussion of the He-Ne laser.) Thus a large number of He* atoms build up during the electrical discharge because they are not allowed to simply decay back to the ground state.

When an excited He atom collides with a Ne atom, it transfers its energy to the Ne atom by resonance energy exchange because, by good fortune, Ne happens to have an empty energy level, corresponding to the $(2p^5 5s^1)$ configuration, matching that of $(1s^1 2s^1)$ of He*. Thus the collision process excites the Ne atom and de-excites He* down to its ground energy, *i.e.*

$$\text{He}^* + \text{Ne} \rightarrow \text{He} + \text{Ne}^*$$

With many He*-Ne collisions in the gaseous discharge we end up with a large number of Ne* atoms and a population inversion between $(2p^5 5s^1)$ and $(2p^5 3p^1)$ states of the Ne atom as indicated in Figure 4.6. A spontaneous emission of a photon from

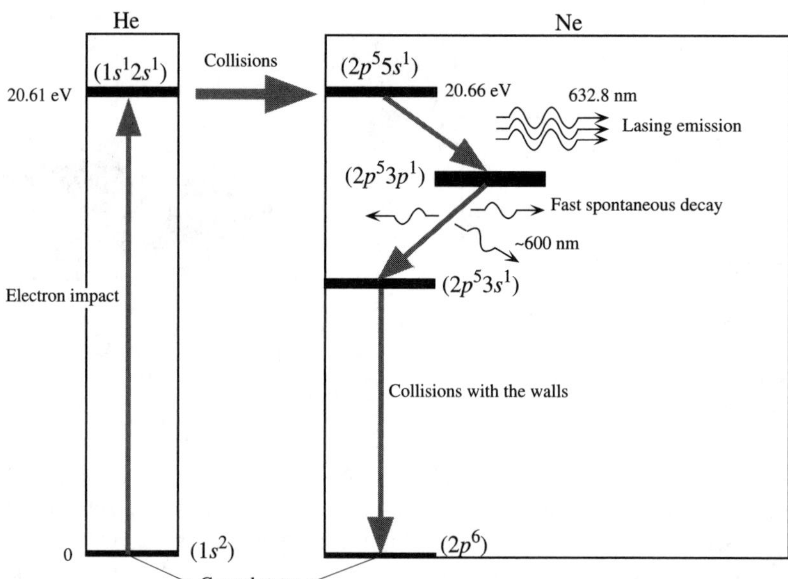

FIGURE 4.6 The principle of operation of the He-Ne laser. He-Ne laser energy levels (for 632.8 nm emission).

one Ne* atom falling from $5s$ to $3p$ gives rise to an avalanche of stimulated emission processes which leads to a lasing emission with a wavelength 632.8 nm in the red.

There are a few interesting facts about the He-Ne laser, some of which are quite subtle. First, the $(2p^5 5s^1)$ and $(2p^5 3p^1)$ electronic configurations of the Ne atom actually have a spread of energies. For example, for Ne$(2p^5 5s^1)$, there are four closely spaced energy levels. Similarly for Ne$(2p^5 3p^1)$ there are ten closely separated energies. We see that we can achieve population inversion with respect to a number of energy levels and, as a result, the lasing emissions from the He-Ne laser contain a variety of wavelengths. The two lasing emissions in the visible spectrum at 632.8 nm and 543 nm can be used to build a red or a green He-Ne laser. Further, we should note that the energy of the state Ne$(2p^5 4p^1)$ (not shown) is above Ne$(2p^5 3p^1)$ but below Ne$(2p^5 5s^1)$. There will therefore also be stimulated transitions from Ne$(2p^5 5s^1)$ to Ne$(2p^5 4p^1)$ and hence a lasing emission at a wavelength of ~3.39 μm (infrared). To suppress lasing emissions at the unwanted wavelengths (*e.g.*, the infrared) and to obtain lasing only at the wavelength of interest, the reflecting mirrors can be made wavelength selective. This way the optical cavity builds up optical oscillations at the selected wavelength.

From the $(2p^5 3p^1)$ energy levels, the Ne atoms decay rapidly to the $(2p^5 3s^1)$ energy levels by spontaneous emission. Most of Ne atoms with the $(2p^5 3s^1)$ configuration, however, cannot simply return to the ground state $2p^6$ by photon emission because the return of the electron in $3s$ requires that its spin is flipped to close the $2p$-subshell. An electromagnetic radiation cannot change the electron spin. Thus the Ne$(2p^5 3s^1)$ energy levels are *metastable* states. The only possible return to the ground state (and for the next repumping act) is by collisions with the walls of the laser tube. We cannot therefore increase the power obtainable from a He-Ne laser by simply increasing the laser tube diameter because that will accumulate more Ne atoms at the metastable $(2p^53s^1)$ states.

A typical He-Ne laser, as illustrated in Figure 4.5, consist of a narrow glass tube which contains the He and Ne gas mixture; typically He to Ne ratio of 5 to 1 and a pressure of several torrs. The lasing emission intensity (optical gain) increases with the tube length since then more Ne atoms are used in stimulated emission. The intensity decreases with increasing tube diameter since Ne atoms in the $(2p^53s^1)$ states can only return to the ground state by collisions with the walls of the tube. The ends of the tube are generally sealed with a flat mirror (99.9% reflecting) at one end and, for easy alignment, a concave mirror (99% reflecting) at the other end to obtain an **optical cavity** within the tube. The outer surface of the concave mirror is ground to behave like a convergent lens to compensate for the divergence in the beam arising from reflections from the concave mirror. The output radiation from the tube is typically a beam of diameter 0.5–1 mm and a divergence of 1 milliradians at a power of few milliwatts. In high power He-Ne lasers, the mirrors are external to the tube. In addition, *Brewster windows* are typically used at the ends of the laser tube to allow only polarized light to be transmitted and amplified within the cavity so that the output radiation is polarized (has electric field oscillations in one plane).

Even though we can try to get as parallel a beam as possible by lining up the mirrors perfectly, we will still be faced with diffraction effects at the output. When the output laser beam hits the end of the laser tube it becomes diffracted so that the emerging beam is necessarily divergent. Simple diffraction theory can readily predict the divergence angle. Further, typically one or both of the reflecting mirrors in many gas lasers are made concave for a more efficient containment of the stimulated photons within the **active medium** and for easier alignment. The beam within the cavity and hence the emerging radiation is approximately a *Gaussian beam*. As mentioned in Chapter 1, a Gaussian beam diverges as it propagates in free space. Optical cavity engineering is an important part of the laser design and there are various advanced texts on the subject.

Due to their relatively simple construction, He-Ne lasers are widely used in numerous applications such interferometry, for example, accurately measuring distances or flatness of an object, laser printing, holography, and various pointing and alignment applications (as in civil engineering).

EXAMPLE 4.4.1 Efficiency of the HeNe laser

A typical low-power 5 mW He-Ne laser tube operates at a dc voltage of 2000 V and carries a current of 7 mA. What is the efficiency of the laser?

Solution From the definition of efficiency,

$$\text{Efficiency} = \frac{\text{Output Light Power}}{\text{Input Electrical Power}} = \frac{5 \times 10^{-3}\ \text{W}}{(7 \times 10^{-3}\ \text{A})(2000\ \text{V})}$$

$$= 0.036\%$$

Typically He-Ne efficiencies are less than 0.1%. What is important is the high concentration of coherent photons. Note that 5 mW over a beam diameter of 1 mm is 6.4 kW m^{-2}.

EXAMPLE 4.4.2 Laser beam divergence

The laser beam emerging from a laser tube has a certain amount of divergence as schematically illustrated in Figure 4.7. A typical He-Ne laser has an output beam with a diameter of 1 mm and a divergence of 1 mrad. What is the diameter of the beam at a distance of 10 m?

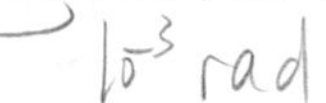

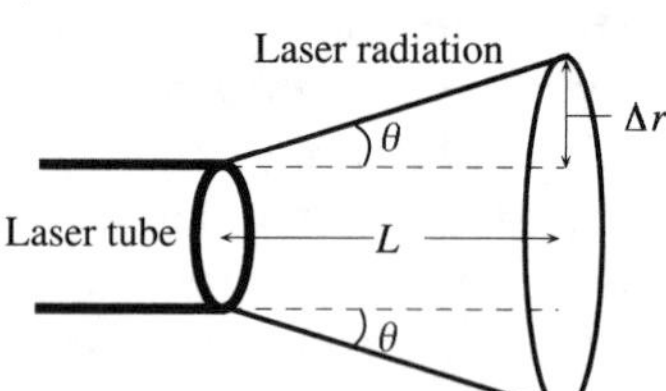

FIGURE 4.7 The output laser beam has a divergence characterized by the angle 2θ (highly exaggerated in the figure).

Solution We can assume that the laser beam emanates like a light-cone, as illustrated in Figure 4.7, with an apex angle 2θ, from the end of the laser tube. The angle 2θ is then the divergence of the beam which is 1 mrad.

If Δr is the increase in the radius of the beam over a distance L then by the definition of divergence,

$$\tan\theta = \Delta r/L$$

where 2θ is the angle of divergence. Thus

$$\Delta r = (10\,\text{m})\tan\left(\tfrac{1}{2}10^{-3}\,\text{rad}\right) = 10(5 \times 10^{-4})\text{m} = 5\text{ mm}$$

so that the diameter is 11 mm.

4.5 THE OUTPUT SPECTRUM OF A GAS LASER

The output radiation from a gas laser is not actually at one single well-defined wavelength corresponding to the lasing transition, but covers a spectrum of wavelengths with a central peak. This is not a simple consequence of the Heisenberg uncertainty principle but a direct result of the broadening of the emitted spectrum by the **Doppler effect**. We recall from the kinetic molecular theory that gas atoms are in random motion with an average kinetic energy of $(3/2)k_BT$. Suppose that these gas atoms emit radiation of frequency υ_o which we label as the source frequency. Then, due to the Doppler effect, when a gas atom is moving *away* from an observer, the latter detects a lower frequency υ_1 given by

$$\upsilon_1 = \upsilon_o\left(1 - \frac{v_x}{c}\right) \qquad \textbf{(1)}$$

where v_x is the relative velocity of the atom along the laser tube (x-axis) with respect to the observer and c is the speed of light. When the atom is moving *towards* the observer, the detected frequency υ_2 is higher and corresponds to

$$\upsilon_2 = \upsilon_o\left(1 + \frac{v_x}{c}\right) \qquad \textbf{(2)}$$

Since the atoms are in random motion the observer will detect a range of frequencies due to this Doppler effect. As a result, the frequency or wavelength of the output radiation from a gas laser will have a "linewidth" $\Delta\upsilon = \upsilon_2 - \upsilon_1$. This is what we mean by a **Doppler broadened linewidth** of a laser radiation. There are other mechanisms which also broaden the output spectrum but we will ignore these in the present case of gas lasers.

From the kinetic molecular theory we know that the velocities of gas atoms obey the Maxwell distribution. Consequently, the stimulated emission wavelengths in the lasing medium must exhibit a distribution about a central wavelength $\lambda_o = c/\upsilon_o$. Stated differently, the lasing medium therefore has an **optical gain** (or a photon gain) that has a distribution around $\lambda_o = c/\upsilon_o$ as shown in Figure 4.8 (a). The variation in the optical gain with the wavelength is called the **optical gain lineshape**. For the Doppler broadening case, this lineshape turns out to be a Gaussian function. For many gas lasers, this spread in the frequencies from υ_1 to υ_2 is 2-5 GHz (for the He-Ne laser the corresponding wavelength spread of ~0.02 Å).

When we consider the Maxwell velocity distribution of the gas atoms in the laser tube, we find that the linewidth $\Delta\upsilon_{1/2}$ between the half-intensity points (*full width at half maximum* FWHM) in the output intensity vs. frequency spectrum is given by,

$$\Delta\upsilon_{1/2} = 2\upsilon_o\sqrt{\frac{2k_BT\ln(2)}{Mc^2}} \qquad (3)$$

Frequency linewidth (FWHM)

where M is the mass of the lasing atom or molecule. The FWHM width $\Delta\upsilon_{1/2}$ has about 18% difference compared to simply taking the difference $\upsilon_2 - \upsilon_1$ from Eqs. (1) and (2) and using a root- mean-square effective velocity along x, that is using v_x in $(1/2)Mv_x^2 = (1/2)k_BT$. Equation (3) can be taken to be the FWHM width $\Delta\upsilon_{1/2}$ of the

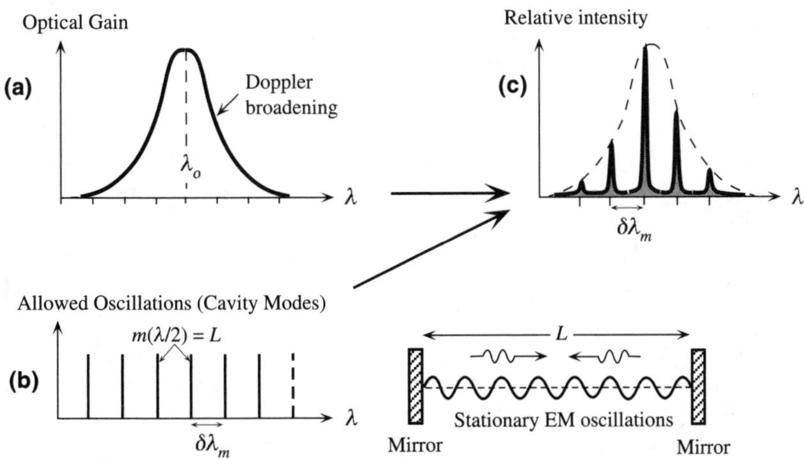

FIGURE 4.8 (a) Optical gain vs. wavelength characteristics (called the optical gain curve) of the lasing medium. (b) Allowed modes and their wavelengths due to stationary EM waves within the optical cavity. (c) The output spectrum (relative intensity vs. wavelength) is determined by satisfying (a) and (b) simultaneously, assuming no cavity losses.

optical gain curve of nearly all gas lasers. It does not apply to solid state lasers in which other broadening mechanisms operate.

Suppose that for simplicity we consider an optical cavity of length L with parallel end mirrors as shown in Figure 4.8 (b). Such an optical cavity is called a **Fabry-Perot optical resonator or etalon.**[10] The reflections from the end mirrors of a laser give rise to traveling waves in opposite directions within the cavity. These oppositely traveling waves interfere constructively to set up a standing wave, that is stationary electromagnetic (EM) oscillations. Some of the energy in these oscillations is tapped out by the 99% reflecting mirror to get an output just like the way we tap out the energy from an oscillating field in an LC circuit by attaching an antenna to it. Only standing waves with certain wavelengths however can be maintained within the optical cavity just as only certain acoustic wavelengths can be obtained from musical instruments. Any standing wave in the cavity must have an integer number of half-wavelengths $\lambda/2$ that fit into the cavity length L,

Laser cavity modes in a gas laser

$$m\left(\frac{\lambda}{2}\right) = L \tag{4}$$

where m is an integer that is called the **mode number** of the standing wave. The wavelength λ in Eq. (4) is that within the cavity medium but for gas lasers the refractive index is nearly unity and λ is the same as the free space wavelength. Each possible standing wave within the laser tube (cavity) satisfying Eq. (4) is called a **cavity mode**. The cavity modes, as determined by Eq. (4), are shown in Figure 4.8 (b). Modes that exist along the cavity axis are called **axial** (or **longitudinal**) **modes**. Other types of modes, that is stationary EM oscillations, are possible when the end mirrors are not flat. An example of an optical cavity formed by confocal spherical mirrors is shown in Figure 1.31 (Chapter 1). The EM radiation within such a cavity is a *Gaussian beam*.

The laser output thus has a broad spectrum with peaks at certain wavelengths corresponding to various cavity modes existing within the Doppler broadened optical gain curve as indicated in Figure 4.8 (c). At wavelengths satisfying Eq. (4), that is representing certain cavity modes, we have spikes of intensity in the output. The net envelope of the output radiation is a Gaussian distribution which is essentially due to the Doppler broadened linewidth. Notice that there is a finite width to the individual intensity spikes within the spectrum which is primarily due to nonidealities of the optical cavity such as acoustic and thermal fluctuations of the cavity length L and nonideal end mirrors (less than 100% reflection). Typically, the frequency width of an individual spike in a He-Ne gas laser is ~1 MHz, though in highly stabilized gas lasers widths as low as ~1 kHz have been reported.

It is important to realize that even if the laser medium has an optical gain, the optical cavity will always have some losses inasmuch as some radiation will be transmitted through the mirrors, and there will be various losses such as scattering within the cavity. Only those modes that have an optical gain that can make up for the radiation losses from the cavity can exist (as discussed later in §4.6).

[10] Question 1.11 in Chapter 1 considers the Fabry-Perot optical resonator. Interested students should try this question.

EXAMPLE 4.5.1 Doppler broadened linewidth

Calculate the Doppler broadened linewidths in frequency and wavelength for the He-Ne laser transition for $\lambda = 632.8$ nm if the gas discharge temperature is about 127°C. The atomic mass of Ne is 20.2 (g mol^{-1}). The laser tube length is 40 cm. What is the linewidth in the output wavelength spectrum? What is mode number m of the central wavelength, the separation between two consecutive modes and how many modes do you expect within the linewidth $\Delta\lambda_{1/2}$ of the optical gain curve?

Solution Due to the Doppler effect arising from the random motions of the gas atoms, the laser radiation from gas-lasers is broadened around a central frequency υ_o. The central υ_o corresponds to the source frequency. Higher frequencies detected will be due to radiations emitted from atoms moving towards the observer whereas lower frequencies will be due to the emissions from atoms moving away from the observer. We will first calculate the frequency width using two approaches, one approximate and the other more accurate. Suppose that v_x is the root-mean-square (rms) velocity along the x-direction. We can intuitively expect the frequency width $\Delta\upsilon_{\text{rms}}$ between rms points of the Gaussian output frequency spectrum to be[11]

$$\Delta\upsilon_{\text{rms}} = \upsilon_o\left(1 + \frac{v_x}{c}\right) - \upsilon_o\left(1 - \frac{v_x}{c}\right) = \frac{2\upsilon_o v_x}{c} \tag{5}$$

We need to know the rms velocity v_x along x which is given by the kinetic molecular theory as $v_x^2 = kT/M$, where M is the mass of the atom. For the He-Ne laser, it is the Ne atoms that lase, so $M = (20.2 \times 10^{-3}\text{ kg mol}^{-1})/(6.02 \times 10^{23}\text{ mol}^{-1}) = 3.35 \times 10^{-26}$ kg. Thus,

$$v_x = [(1.38 \times 10^{-23}\text{ J K}^{-1})(127 + 273\text{ K})/(3.35 \times 10^{-26}\text{ kg})]^{1/2} = 405.8\text{ m s}^{-1}$$

The central frequency is

$$\upsilon_o = c/\lambda_o = (3 \times 10^8\text{ m s}^{-1})/(632.8 \times 10^{-9}\text{ m}) = 4.74 \times 10^{14}\text{ s}^{-1}.$$

The rms frequency linewidth is approximately,

$$\Delta\upsilon_{\text{rms}} \approx (2\upsilon_o v_x)/c$$
$$= 2(4.74 \times 10^{14}\text{ s}^{-1})(405.8\text{ m s}^{-1})/(3 \times 10^8\text{ m s}^{-1}) = 1.282\text{ GHz.}$$

The observed FWHM width of the frequencies $\Delta\upsilon_{1/2}$ will be given by Eq. (3)

$$\Delta\upsilon_{1/2} = 2\upsilon_o\sqrt{\frac{2k_BT\ln(2)}{Mc^2}} = 2(4.748 \times 10^{14})\sqrt{\frac{2(1.38 \times 10^{-23})(400)\ln(2)}{(3.35 \times 10^{-26})(3 \times 10^8)^2}}$$
$$= 1.51\text{ GHz,}$$

which is about 18% wider.

To get FWHM wavelength width $\Delta\lambda_{1/2}$, differentiate $\lambda = c/\upsilon$

$$\frac{d\lambda}{d\upsilon} = -\frac{c}{\upsilon^2} = -\frac{\lambda}{\upsilon} \tag{6}$$

so that

$$\Delta\lambda_{1/2} \approx \Delta\upsilon_{1/2}|-\lambda/\upsilon| = (1.51 \times 10^9\text{ Hz})(632.8 \times 10^{-9}\text{ m})/(4.74 \times 10^{14}\text{ s}^{-1})$$

or

$$\Delta\lambda_{1/2} \approx 2.02 \times 10^{-12}\text{ m or 0.0020 nm.}$$

---

[11] The fact that this is the width between the rms points of the Gaussian output spectrum can be shown from detailed mathematics.

This width is between the half-points of the spectrum. The rms linewidth would be 0.0017 nm. Each mode in the cavity satisfies $m(\lambda/2) = L$ and since L is some 6.3×10^5 times greater than λ, the mode number m must be very large. For $\lambda = \lambda_o = 632.8$ nm, the corresponding mode number m_o is,

$$m_o = 2L/\lambda_o = (2 \times 0.3\ \text{m})/(632.8 \times 10^{-9}\ \text{m}) = 9.4817 \times 10^5$$

and actual m_o has to be the closest integer value to 9.4817×10^5.

The separation $\Delta\lambda_m$ between two consecutive modes (m and $m + 1$) is

$$\delta\lambda_m = \lambda_m - \lambda_{m+1} = \frac{2L}{m} - \frac{2L}{m+1} \approx \frac{2L}{m^2}$$

or

Separation between modes

$$\delta\lambda_m \approx \frac{\lambda_o^2}{2L}$$

Substituting the values we find $\delta\lambda_m = (632.8 \times 10^{-9})^2/(2 \times 0.4) = 5.006 \times 10^{-13}$ m or 0.501 pm.

The number modes, that is the number of m values, within the linewidth, that is, between the half-intensity points will depend on how the cavity modes and the optical gain curve coincide, for example, whether there is a cavity mode right at the peak of the optical gain curve as illustrated in Figure 4.9. Suppose that we use,

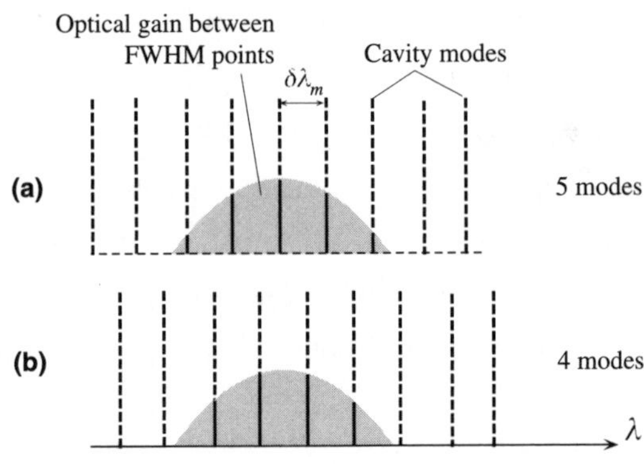

FIGURE 4.9 Number of laser modes depends on how the cavity modes intersect the optical gain curve. In this case we are looking at modes within the linewidth $\Delta\lambda_{1/2}$.

$$\text{Modes} = \frac{\text{Linewidth of spectrum}}{\text{Separation of two modes}} \approx \frac{\Delta\lambda_{1/2}}{\delta\lambda_m} = \frac{2.02\ \text{pm}}{0.501\ \text{pm}} = 4.03$$

We can expect at most 4 to 5 modes within the linewidth of the output as shown in Figure 4.9. We neglected the cavity losses.

4.6 LASER OSCILLATION CONDITIONS

A. Optical Gain Coefficient *g*

Consider a general laser medium which has an optical gain for coherent radiation along some direction x as shown Figure 4.10 (a). This means that the medium is appropriately pumped. Consider an electromagnetic wave propagating in the medium along the

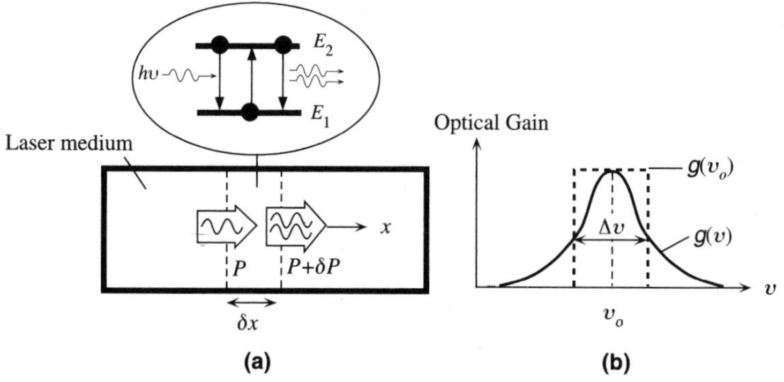

FIGURE 4.10 (a) A laser medium with an optical gain. (b) The optical gain curve of the medium. The dashed line is the approximate derivation in the text.

x-direction. As it propagates its power (energy flow per unit time) increases due to greater stimulated emissions over spontaneous emissions and absorption across the same two energy levels $E_2 - E_1$ as in Figure 4.10 (a). If the light intensity were decreasing, we would have used a factor $\exp(-\alpha x)$, where α is the absorption coefficient, to represent the power loss along the distance x. Similarly, we represent the power increase as $\exp(g x)$ where g is the optical gain per unit length and is called the **optical gain coefficient** of the medium. The gain coefficient g is defined as the fractional change in the light power (or intensity) per unit distance. Optical power P along x at any point is proportional to the concentration of coherent photons N_{ph} and their energy $h\upsilon$. These coherent photons travel with a velocity c/n, where n is the refractive index.[12] Thus in time δt they travel a distance $\delta x = (c/n)\delta t$ in the tube. Then,

$$g = \frac{\delta P}{P\delta x} = \frac{\delta N_{ph}}{N_{ph}\delta x} = \frac{n}{cN_{ph}}\frac{\delta N_{ph}}{\delta t} \quad (1)$$

Optical gain coefficient

The gain coefficient g describes the increase in intensity of the lasing radiation in the cavity per unit length due to stimulated emission transitions from E_2 to E_1 exceeding photon absorption across the same two energy levels. We know that the difference between stimulated emission and absorption rates (see Eqs (1) and (2) in §4.2) gives the *net* rate of change in the coherent photon concentration, that is

$$\begin{aligned} \frac{dN_{ph}}{dt} &= \text{Net rate of stimulated photon emission} \\ &= N_2 B_{21}\rho(h\upsilon) - N_1 B_{21}\rho(h\upsilon) \\ &= (N_2 - N_1)B_{21}\rho(h\upsilon) \end{aligned} \quad (2)$$

It is now straightforward to obtain the optical gain by using Eq. (2) in Eq. (1) with certain assumptions. As we are interested in the amplification of a coherent wave trav-

[12] In semiconductor related chapters, n is the electron concentration and n is the refractive index; E is the energy and E is the electric field.

eling along a defined direction (x) in Figure 4.10 we can neglect spontaneous emissions which are in random directions and do not, on average, contribute to the directional wave.

Normally, the emission and absorption processes occur not at a discrete photon energy hv but they would be distributed in photon energy or frequency over some frequency interval Δv. The spread Δv, for example, can be due to Doppler broadening or broadening of the energy levels E_2 and E_1. In any event, this means that the optical gain will reflect this distribution, that is $g = g(v)$ as depicted in Figure 4.10 (b). The spectral shape of the gain curve is called the **lineshape function**.

We can express $\rho(hv)$ in terms of N_{ph} by noting that $\rho(hv)$ is the radiation energy density per unit frequency so that at hv_o,

$$\rho(hv_o) \approx \frac{N_{ph}hv_o}{\Delta v} \tag{3}$$

We can now substitute for dN_{ph}/dt in Eq. (1) from Eq. (2) and use Eq. (3) to obtain the optical gain coefficient,

General optical gain coefficient

$$g(v_o) \approx (N_2 - N_1)\frac{B_{21}nhv_o}{c\Delta v} \tag{4}$$

Equation (4) gives the optical gain of the medium at the center frequency v_o. A more rigorous derivation would have found the optical gain curve as a function of frequency, shown as $g(v)$ in Figure 4.10 (b), and would derive $g(v_o)$ from this lineshape.[13]

B. Threshold Gain g_{th}

Consider an optical cavity with mirrors at the ends, such as the Fabry-Perot optical cavity shown in Figure 4.11. The cavity contains a laser medium so that lasing emissions build up to a steady state, that is we have continuous operation. We effectively assume that we have stationary electromagnetic (EM) oscillations in the cavity and that we have reached steady state. The optical cavity acts as an optical resonator. Consider an EM wave with an initial optical power P_i starting at some point in the cavity and traveling towards the cavity face 1 as shown in Figure 4.11. It will travel the length of the cavity, become reflected at face 1, travel back the length of the cavity to face 2, become reflected at 2 and arrive at the starting point with a final optical power P_f. Under steady state conditions, oscillations do not build up and do not die out which means that P_f must be the same as P_i. Thus there should be no optical power loss in the round trip which means that the **net round-trip optical gain** G_{op} must be unity,

Power condition for maintaining oscillations

$$G_{op} = P_f/P_i = 1 \tag{5}$$

Reflections at the faces 1 and 2 reduce the optical power in the cavity by the reflectances R_1 and R_2 of the faces. There are other losses such as some absorption and scattering during propagation in the medium. All these losses have to be made up by stimulated emissions in the optical cavity which effectively provides an optical gain in

---

[13] Commonly known as Füchtbauer-Ladenburg relation.

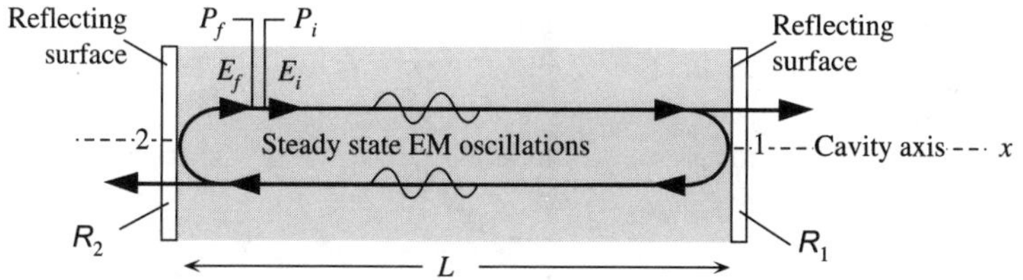

FIGURE 4.11 Optical cavity resonator

the medium. As the wave propagates, its power increases as $\exp(gx)$. However, there are a number of losses in the cavity medium acting against the stimulated emission gain such as light scattering at defects and inhomogenities, absorption by impurities, absorption by free carriers (important in semiconductors) and other loss phenomena. These losses decrease the power as $\exp(-\gamma x)$ where γ is the **attenuation** or **loss coefficient of the medium.** γ represents all losses in the cavity and its walls, *except* light transmission losses though the end mirrors and absorption across the energy levels involved in stimulated emissions (which is incorporated into g).[14]

The power P_f of the EM radiation after one round trip of path length $2L$ (Figure 4.11) is given by

$$P_f = P_i R_1 R_2 \exp[g(2L)] \exp[-\gamma(2L)] \tag{6}$$

For steady state oscillations Eq. (5) must be satisfied, and the value of the gain coefficient g that makes $P_f/P_i = 1$ is called the **threshold gain** g_{th}. From Eq. (6),

$$g_{th} = \gamma + \frac{1}{2L}\ln\left(\frac{1}{R_1 R_2}\right) \tag{7}$$

Threshold optical gain

Equation (7) gives the optical gain needed in the medium to achieve a continuous wave lasing emission. The necessary g_{th} as required by Eq. (3) has to be obtained by suitably pumping the medium so that N_2 is sufficiently greater than N_1. This corresponds to a **threshold population inversion** or $N_2 - N_1 = (N_2 - N_1)_{th}$. From Eq. (4),

$$(N_2 - N_1)_{th} \approx g_{th}\frac{c\Delta\upsilon}{B_{21} n h \upsilon_o} \tag{8}$$

Threshold population inversion

Initially the medium must have a gain coefficient g greater than g_{th}. This allows the oscillations to build-up in the cavity until a steady state is reached when $g = g_{th}$. By analogy, an electrical oscillator circuit has an overall gain (loop gain) of unity once a steady state is reached and oscillations are maintained. Initially, however, when the circuit is just switched on, the overall gain is greater then unity. The oscillations start from a small noise voltage, become amplified, that is built-up, until the overall gain becomes unity and a steady state operation is reached. The reflectance of the mirrors R_1 and R_2 are important in determining the threshold population inversion as they control g_{th} in

[14] γ should *not* be confused with the natural absorption coefficient α.

FIGURE 4.12 Simplified description of a laser oscillator. $(N_2 - N_1)$ and coherent output power (P_o) vs. pump rate under continuous wave steady state operation.

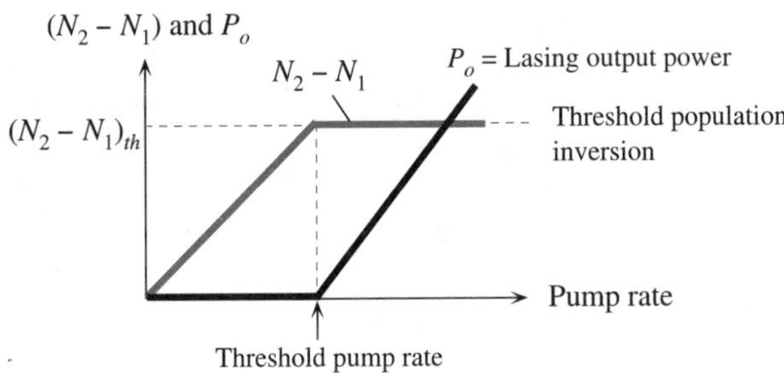

Eq. (8). It should be apparent that the laser device emitting coherent emission is actually a **laser oscillator**.

The examination of the steady state continuous wave (cw) coherent radiation output power P_o and the population difference $(N_2 - N_1)$ in a laser as a function of the pump rate would reveal the simplified behavior shown in Figure 4.12. Until the pump rate can bring $(N_2 - N_1)$ to the threshold value $(N_2 - N_1)_{th}$, there would be no coherent radiation output. When the pumping rate exceeds the threshold value, then $(N_2 - N_1)$ remains clamped at $(N_2 - N_1)_{th}$ because this controls the optical gain g which must remain at g_{th}. Additional pumping increases the *rate of stimulated transitions* and hence increases the optical output power P_o. Also note that we have not considered how pumping actually modifies N_1 and N_2 except that $(N_2 - N_1)$ is proportional to the pumping rate as in Figure 4.12.[15]

C. Phase Condition and Laser Modes

The laser oscillation condition stated in Eq. (5) which leads to the threshold gain g_{th} in Eq. (7) considers only the intensity of the radiation inside the cavity. Examination of Figure 4.11 reveals that the initial wave E_i with power P_i attains a power P_f after *one round trip* when the wave has arrived back exactly at the same position as E_f as shown in Figure 4.11. Unless the total phase change after one round trip from E_i to E_f is a multiple of 2π, the wave E_f cannot be identical to the initial wave E_i. We therefore need the additional condition that the round-trip phase change $\Delta\phi_{\text{round-trip}}$ must be a multiple of 360°,

Phase condition for laser oscillations

$$\Delta\phi_{\text{round-trip}} = m(2\pi) \tag{9}$$

where m is an integer, 1, 2, ... This condition ensures *self-replication* rather than self-destruction. There are various factors that complicate any calculation from the phase condition in Eq. (9). The refractive index n of the medium in general will depend on the pumping (especially so in semiconductors), and the end-reflectors can also introduce

[15] To relate N_1 and N_2 to the pumping rate we have to consider the actual energy levels involved in the laser operation (three or four levels) and develop *rate equations* to describe the transitions in the system; see for example J. Wilson and J.F.B. Hawkes, *Optoelectronics, An Introduction, Third Ed.* (Prentice-Hall, 1998), Ch. 5.

phase changes. In the simplest case, we can assume that n is constant and neglect phase changes at the mirrors. If $k = 2\pi/\lambda$ is the free space wavevector, only those special k-values, denoted as k_m, that satisfy Eq. (9) can exits as radiation in the cavity, *i.e.* for propagation along the cavity axis,

$$nk_m(2L) = m(2\pi) \quad \textbf{(10)}$$

Approximate laser cavity modes

which leads to the usual mode condition,

$$m\left(\frac{\lambda_m}{2n}\right) = L \quad \textbf{(11)}$$

Approximate laser cavity modes

Thus, our earlier intuitive representation of modes as standing waves described by Eq. (11) is a simplified conclusion from the general phase condition in Eq. (9). Furthermore, the above modes in Eq. (11) are controlled by the length L of the optical cavity along its axis and are called **longitudinal axial modes.**

In the discussions of threshold gain and phase conditions we referred to Figure 4.11 and tacitly assumed plane EM waves traveling inside the cavity between two perfectly flat and aligned mirrors. A plane wave is an idealization as it has an infinite extent over the plane normal to the direction of propagation. All practical laser cavities have a finite transverse size, a size perpendicular to the cavity axis. Furthermore, not all cavities have flat reflectors at the ends. In gas lasers, one or both mirrors at the tube ends may be spherical to allow a better mirror alignment as illustrated in Figure 4.13 (a) and (b). One can easily visualize off-axis self-replicating rays that can travel off the axis as shown in one example in Figure 4.13 (a). Such a mode would be non-axial. Its properties would be determined not only by the off-axis round-trip distance, but also by the transverse size of the cavity. The greater the transverse size, the more of these off-axis modes can exist.

A better way of thinking about modes is to realize that a mode represents a particular electric field pattern in the cavity that can replicate itself after one round trip.

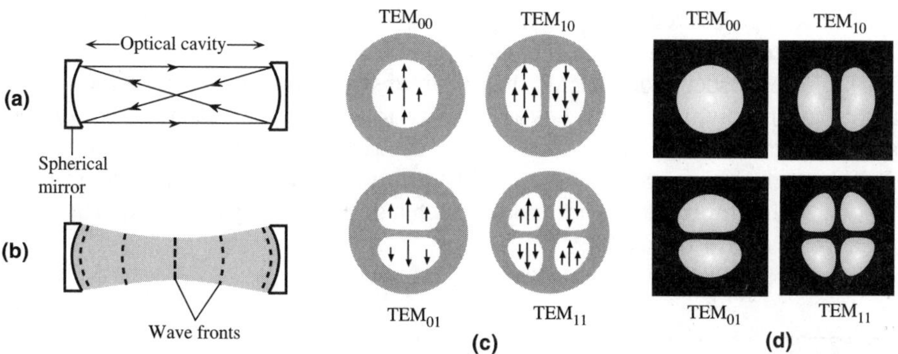

FIGURE 4.13 Laser Modes (a) An off-axis transverse mode is able to self-replicate after one round trip. (b) Wavefronts in a self-replicating wave (c) Four possible modes low order transverse cavity modes and their fields. (d) Intensity patterns in the modes of (c). (For rectangular symmetry.)

Figure 4.13 (b) shows how a wavefront of a particular mode starts parallel to the surface of one of the mirrors, and after one round trip, it replicates it self. The wavefront curvature changes as the radiation propagates in the cavity and it is parallel to mirror surfaces at the end-mirrors. Such a mode has similarities to the Gaussian beam discussed in Ch. 1.

More generally, whether we have flat or spherical end-mirrors, we can find all possible allowed modes by considering what spatial field patterns at one mirror can self-replicate itself after one round trip[16] through the cavity to the other mirror and back as in the example in Figure 4.13 (b). *A mode with a certain field pattern at a reflector can propagate to the other reflector and back again and return the same field pattern.* All these modes, can be represented by fields (**E** and **B**) that are nearly normal to the cavity axis; they are referred to as **transverse modes** or **transverse electric and magnetic (TEM) modes.**[17] Each allowed mode corresponds to a distinct spatial field distribution at a reflector. These modal field patterns at a reflector can be described by three integers p, q, m and designated by TEM_{pqm}. The integers p, q represent the number of nodes in the field distribution along the transverse directions y and z to the cavity axis x (put differently across the beam cross section). The integer m is the number of nodes along the cavity axis x and is the usual longitudinal mode number. Figure 4.13 (c) and (d) show the field patterns for four TEM modes and the corresponding intensity patterns for four example TEM modes. Each transverse mode with a given p, q has a set of longitudinal modes (m values) but usually m is very large ($\sim 10^6$ in gas lasers) and is not written, though understood. Thus, transverse modes are written as TEM_{pq} and each has a set of longitudinal modes ($m = 1, 2, \ldots$). Moreover, two different transverse modes may not necessarily have the same longitudinal frequencies implied by Eq. (11). (For example, n may be not be spatially uniform and different TEM modes have different spatial field distributions.)

Transverse modes depend on the optical cavity dimensions, reflector sizes, and other size limiting apertures that mat be present in the cavity. The modes either have **Cartesian (rectangular)** or **polar (circular)** symmetry about the cavity axis. Cartesian symmetry arises whenever a feature of the optical cavity imposes a more favorable field direction; otherwise, the patterns exhibit circular symmetry. The examples in Figure 4.13 (c) and (d) posses rectangular symmetry and would arise, for example, if polarizing Brewster windows are present at the ends of the cavity.

The lowest order mode TEM_{00} has an intensity distribution that is radially symmetric about the cavity axis and has a Gaussian intensity distribution across the beam cross section everywhere inside and outside cavity. It also has the lowest divergence angle. These properties render TEM_{00} highly desirable and many laser designs optimize on TEM_{00} while suppressing other modes. Such a design usually requires restrictions in the transverse size of the cavity.

[16] We actually have to solve Maxwell's equations with the boundary conditions of the cavity to determine what EM wave patterns that are allowed. Further we have to incorporate optical gain into these equations (not a trivial task).

[17] Or, *transverse electromagnetic modes*.

EXAMPLE 4.6.1 Threshold population inversion for the He-Ne laser

Show that the threshold population inversion $\Delta N_{th} = (N_2 - N_1)_{th}$ can be written as,

$$\Delta N_{th} \approx g_{th} \frac{8\pi n^2 \upsilon_o^2 \tau_{sp} \Delta\upsilon}{c^2} \tag{12}$$

Threshold population inversion

where υ_o = peak emission frequency (at peak of output spectrum), n = refractive index, $\tau_{sp} = 1/A_{21}$ = mean time for spontaneous transition and $\Delta\upsilon$ = optical gain bandwidth (frequency-linewidth of the optical gain lineshape).

Consider a He-Ne gas laser operating at 623.8 nm. The tube length $L = 50$ cm, tube diameter is 1.5 mm and mirror reflectances are approximately 100% and 90%. The linewidth $\Delta\upsilon = 1.5$ GHz, the loss coefficient is $\gamma \approx 0.05$ m^{-1}, spontaneous decay time constant $\tau_{sp} = 1/A_{21} \approx 300$ ns, $n \approx 1$. What is the threshold population inversion?

Solution The B_{21} coefficient in Eq. (8) can be replaced in terms of A_{21} (which can be determined experimentally), $A_{21}/B_{21} = 8\pi h\upsilon^3/c^3$,

$$(N_2 - N_1)_{th} \approx g_{th} \frac{c\Delta\upsilon}{\dfrac{A_{21}c^3}{8\pi h n^3 \upsilon_o^3} nh\upsilon_o} = g_{th} \frac{8\pi n^2 \upsilon_o^2 \tau_{sp} \Delta\upsilon}{c^2}$$

The emission frequency $\upsilon_o = c/\lambda_o = (3 \times 10^8 \text{ m s}^{-1})/(632.8 \times 10^{-9} \text{ m}) = 4.74 \times 10^{15}$ Hz. Given the laser characteristics,

$$g_{th} = \gamma + \frac{1}{2L} \ln\left(\frac{1}{R_1 R_2}\right) = 0.05 \text{ m}^{-1} + \frac{1}{2(0.5 \text{ m})} \ln\left(\frac{1}{1 \times 0.9}\right) = 0.155 \text{ m}^{-1}.$$

$$\Delta N_{th} \approx g_{th} \frac{8\pi \upsilon_o^2 n^2 \tau_{sp} \Delta\upsilon}{c^2}$$

and

$$= (0.155 \text{ m}^{-1}) \frac{8\pi (4.74 \times 10^{14} \text{ s}^{-1})^2 (1)^2 (300 \times 10^{-9} \text{ s})(1.5 \times 10^9 \text{ s}^{-1})}{(3 \times 10^8 \text{ m s}^{-1})^2}$$

$$= 4.4 \times 10^{15} \text{ m}^{-3}.$$

Note that this is the threshold population inversion for Ne atoms in configurations $2p^5 5s^1$ and $2p^5 3p^1$.

4.7 PRINCIPLE OF THE LASER DIODE[18]

Consider a degenerately doped direct bandgap semiconductor *pn* junction whose band diagram is shown in Figure 4.14 (a). By degenerate doping we mean that the Fermi level E_{Fp} in the *p*-side is in the valence band (VB) and that E_{Fn} in the *n*-side is in the conduction band (CB). All energy levels up to the Fermi level can be taken to be occupied

[18] The first semiconductor lasers used GaAs *pn* junctions, and were reported by the American researchers R.N. Hall *et al.* (General Electric Research, Schenectady) and and Marshall I. Nathan *et al.* (IBM, Thomas J. Watson Research Center) in 1962.

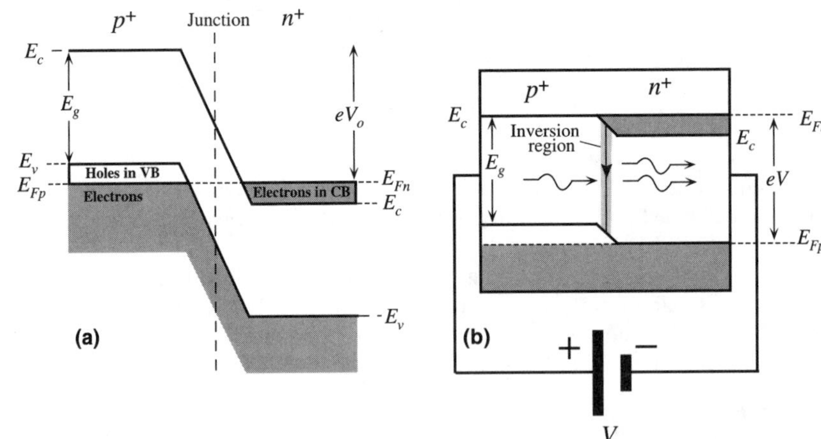

FIGURE 4.14 The energy band diagram of a degenerately doped pn junction with no bias. (b) Band diagram with a sufficiently large forward bias to cause population inversion and hence stimulated emission.

by electrons as in Figure 4.14 (a). In the absence of an applied voltage, the Fermi level is continuous across the diode, $E_{Fp} = E_{Fn}$. The depletion region or the space charge layer (SCL) in such a pn junction is very narrow. There is a built-in voltage V_o that gives rise to a potential energy barrier eV_o that prevents electrons in the CB of n^+-side diffusing into the CB of the p^+-side. There is a similar barrier stopping hole diffusion from p^+-side to n^+-side.

Recall that when a voltage is applied to a pn junction device, the change in the Fermi level from end-to-end is the electrical work done by the applied voltage[19], that is $\Delta E_F = eV$. Suppose that this degenerately doped pn junction is forward biased by a voltage V greater than the bandgap voltage; $eV > E_g$ as shown in Figure 4.14 (b). The separation between E_{Fn} and E_{Fp} is now the applied potential energy or eV. The applied voltage diminishes the built-in potential barrier to almost zero which means that electrons flow into the SCL and flow over to the p^+-side to constitute the diode current. There is a similar reduction in the potential barrier for holes from p^+ to n^+-side. The final result is that electrons from n^+ side and holes from p^+ side flow into the SCL, and this SCL region is no longer depleted, as apparent in Figure 4.14 (b). If we draw the energy band diagram with $E_{Fn} - E_{Fp} = eV > E_g$ this conclusion is apparent. In this region, there are *more* electrons in the conduction band at energies near E_c than electrons in the valence band near E_v as illustrated by density of states diagram for the junction region in Figure 4.15 (a). In other words, there is a **population inversion** between energies near E_c and those near E_v around the junction.

[19] There is a useful theorem in Thermodynamics that states that any change in the Gibbs free energy of a system corresponds to an external electrical work done by, or on, the system. Fermi energy is simply the Gibbs free energy per electron and $\Delta E_F = eV$.

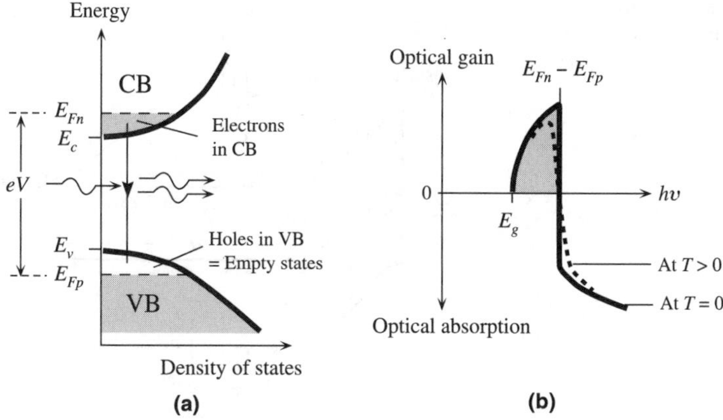

FIGURE 4.15 (a) The density of states and energy distribution of electrons and holes in the conduction and valence bands respectively at $T \approx 0$ in the SCL under forward bias such that $E_{Fn} - E_{Fp} > E_g$. Holes in the VB are empty states. (b) Gain vs. photon energy.

This population inversion region is a layer along the junction and is called the **inversion layer** or the **active region**. An incoming photon with an energy of $(E_c - E_v)$ cannot excite an electron from E_v to E_c as there are almost none near E_v. It can, however, stimulate an electron to fall down from E_c to E_v as shown in Figure 4.14 (b). Put differently, the incoming photon stimulates direct recombination. The region where there is population inversion and hence more stimulated emission than absorption, or the active region, has an **optical gain** because an incoming photon is more likely to cause stimulated emission than being absorbed. The optical gain depends on the photon energy (and hence on the wavelength) as apparent by the energy distributions of electrons and holes in the conduction and valence bands in the active layer in Figure 4.15 (a). At low temperatures ($T \approx 0\ K$), the states between E_c and E_{Fn} are filled with electrons and those between E_{FP} and E_v are empty. Photons with energy greater than E_g but less than $E_{Fn} - E_{fp}$ cause stimulated emissions whereas those photons with energies greater than $E_{Fn} - E_{Fp}$ become absorbed. Figure 4.15 (b) shows the expected dependence of optical gain and absorption on the photon energy at low temperatures ($T \approx 0\ K$). As the temperature increases, the Fermi-Dirac function spreads the energy distributions of electrons in the CB to above E_{Fn} and holes below E_{Fp} in the VB. The result is a reduction in optical gain as indicated in Figure 4.15 (b). The optical gain depends on $E_{Fn} - E_{Fp}$ which depends on the applied voltage and hence on the diode current.

It is apparent that population inversion between energies near E_c and those near E_v is achieved by the injection of carriers across the junction under a sufficiently large forward bias. The pumping mechanism is therefore the forward diode current and the pumping energy is supplied by the external battery. This type of pumping is called **injection pumping**.

In addition to population inversion we also need to have an *optical cavity* to implement a laser oscillator, that is, to build up the intensity of stimulated emissions by

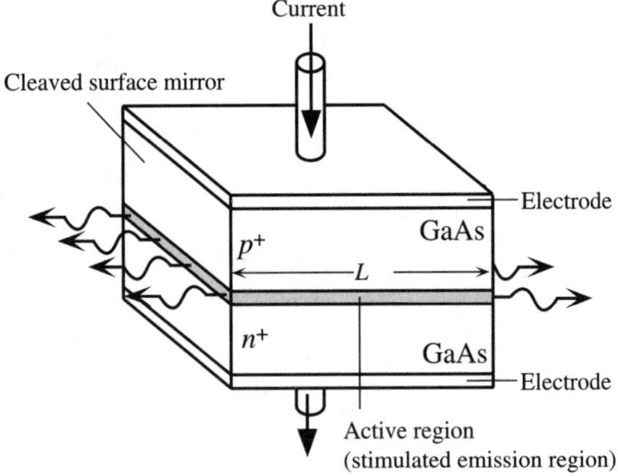

FIGURE 4.16 A schematic illustration of a GaAs homojunction laser diode. The cleaved surfaces act as reflecting mirrors.

means of an optical resonator. This would provide a continuous coherent radiation as output from the device. Figure 4.16 shows schematically the structure of a **homojunction laser diode**. The *pn* junction uses the same direct bandgap semiconductor material throughout, for example GaAs, and hence has the name homojunction. The ends of the crystal are cleaved to be flat and optically polished to provide reflection and hence form an optical cavity. Photons that are reflected from the cleaved surfaces stimulate more photons of the same frequency and so on. This process builds up the intensity of the radiation in the cavity. The wavelength of the radiation that can build up in the cavity is determined by the length L of the cavity because only multiples of the half-wavelength can exists in such an optical cavity as explained above, *i.e.*

Modes in an optical cavity

$$m\frac{\lambda}{2n} = L \tag{1}$$

where m is an integer, n is the refractive index of the semiconductor and λ is the free space wavelength. Each radiation satisfying the above relationship is essentially a **resonant frequency** of the cavity, that is, a **mode** of the cavity. The separation between possible modes of the cavity (or separation between allowed wavelengths) $\Delta\lambda_m$ can be readily found from Eq. (1) as in the case of the He-Ne gas laser previously.

The dependence of the optical gain of the medium on the wavelength of radiation can be deduced from the energy distribution of the electrons in the CB and holes in the VB around the junction as in Figure 4.15. The exact output spectrum from the laser diode depends both on the nature of the optical cavity and the optical gain vs. wavelength characteristics. Lasing radiation is only obtained when the optical gain in the medium can overcome the photon losses from the cavity, which requires the diode current I to exceed a **threshold value** I_{th}. Below I_{th}, the light from the device is due to spontaneous emission and not stimulated emission. The light output is then composed of incoherent photons that are emitted randomly and the device behaves like an LED.

We can identify two critical diode currents. First is the diode current that provides just sufficient injection to lead to stimulated emissions just balancing absorption. This is called the **transparency current** I_{trans} inasmuch as there is then no net photon absorption; the medium is *transparent*. Above I_{trans} there is optical gain in the medium though the op-

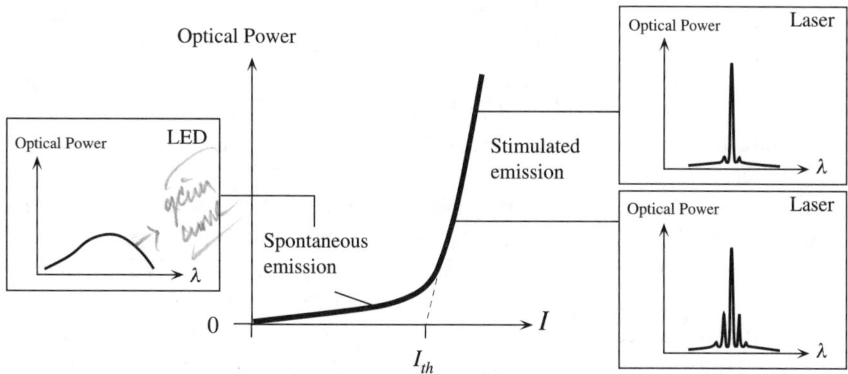

FIGURE 4.17 Typical output optical power vs. diode current (I) characteristics and the corresponding output spectrum of a laser diode.

tical output is not yet a continuous wave coherent radiation. Lasing oscillations occur only when the optical gain in the medium can overcome the photon losses from the cavity, that is when the optical gain g reaches the threshold gain g_{th}. This occurs at the **threshold current** I_{th}. Those cavity resonant frequencies that experience the threshold optical gain can resonate within the cavity. Some of this cavity radiation is transmitted out from the cleaved ends as these are not perfectly reflecting (typically about 32% reflecting). Figure 4.17 shows the output light intensity as a function of diode current. Above I_{th}, the light intensity becomes coherent radiation consisting of cavity wavelengths (or modes) and increases steeply with the current. The number of modes in the output spectrum and their relative strengths depend on the diode current as depicted in Figure 4.17.

The main problem with the homojunction laser diode is that the threshold current density J_{th} is too high for practical uses. For example, the threshold current density is of the order of ~ 500 A mm^{-2} for GaAs at room temperature which means that the GaAs homojunction laser can only be operated continuously at very low temperatures. However J_{th} can be reduced by orders of magnitude by using **heterostructured** semiconductor laser diodes.

4.8 HETEROSTRUCTURE LASER DIODES

The reduction of the threshold current I_{th} to a practical value requires improving the rate of stimulated emission and also improving the efficiency of the optical cavity. First we can confine the injected electrons and holes to a narrow region around the junction. This narrowing of the active region means that less current is needed to establish the necessary concentration of carriers for population inversion. Secondly, we can build a dielectric waveguide around the optical gain region to increase the photon concentration and hence the probability of stimulated emission. This way we can reduce the loss of photons traveling off the cavity axis. We therefore need both **carrier confinement** and **photon confinement.** Both of these requirements are readily achieved in modern laser diodes by the use of heterostructured devices as in the case of high-intensity double heterostructure LEDs. However, in the case of laser diodes, there is an additional requirement for maintaining a good optical cavity that will increase stimulated emissions over spontaneous emissions.

Figure 4.18 (a) shows a **double heterostructure (DH)** device based on two junctions between different semiconductor materials with different bandgaps. In this case the semiconductors are AlGaAs with $E_g \approx 2$ eV and GaAs with $E_g \approx 1.4$ eV. The *p*-GaAs region is a thin layer, typically 0.1 – 0.2 μm, and constitutes the **active layer** in which lasing recombination takes place. Both *p*-GaAs and *p*-AlGaAs are heavily *p*-type doped and are degenerate with E_F in the valence band. When a sufficiently large forward bias is applied, E_c of *n*-AlGaAs moves above E_c of *p*-GaAs which leads to a large injection of electrons in the CB of *n*-AlGaAs into *p*-GaAs as shown in Figure 4.18 (b). These electrons, however, are *confined* to the CB of *p*-GaAs since there is a barrier ΔE_c between *p*-GaAs and *p*-AlGaAs due to the change in the bandgap (there is also a small change in E_v but we ignore this). Inasmuch as *p*-GaAs is a thin layer, the concentration of injected electrons in the *p*-GaAs layer can be increased quickly even with moderate increases in forward current. This effectively reduces the threshold current for population inversion or optical gain. Thus even moderate forward currents can inject sufficient number of electrons into the CB of *p*-GaAs to establish the necessary electron concentration for population inversion in this layer.

A wider bandgap semiconductor generally has a lower refractive index. AlGaAs has a lower refractive index than that of GaAs. The change in the refractive index defines an optical dielectric waveguide, as depicted in Figure 4.18 (c), that confines the

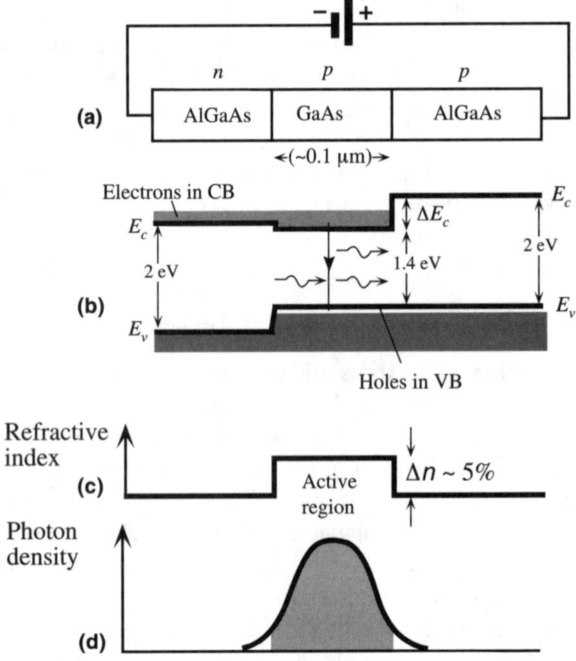

FIGURE 4.18 (a) A double heterostructure diode has two junctions which are between two different bandgap semiconductors (GaAs and AlGaAs). (b) Simplified energy band diagram under a large forward bias. Lasing recombination takes place in the *p*-GaAs layer, the *active layer*. (c) Higher bandgap materials have a lower refractive index. (d) AlGaAs layers provide lateral optical confinement.

photons to the active region of the optical cavity and thereby reduces photon losses and increases the photon concentration. The photon concentration across the device is shown in Figure 4.18 (d). This increase in the photon concentration increases the rate of stimulated emissions. Thus both carrier and optical confinement lead to a reduction in the threshold current density. Without double-heterostructure devices we would not have practical solid state lasers that can be operated continuously at room temperature.

A typical structure of a **double heterostructure laser diode** is similar to a double heterostructure LED and is shown schematically in Figure 4.19. The doped layers are grown epitaxially on a crystalline substrate which in this case is *n*-GaAs. The double heterostructure described above consists of the first layer on the substrate, *n*-AlGaAs, the active *p*-GaAs layer and the *p*-AlGaAs layer. There is an additional *p*-GaAs layer, called **contacting layer**, next to *p*-AlGaAs. It can be seen that the electrodes are attached to the GaAs semiconductor materials rather than AlGaAs. This choice allows for better contacting and avoids Schottky junctions which would limit the current. The *p* and *n*-AlGaAs layers provide carrier and optical confinement in the vertical direction by forming heterojunctions with *p*-GaAs. The active layer is *p*-GaAs which means that the lasing emission will be in the range 870–900 nm depending on the doping level. This layer can also be made to be $Al_yGa_{1-y}As$ but of different composition than the confining $Al_xGa_{1-x}As$ layers and still preserve heterojunction properties. This allows the lasing wavelength to be controlled by the composition of the active layer. The advantage of the AlGaAs/GaAs heterojunction is that there is only a small lattice mismatch between the two crystal structures and hence negligible strain induced interfacial defects (*e.g.* dislocations) in the device.

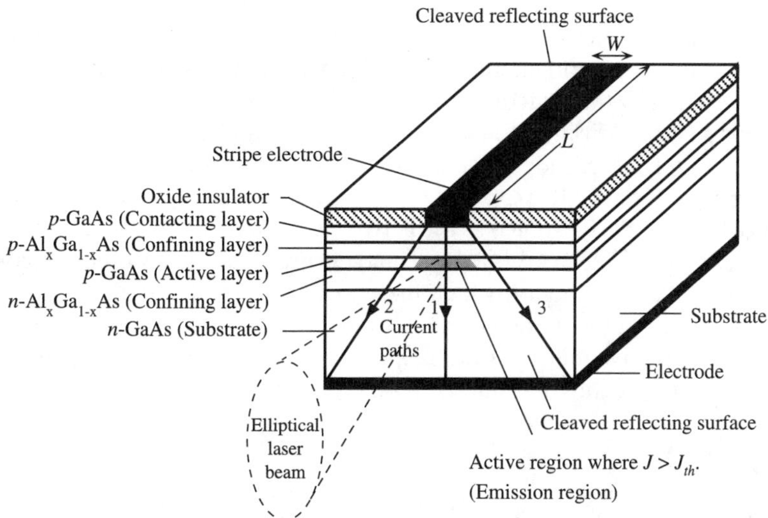

FIGURE 4.19 Schematic illustration of the structure of a double heterojunction stripe contact laser diode

Such defects invariably act as non-radiative recombination centers and hence reduce the rate of radiative transitions.

An important feature of this laser diode is the **stripe geometry**, or stripe contact on p-GaAs. The current density J from the stripe contact is not uniform laterally. J is greatest along the central path, 1, and decreases away from path 1, towards 2 or 3. The current is confined to flow within paths 2 and 3. The current density paths through the active layer where J is greater than the threshold value J_{th} as shown in Figure 4.19, define the **active region** where population inversion and hence optical gain takes place. The lasing emission emerges from this active region. The width of the active region, or the optical gain region, is therefore defined by the current density from the stripe contact. Optical gain is highest where the current density is greatest. Such lasers are called **gain guided**. There are two advantages to using a stripe geometry. First, the reduced contact area also reduces the threshold current I_{th}. Secondly, the reduced emission area makes light coupling to optical fibers easier. Typical stripe widths (W) may be as small as a few microns leading to typical threshold currents that may be tens of milliamperes.

The laser efficiency can be further improved by reducing the reflection losses from the rear crystal facet. Since the refractive index of GaAs is about 3.7, the reflectance is 0.33. However, by fabricating a dielectric mirror (Chapter 1) at the rear facet, that is a mirror consisting of a number of quarter wavelength semiconductor layers of different refractive index, it is possible to bring the reflectance close to unity and thereby improve the optical gain of the cavity. This corresponds to a reduction in the threshold current.

The width, or the lateral extent, of the optical gain region in the stripe geometry DH laser in Figure 4.19 is defined by the current density and changes with the current. More importantly, the lateral optical confinement of photons to the active region is poor because there is no marked change in the refractive index laterally. It would be advantageous to laterally confine the photons to the active region to increase the rate of stimulated emissions. This can be achieved by shaping the refractive index profile in the same way the vertical confinement was defined by the heterostructure. Figure 4.20 illustrates schematically the structure of such a DH laser diode where the active layer, p-GaAs, is bound both vertically and laterally by a wider bandgap semiconductor, AlGaAs, which has lower refractive index. The active layer (GaAs) is effectively *buried* within a wider bandgap material (AlGaAs) and the structure is hence called **buried double heterostructure** laser diode. Inasmuch as the active layer is surrounded by a lower index AlGaAs it behaves as a dielectric waveguide and ensures that the photons are confined to the active or optical gain region which increases the rate of stimulated emission and hence the efficiency of the diode. Since the optical power is confined to the waveguide defined by the refractive index variation, these diodes are called **index guided**. Further, if the buried heterostructure has the right dimensions compared with the wavelength of

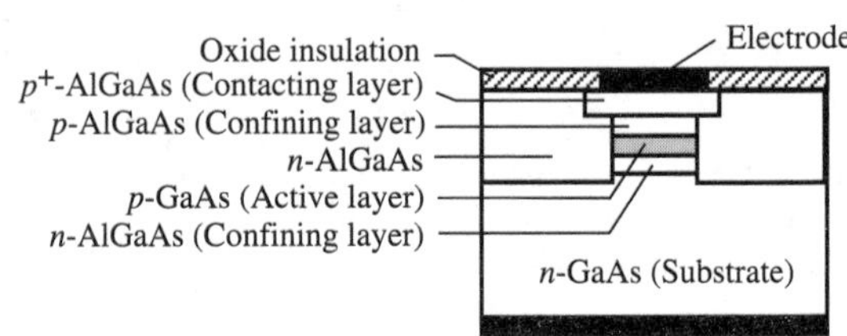

FIGURE 4.20 Schematic illustration of the cross sectional structure of a buried heterostructure laser diode.

the radiation then only the fundamental mode can exist in this waveguide structure as in the case of dielectric waveguides. This would be the case in a **single mode laser diode.**

The laser diode heterostructures based on GaAs and AlGaAs are suitable for emissions around 900 nm. For operation in the optical communication wavelengths of 1.3 and 1.55 μm, typical heterostructures are based on InP (substrate) and quarternary alloys InGaAsP where InGaAsP alloys have a narrower bandgap than that of InP and a greater refractive index. The composition of the InGaAsP alloy is adjusted to obtain the required bandgap for the active and confining layers (see Example 3.8.3).

EXAMPLE 4.8.1 Modes in a laser and the optical cavity length

Consider an AlGaAs based heterostructure laser diode which has an optical cavity of length 200 microns. The peak radiation is at 870 nm and the refractive index of GaAs is about 3.7. What is the mode integer m of the peak radiation and the separation between the modes of the cavity? If the optical gain vs. wavelength characteristics has a FWHM wavelength width of about 6 nm how many modes are there within this bandwidth? How many modes are there if the cavity length is 20 μm?

Solution Figure 4.8 schematically illustrates the cavity modes, the optical gain characteristics and a typical output spectrum from a laser. The free-space wavelength λ of a cavity mode and length L are related by

$$m\frac{\lambda}{2n} = L$$

so that

$$m = \frac{2nL}{\lambda} = \frac{2(3.7)(200 \times 10^{-6})}{(900 \times 10^{-9})} = 1644.4 \text{ or } 1644.$$

The wavelength separation $\delta\lambda_m$ between the adjacent cavity modes m and $(m + 1)$ in Figure 4.8 is

$$\delta\lambda_m = \frac{2nL}{m} - \frac{2nL}{m+1} \approx \frac{2nL}{m^2} = \frac{\lambda^2}{2nL}$$

Thus the separation between the modes for a given peak wavelength increases with decreasing L. When $L = 200$ μm,

$$\delta\lambda_m = \frac{(900 \times 10^{-9})^2}{2(3.7)(200 \times 10^{-6})} = 5.47 \times 10^{-10} \text{ m or } 0.547 \text{ nm}$$

If the optical gain has a bandwidth of $\Delta\lambda_{1/2}$ as *in* Figure 4.8 then there will be $\Delta\lambda_{1/2}/\Delta\lambda_m$ number of modes, or (6 nm)/(0.547 nm), that is 10 modes.

When $L = 20$ μm, the separation between the modes becomes,

$$\delta\lambda_m = \frac{(900 \times 10^{-9})^2}{2(3.7)(20 \times 10^{-6})} = 5.47 \text{ nm}$$

Then $(\Delta\lambda_{1/2})/\delta\lambda_m = 1.1$ and there will be one mode that corresponds to about 900 nm. In fact m must be an integer and when $m = 1644$, $\lambda = 902.4$ nm. It is apparent that reducing the cavity length suppresses higher modes. Note that the optical bandwidth depends on the diode current.

4.9 ELEMENTARY LASER DIODE CHARACTERISTICS

The output spectrum from a laser diode (LD) depends on two factors: the nature of the optical resonator used to build the laser oscillations and the optical gain curve (lineshape) of the active medium. The optical resonator is essentially a Fabry-Perot cavity as depicted in Figure 4.21 which can be assigned a length (L), width (W) and height (H). The length L determines the **longitudinal mode** separation whereas the width W and height H determine the transverse modes, or **lateral modes** in LD nomenclature. If the transverse dimensions (W and H) are sufficiently small, only the lowest transverse mode, TEM_{00} mode, will exit. This TEM_{00} mode however will have longitudinal modes whose separation depends on L. Figure 4.21 also shows that the emerging laser beam exhibits divergence. This is due to diffraction of the waves at the cavity ends. The smallest aperture (H in the figure) causes the greatest diffraction.

The actual modes that exist in the output spectrum of a LD will depend on the optical gain these modes will experience. The spectrum, that is optical power density vs. wavelength characteristics, is either multimode or single mode depending on the optical resonator structure and the pumping current level. Figure 4.22 shows the output spectrum from an index guided LD at various output power levels. The multimode spectrum at low output power becomes single mode at high output powers. In contrast, the output spectrum of most gain guided LDs tend to remain multimode even at high diode currents.

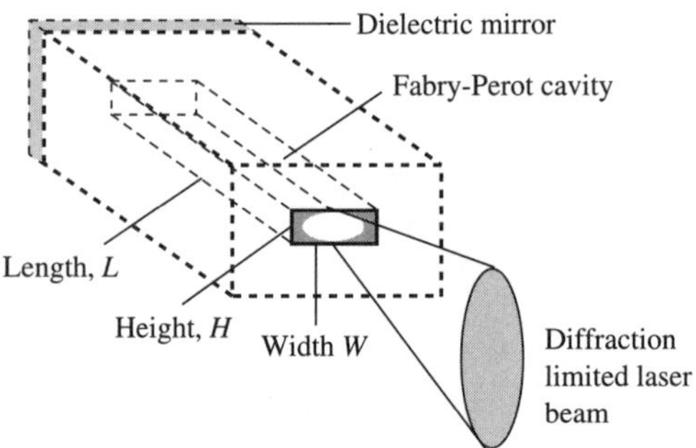

FIGURE 4.21 The laser cavity definitions and the output laser beam characteristics.

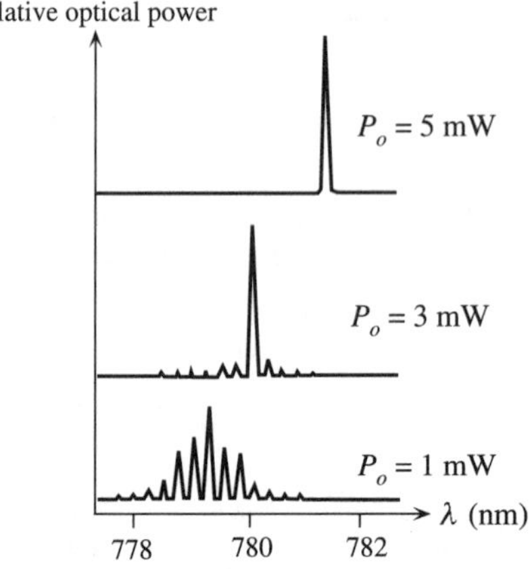

FIGURE 4.22 Output spectra of lasing emission from an index guided LD. At sufficiently high diode currents corresponding to high optical power, the operation becomes single mode. (Note: Relative power scale applies to each spectrum individually and not between spectra)

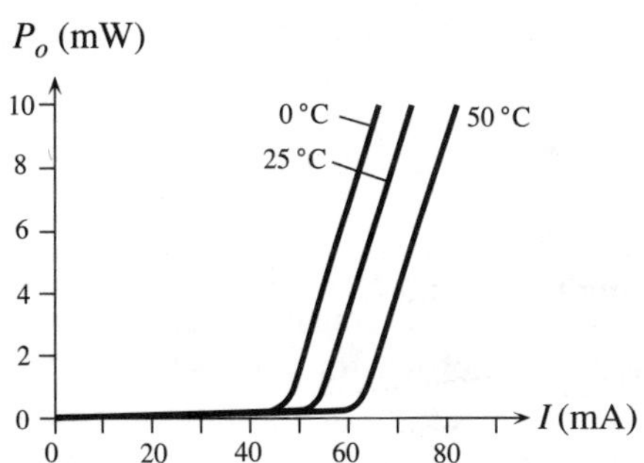

FIGURE 4.23 Output optical power vs. diode current at three different temperatures. The threshold current shifts to higher temperatures.

The LD output characteristics tend to be temperature sensitive. Figure 4.23 shows the changes in the optical output power vs. diode current characteristics with the case temperature. As the temperature increases, the threshold current increases steeply, typically as the exponential of the absolute temperature. The output spectrum also changes with the temperature. In the case of a single mode LD, the peak emission wavelength λ_o exhibits "jumps" at certain temperatures as apparent in Figure 4.24 (a) and (b). A jump corresponds to a **mode hop** in the output. That is, at the new operating temperature, another mode fulfills the laser oscillation condition which means a discrete change in the laser oscillation wavelength. Between mode hops, λ_o increases slowly with the temperature due to the slight increase in the refractive index n (and the cavity length) with temperature. If mode hops are undesirable, then the device structure must be such to keep the modes sufficiently separated. In contrast, the output spectrum of a gain guided laser has many modes so that λ_o vs. T behavior tends to follow the changes in the bandgap (the optical gain curve) rather than the cavity properties. Highly stabilized LDs are usually marketed with thermoelectric coolers integrated into the diode package to control the device temperature.

One commonly stated important and useful laser diode parameter is the **slope efficiency** which determines the optical power P_o of the output coherent radiation in

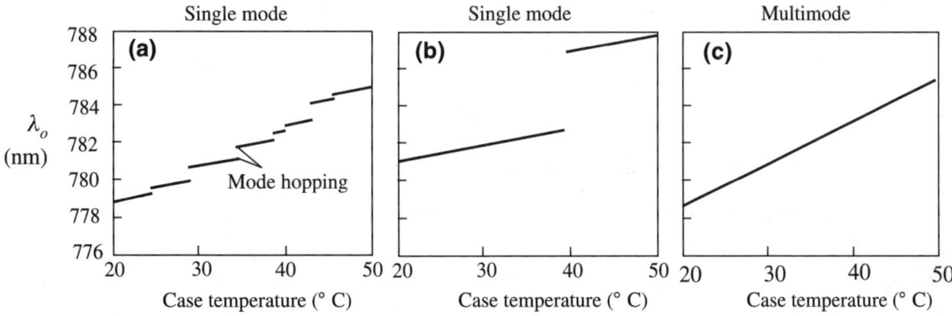

FIGURE 4.24 Peak wavelength vs. case temperature characteristics. (a) Mode hops in the output spectrum of a single mode LD. (b) Restricted mode hops and none over the temperature range of interest (20–40°C). (c) Output spectrum from a multimode LD.

terms of the diode current above the threshold current I_{th}. If I is the diode current, the slope efficiency η_{slope} is

Slope efficiency

$$\eta_{\text{slope}} = \frac{P_o}{I - I_{th}} \qquad \textbf{(1)}$$

and is measured in W/A or W/mA. The slope efficiency depends on the LD structure as well as the semiconductor packaging, and typically values for commonly available LDs are less than 1W/A. The **conversion efficiency** gauges the overall efficiency of the conversion from the input of electrical power to the output of optical power. Although this is not generally quoted in data sheets, it can be easily determined from the output power at the operating diode current and voltage. In some modern LDs this may be as high as 30-40%.

EXAMPLE 4.9.1 Laser output wavelength variations

Given that the refractive index n of GaAs has a temperature dependence $dn/dT \approx 1.5 \times 10^{-4}\,\text{K}^{-1}$ estimate the change in the emitted wavelength 870 nm per degree change in the temperature between mode hops.

Solution Consider a particular given mode with wavelength λ_m,

$$m\left(\frac{\lambda_m}{2n}\right) = L$$

Then,

$$\frac{d\lambda_m}{dT} = \frac{d}{dT}\left[\frac{2}{m}nL\right] \approx \frac{2L}{m}\frac{dn}{dT}$$

Substituting for L/m in terms of λ_m,

$$\frac{d\lambda_m}{dT} \approx \frac{\lambda_m}{n}\frac{dn}{dT} = \frac{870\text{ nm}}{(3.7)}\left(1.5 \times 10^{-4}\,\text{K}^{-1}\right) = 0.035\text{ nm K}^{-1}.$$

Note that we have used n for a passive cavity whereas n above should be the effective refractive index of the *active* cavity which will also depend on the optical gain of the medium, and hence its temperature dependence is likely to be somewhat higher than the dn/dT value we used.

4.10 STEADY STATE SEMICONDUCTOR RATE EQUATIONS

Consider a double heterostructure laser diode under forward bias as in Figure 4.19. The current carries the electrons into the active layer where they recombine with holes radiatively. If d is the thickness, L is the length and W is the width of the active layer then, under steady state operation, the rate of electron injection into the active layer by the current I is equal to their rate of recombination by spontaneous and stimulated emissions (neglecting nonradiative recombinations).

Rate of electron injection by current
= Rate of spontaneous emissions + Rate of stimulated emissions

that is,

Active layer rate equation

$$\frac{I}{edLW} = \frac{n}{\tau_{sp}} + CnN_{ph} \quad \textbf{(1)}$$

where n is the injected electron concentration and N_{ph} is the coherent photon concentration in the active layer, τ_{sp} is the average time a for spontaneous recombination and C is a constant (depends on B_{21}). The second term represents the stimulated emission rate which depends on the concentration of available electrons in the conduction band n and the coherent photon concentration N_{ph} in the active layer. N_{ph} considers only those coherent photons encouraged by the optical cavity, that is, a mode of the cavity. As the current increases and provides more pumping, N_{ph} increases (helped by the optical cavity), and eventually the stimulated term dominates the spontaneous term (as in Figure 4.17). The output light power P_o is proportional to N_{ph}.

Consider the coherent photon concentration N_{ph} in the cavity. Under steady state conditions:

Rate of coherent photon loss in the cavity = Rate of stimulated emissions

that is,

$$\frac{N_{ph}}{\tau_{ph}} = CnN_{ph} \quad \textbf{(2)}$$

where τ_{ph} is the average time for a photon to be lost from the cavity due to transmission through the end-faces, scattering and absorption in the semiconductor. If α_t is the total attenuation coefficient representing all these loss mechanisms, then the power in a light wave, in the absence of amplification, decreases as $\exp(-\alpha_t x)$ which is equivalent to a decay in time as $\exp(-t/\tau_{ph})$ where $\tau_{ph} = n/(c\alpha_t)$ and n is the refractive index.

In semiconductor laser science, the threshold electron concentration n_{th} and threshold current I_{th} refers to that condition when the stimulated emission just overcomes the spontaneous emission and the total loss mechanisms in τ_{ph}. This occurs when the injected n reaches n_{th}, the threshold concentration. From Eq. (2), this is when

Threshold concentration

$$n_{th} = \frac{1}{C\tau_{ph}} \quad \textbf{(3)}$$

This is the point when coherent radiation gain in the active layer by stimulated emission just balances all the cavity losses (represented by τ_{ph}) plus losses by spontaneous emission which is random. When the current exceeds I_{th}, the output optical power increases sharply with the current (Figure 4.17) so we can just as well take $N_{ph} = 0$ when $I = I_{th}$ in Eq. (1) which gives

Threshold current

$$I_{th} = \frac{n_{th}edLW}{\tau_{sp}} \quad \textbf{(4)}$$

Clearly the threshold current decreases with d, L and W which explains the reasons for the heterostructure and stripe geometry lasers and the avoidance of the homojunction laser.

When the current exceeds the threshold current, the excess carriers above n_{th} brought in by the current recombine by stimulated emission. The reason is that above threshold, the active layer has optical gain and therefore builds up coherent radiation quickly and stimulated emission depends on N_{ph}. The steady state electron concentration remains constant at n_{th} though the rates of carrier injection and stimulated recombination have increased. Above threshold, from Eq. (1) with n clamped at n_{th},

$$\frac{I - I_{th}}{edLW} = Cn_{th}N_{ph} \tag{5}$$

so that using Eq. (3) and defining $J = I/WL$ we can find N_{ph}

Coherent photon concentration

$$N_{ph} = \frac{\tau_{ph}}{ed}(J - J_{th}) \tag{6}$$

To find the optical output power P_o consider the following. It takes $\Delta t = nL/c$ seconds for photons to cross the laser cavity length L. Only half of the photons, $(1/2)\,N_{ph}$, in the cavity would be moving towards the output face of the crystal at any instant. Only a fraction $(1 - R)$ of the radiation power will escape. Thus, the output optical power P_o is

$$P_o = \frac{(\frac{1}{2}N_{ph})(\text{Cavity Volume})(\text{Photon energy})}{\Delta t}(1 - R)$$

and using N_{ph} from Eq. (6) we obtain the **laser diode equation**

Laser diode equation

$$P_o = \left[\frac{hc^2\tau_{ph}W(1 - R)}{2en\lambda}\right](J - J_{th}) \tag{7}$$

The above semiquantitative steady state approach is a special case of the more general semiconductor rate equations described in more advanced textbooks where the time response of the laser diode is also analyzed. The key conclusions are embedded in Eqs. (3), (4) and (7), which determine I_{th} and the theoretical coherent light output power vs. diode current behavior as illustrated in Figure 4.25.

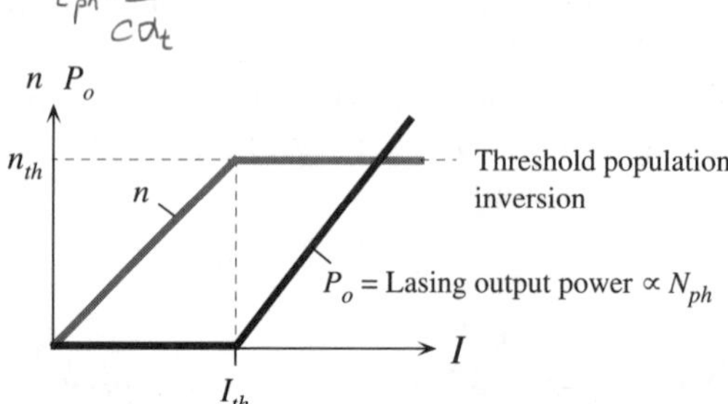

FIGURE 4.25 Simplified and idealized description of a semiconductor laser diode based on rate equations. Injected electron concentration n and coherent radiation output power P_o vs. diode current I.

4.11 LIGHT EMITTERS FOR OPTICAL FIBER COMMUNICATIONS

The type of light source suitable for optical communications depends not only on the communication distance but also on the bandwidth requirement. For short haul applications, for example, local networks, LEDs are preferred as they are simpler to drive, more economic, have a longer lifetime and provide the necessary output power even though their output spectrum is much wider than that of a laser diode. LEDs are typically used with multimode and graded index fibers because the dispersion arising from the finite linewidth $\Delta\lambda$ of the output spectrum is not a major concern with these fibers. For long-haul and wide bandwidth communications invariably laser diodes are used because of their narrow linewidth and high output power.

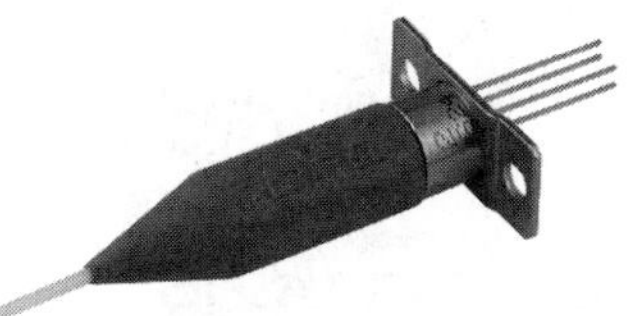

A laser diode pigtailed to a fiber. Two of the leads are for a back-facet photodetector to allow the monitoring of the laser output power. *(Courtesy of Alcatel)*

Figure 4.26 and Table 4.1 compare the characteristics of typical LED and laser diode sources. The linewidth $\Delta\lambda$ of the output spectrum is obviously narrowest for laser diodes since the optical cavity and the optical gain characteristics of the laser diode structure define the resonant wavelengths that lead to lasing emission. By allowing only one mode to exist by suitably suppressing the unwanted modes through design, the output spectrum of the laser diode can be made to be very narrow, *e.g.* 0.01–0.1nm. The present technological drive in solid state laser design is to achieve what is called single frequency operation in which the emitted radiation has a very narrow bandwidth $\Delta\lambda$, typically less than 0.01 nm. The description of these lasers is beyond the scope of this book. The speed of response of an emitter is generally described by a rise time. If the driving

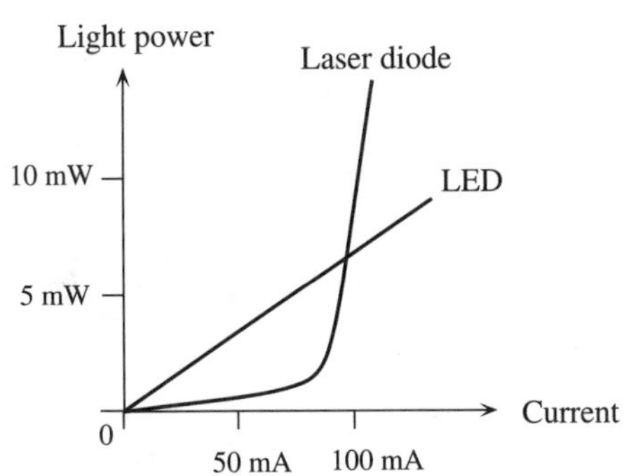

FIGURE 4.26 Typical optical power output vs. forward current for an LED and a laser diode.

TABLE 4.1 Typical characteristics of LEDs and Laser diodes for 1.3 μm emission. Rise time is the time it takes for the output optical power to rise from 10% to 90% in response to a step current input.

	LED	Laser diode
Structure	Double heterojunction	Double heterojunction
Material	InGaAsP on InP	InGaAsP on InP
Output radiation	Incoherent (Spontaneous emission)	Coherent (Stimulated emission)
Typical spectral linewidth, $\Delta\lambda$	100 nm	2–4 nm (multimode laser) < 0.1 nm (single mode laser)
Rise time	5–20 ns	< 1 ns

current is applied suddenly as a step to the diode, then the **rise time** is the time it takes for the light output to rise from 10% to 90% of the final value. Laser diodes have shorter rise times and hence are used whenever wide bandwidths are required.

4.12 SINGLE FREQUENCY SOLID STATE LASERS

Ideally the output spectrum from a laser device should be as narrow as possible which generally means that we have to allow only a single mode to exist. There are a number of device structures that operate with an output spectrum that has high modal purity.

One method of ensuring a single mode of radiation in the laser cavity is to use frequency selective dielectric mirrors at the cleaved surfaces of the semiconductor. The **distributed Bragg reflector**, as shown in Figure 4.27 (a), is a mirror that has been designed like a reflection type diffraction grating; it has a periodic corrugated structure. Intuitively, partial reflections of waves from the corrugations interfere constructively (that is reinforce each other) to give a reflected wave only when the wavelength corresponds to twice the corrugation periodicity as illustrated in Figure 4.27 (b). For example, two partially reflected waves such as A and B have an optical path difference of 2Λ where Λ is the corrugation period. They can only interfere constructively if 2Λ is a multiple of the wavelength within the medium. Each of these wavelengths is called a **Bragg wavelength** λ_B and given by the condition for in-phase interference,

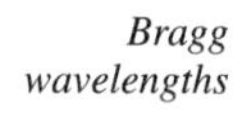
Bragg wavelengths

$$q\frac{\lambda_B}{n} = 2\Lambda \tag{1}$$

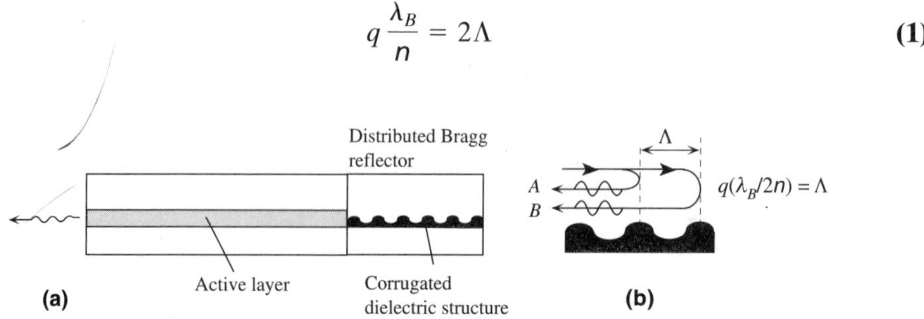

FIGURE 4.27 (a) Distributed Bragg reflection (DBR) laser principle. (b) Partially reflected waves at the corrugations can only constitute a reflected wave when the wavelength satisfies the Bragg condition. Reflected waves A and B interfere constructively when $q(\lambda_B/2n) = \Lambda$.

where n is the refractive index of the corrugated material and $q = 1, 2, \ldots$ is an integer called the **diffraction order**. The DBR therefore has a high reflectance around λ_B but low reflectance away from λ_B. The result is that only that particular Fabry-Perot cavity mode, within the optical gain curve, that is close to λ_B can lase and exist in the output. (The exact calculation of the mode frequency is beyond the scope of this book.)

In a normal laser, the crystal faces provide the necessary optical feedback into the cavity to build up the photon concentration. In the **distributed feedback (DFB)** laser, as shown in Figure 4.28 (a), there is a corrugated layer, called the **guiding layer**, next to the active layer; radiation spreads from the active layer to the guiding layer. These corrugations in the refractive index act as optical feedback over the length of the cavity by producing partial reflections. Thus optical feedback is *distributed* over the cavity length. Intuitively, we might infer that only those Bragg wavelengths λ_B related to the corrugation periodicity Λ as in Eq. (1) can interfere constructively and thereby exist in the cavity in a similar fashion to the depiction in Figure 4.27 (b). However, the operating principle of the DFB laser is totally different. Radiation is fed from the active into the guiding layer along the whole cavity length so that the corrugated medium can be thought of as possessing an optical gain. Partially reflected waves experience gain and we cannot simply add these without considering the optical gain and also possible phase changes (Eq. (1) assumes normal incidence and ignores any phase change on reflection). A left-traveling wave in the guiding layer experiences partial reflections and these reflected waves are optically amplified by the medium to constitute a right-going wave.

In a Fabry-Perot cavity, a wave that is traveling towards the right, becomes reflected to travel towards the left. At any point in the cavity, as a result of end-reflections, we therefore have these right and left traveling waves interfering, or being "coupled". These oppositely traveling waves, assuming equal amplitudes, can only set up a standing wave, a mode, if they are *coherently coupled*, which requires that the round-trip phase change is 2π. In the DFB structure, traveling waves are reflected partially and periodically as they propagate. The left and right traveling waves can only coherently couple to set up a mode if their frequency is related to the corrugation periodicity Λ, taking into account that the medium alters the wave-amplitudes via optical gain.[20]

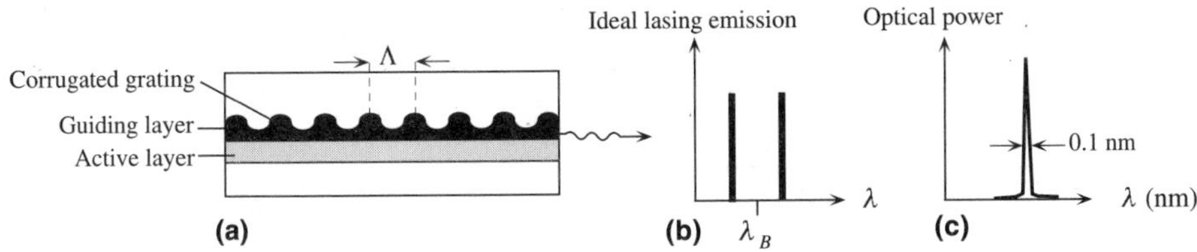

FIGURE 4.28 (a) Distributed feedback (DFB) laser structure. (b) Ideal lasing emission output. (c) Typical output spectrum from a DFB laser.

[20]These partially reflected waves travel in a medium that has the refractive index modulated periodically (with periodicity Λ). We have to consider how left and right propagating waves in this corrugated structure are coupled. The rigorous theory is beyond the scope of this book.

The allowed DFB modes are not exactly at Bragg wavelengths but are symmetrically placed about λ_B. If λ_m is an allowed DFB lasing mode then

DFB Laser wavelengths

$$\lambda_m = \lambda_B \pm \frac{\lambda_B^2}{2nL}(m+1) \qquad (2)$$

where m is a mode integer, 0, 1, 2, ..., and L is the effective length of the diffraction grating (corrugation length). The relative threshold gain for higher modes is so large that only the $m = 0$ mode effectively lases. A perfectly symmetric device has two equally spaced modes placed around λ_B as in Figure 4.28 (b). In reality, either inevitable asymmetry introduced by the fabrication process, or asymmetry introduced on purpose, leads to only one of the modes to appear as shown in Figure 4.28 (c). Further, typically the corrugation length L is so much larger than the period Λ that the second term in Eq. (2) is very small and the emission is very close to λ_B. There are various commercially available single mode DFB lasers in the market with spectral widths of ~0.1 nm at the communications channel of 1.55 μm.

In the **cleaved-coupled-cavity** (C^3) laser device, two different laser optical cavities L and D (different lengths) are coupled as shown in Figure 4.29 (a). The two lasers are pumped by different currents. Only those waves that can exist as modes in *both* cavities are now allowed because the system has been coupled. In this example, the modes in L are more closely spaced than modes in D. These two different set of modes coincide only at far spaced intervals as indicated in Figure 4.29 (b). This restriction in possible modes in the combined cavity and the wide separation between the modes, results in a single mode operation. (Why do you need to pump both cavities?)

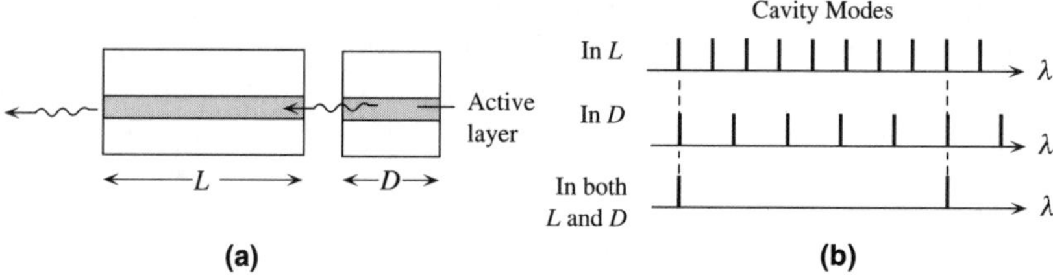

FIGURE 4.29 Cleaved-coupled-cavity (C^3) laser

EXAMPLE 4.12.1 DFB Laser

Consider a DFB laser that has a corrugation period Λ of 0.22 μm and a grating length of 400 μm. Suppose that the effective refractive index of the medium is 3.5. Assuming a first order grating, calculate the Bragg wavelength, the mode wavelengths and their separation.

Solution The Bragg wavelength is

$$\lambda_B = \frac{2\Lambda n}{q} = \frac{2(0.22\ \mu\text{m})(3.5)}{1} = 1.540\ \mu\text{m}.$$

and the symmetric mode wavelengths about λ_B are

$$\lambda_m = \lambda_B \pm \frac{\lambda_B^2}{2nL}(m+1) = 1.54 \pm \frac{(1.54\ \mu\text{m})^2}{2(3.5)(400\ \mu\text{m})}(0+1)$$

so that the $m = 0$ mode wavelengths are

$$\lambda_0 = 1.539 \text{ or } 1.508\ \mu\text{m}.$$

The two are separated by 0.0017 μm, or 1.7 nm. Due some asymmetry, only one mode will appear in the output and for most practical purposes the mode wavelength can be taken as λ_B.

4.13 QUANTUM WELL DEVICES

A typical **quantum well** device has an ultra thin, typically less than 50 nm, narrow bandgap semiconductor, such as GaAs, sandwiched between two wider bandgap semiconductors, such as AlGaAs, as depicted in Figure 4.30 (a) which is a **heterostructure device**. We assume that the two semiconductors are lattice matched in the sense that they have the same lattice parameter a. This means that interface defects due to mismatch of crystal dimensions between the two semiconductor crystals are minimal. Since the bandgap, E_g, changes at the interface, there are discontinuities in E_c and E_v at the interfaces. These discontinuities, ΔE_c and ΔE_v, depend on the semiconductor materials and their doping. In the case of GaAs/AlGaAs heterostructure shown in the figure, ΔE_c is greater than ΔE_v. Very approximately, the change from the wider E_{g2} to narrower E_{g1} is proportioned 60% to ΔE_c and 40% to ΔE_v as shown in Figure 4.30 (b). Because of the potential energy barrier, ΔE_c, conduction electrons in the thin GaAs layer are confined in the x-direction. This confinement length d is so small that we can treat the electron as in a one-dimensional potential energy (PE) well in the x-direction but as if it were free in the yz plane.

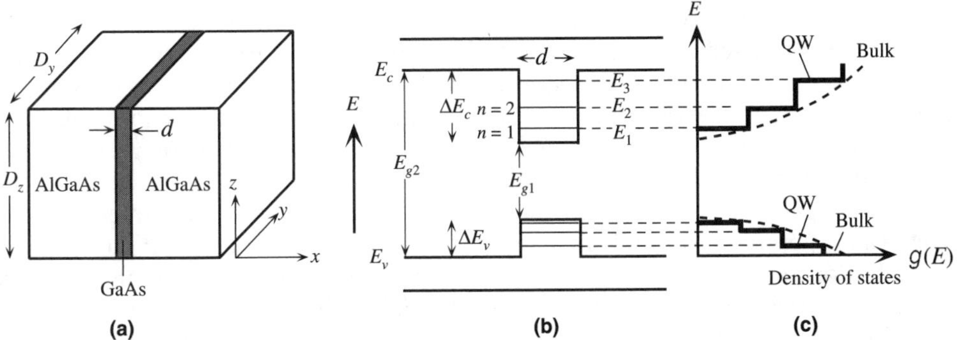

FIGURE 4.30 A quantum well (QW) device. (a) Schematic illustration of a quantum well (QW) structure in which a thin layer of GaAs is sandwiched between two wider bandgap semiconductors (AlGaAs). (b) The conduction electrons in the GaAs layer are confined (by ΔE_c) in the x-direction to a small length d so that their energy is quantized. (c) The density of states of a two-dimensional QW. The density of states is constant at each quantized energy level.

We can appreciate the confinement effect by considering the energy of the conduction electron that is bound by the size of the GaAs layer which is d along x and D_y and D_z along y and z as shown in Figure 4.30 (a). The energy of the conduction electron will be the same as that in a three dimensional PE well of size d, D_y and D_z and given by

Energy in a quantum well

$$E = E_c + \frac{h^2 n^2}{8m_e^* d^2} + \frac{h^2 n_y^2}{8m_e^* D_y^2} + \frac{h^2 n_z^2}{8m_e^* D_z^2} \quad \textbf{(1)}$$

where n, n_y and n_z are quantum numbers having the values $1, 2, 3, \ldots$. The reason for the E_c in Eq. (1) is that the potential energy barriers are defined with respect to E_c. These PE barriers are ΔE_c along x and electron affinity (energy required to take the electron from E_c to vacuum) along y and z. But D_y and D_z are orders of magnitude greater than d so that the minimum energy, denoted as E_1, is determined by the term with n and d, the energy associated with motion along x. The minimum energy E_1 corresponds to $n = 1$ and is above E_c of GaAs as shown in Figure 4.30 (b). The separation between the energy levels identified by n_y and n_z and associated with motion in the yz plane is so small that the electron is free to move in the yz plane as if it were in the bulk semiconductor. We therefore have a two-dimensional electron gas which is confined in the x-direction. The holes in the valence band are confined by the potential energy barrier ΔE_v (hole energy is in the opposite direction to electron energy) and behave similarly as indicated in Figure 4.30 (b).

The density of electronic states for the two dimensional electron system is not the same as that for the bulk semiconductor. For a given electron concentration n, the density of states $g(E)$, number of quantum states per unit energy per unit volume, is constant and does not depend on the energy. The density of states for the confined electron and that in the bulk semiconductor are shown schematically in Figure 4.30 (c). $g(E)$ is constant at E_1 until E_2 where it increases as a step and remains constant until E_3 where again it increases as a step by the same amount and at every value of E_n. Density of states in the valence band behaves similarly as also shown in Figure 4.30 (c).

Since at E_1 there is a finite and substantial density of states, the electrons in the conduction band do not have to spread far in energy to find states. In the bulk semiconductor, on the other hand, the density of states at E_c is zero and increases slowly with energy (as $E^{1/2}$) which means that the electrons are spread more deeply into the conduction band in search for states. A large concentration of electrons can easily occur at E_1 whereas this is not the case in the bulk semiconductor. Similarly, the majority of holes in the valence band will be at the minimum hole energy E_1' since there are sufficient states at this energy; see Figure 4.31. Under a forward bias electrons are injected into the conduction band of the GaAs layer which serves as the active layer. The injected electrons readily populate the ample number of states at E_1 which means that the electron concentration at E_1 increases rapidly with the current and hence population inversion occurs quickly without the need for a large current to bring in a great number of electrons. Stimulated transitions of electrons from E_1 to E_1' leads to a lasing emission as depicted in Figure 4.31. There are two distinct advantages. First is that the threshold current for population inversion and hence lasing emission is markedly reduced with respect to that for bulk semiconductor devices. For example in a **single quantum well** (SQW) laser this is typically in the range 0.5–1 mA whereas in a double heterostructure laser the threshold current is in the range 10–50 mA. Secondly, since majority of the electrons

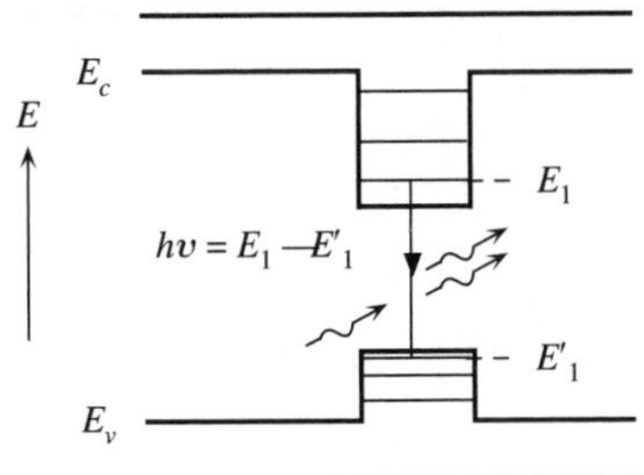

FIGURE 4.31 In single quantum well (SQW) lasers electrons are injected by the forward current into the thin GaAs layer which serves as the active layer. Population inversion between E_1 and E'_1 is reached even with a small forward current which results in stimulated emissions.

are at and near E_1 and holes are at and near E'_1, the range of emitted photon energies are very close to $E_1 - E'_1$. Consequently the spread in the wavelength, the linewidth, in the output spectrum is substantially narrower than that in bulk semiconductor lasers.

The advantages of the SQW can be extended to a larger volume of the crystal by using multiple quantum wells. In **multiple quantum well** (MQW) lasers, the structure has alternating ultrathin layers of wide and narrow bandgap semiconductors as schematically sketched in Figure 4.32. The smaller bandgap layers are the active layers where electron confinement and lasing transition take place whereas the wider bandgap layers are the barrier layers.

Although the optical gain curve is narrower than the corresponding bulk device, the output spectrum from a quantum well device is not necessarily a single mode. The number of modes depends on the indiviual widths of the quantum wells. It is, of course possible, to combine a MQW design with a distribued feedback structure to obtain a single mode operation. Many commercially available LDs are currently MQW devices.

A 1550 nm MQW-DFB InGaAsP laser diode pigtail-coupled to a fiber. *(Courtesy of Alcatel.)*

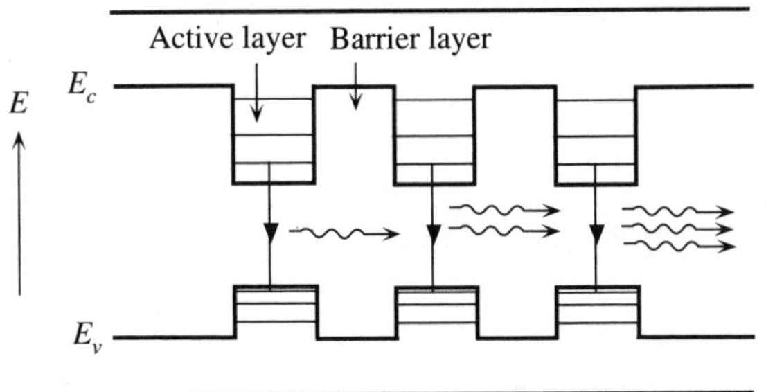

FIGURE 4.32 A multiple quantum well (MQW) structure. Electrons are injected by the forward current into active layers which are quantum wells.

EXAMPLE 4.13.1 A GaAs quantum well

Consider a GaAs quantum well. Effective mass of a conduction electron in GaAs is $0.07m_e$ where m_e is the electron mass in vacuum. Calculate the first two electron energy levels for a quantum well of thickness 10 nm. What is the hole energy, below E_v of GaAs, if the effective mass of the hole is about $0.50m_e$?

What is the change in the emission wavelength with respect to bulk GaAs which has an energy bandgap of 1.42 eV?

Solution The lowest energy levels with respect to the CB edge E_c in GaAs are determined by the energy of an electron in a one-dimensional potential energy well[21]

$$\varepsilon_n = \frac{h^2 n^2}{8m_e^* d^2}$$

where n is a quantum number $1, 2, \ldots$, ε_n is the electron energy with respect to E_c in GaAs, or $\varepsilon_n = E_n - E_c$ in Figure 4.30 (b). Using $d = 10 \times 10^{-9}$ m, $m_e^* = 0.07m_e$ and $n = 1$ and 2, we find, $\varepsilon_1 = 0.0537$ eV and $\varepsilon_2 = 0.215$ eV respectively.

The hole energy levels below E_v in Figure 4.30 (b) are given by

$$\varepsilon'_n = \frac{h^2 n^2}{8m_h^* d^2}$$

Using $d = 10 \times 10^{-9}$ m, $m_h^* \approx 0.5m_e$ and $n = 1$, we find, $\varepsilon'_1 = 0.0075$ eV. The wavelength of emission from bulk GaAs with $E_g = 1.42$ eV is

$$\lambda_g = \frac{hc}{E_g} = \frac{(6.626 \times 10^{-34})(3 \times 10^8)}{(1.42)(1.602 \times 10^{-19})} = 874 \times 10^{-9}\ \text{m}\ (874\ \text{nm})$$

Whereas from the GaAs QW, the wavelength is,

$$\lambda_{QW} = \frac{hc}{E_g + \varepsilon_1 + \varepsilon'_1} = \frac{(6.626 \times 10^{-34})(3 \times 10^8)}{(1.42 + 0.0537 + 0.0075)(1.602 \times 10^{-19})}$$
$$= 839 \times 10^{-9}\ \text{m}\ (839\ \text{nm})$$

The difference is $\lambda_g - \lambda_{QW} = 35$ nm.

---

[21] See, for example, *Principles of Electronic Materials and Devices, Second Edition*, S. O. Kasap (McGraw-Hill, 2001), Ch.3.

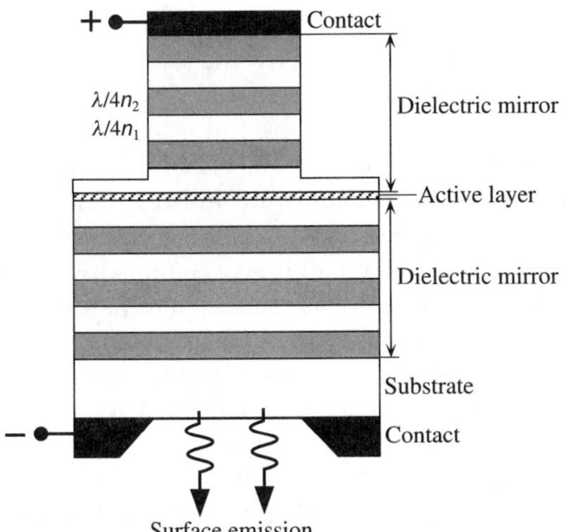

FIGURE 4.33 A simplified schematic illustration of a vertical cavity surface emitting laser (VCSEL).

4.14 VERTICAL CAVITY SURFACE EMITTING LASERS (VCSELs)

Figure 4.33 shows the basic concept of the **vertical cavity surface emitting laser** (VCSEL). A VCSEL has the optical cavity axis along the direction of current flow rather than perpendicular to the current flow as in conventional laser diodes. The active region length is very short compared with the lateral dimensions so that the radiation emerges from the "surface" of the cavity rather than from its edge. The reflectors at the ends of the cavity are **dielectric mirrors** made from alternating high and low refractive index quarter-wave thick multilayers. Such dielectric mirrors provide a high degree of wavelength selective reflectance at the required free-space wavelength λ if the thicknesses of alternating layers d_1 and d_2 with refractive indices n_1 and n_2 are such that

$$n_1 d_1 + n_2 d_2 = \tfrac{1}{2}\lambda \tag{1}$$

which then leads to the constructive interference of all partially reflected waves at the interfaces.[22] Since the wave is reflected because of a periodic variation in the refractive index as in a grating, the dielectric mirror is essentially a **distributed Bragg reflector** (DBR). The wavelength in Eq. (1) is chosen to coincide with the optical gain of the active layer. High reflectance end mirrors are needed because the short cavity length L reduces the optical gain of the active layer inasmuch as the optical gain is proportional to $\exp(gL)$ where g is the optical gain coefficient. There may be 20–30 or so layers in the

[22] Such dielectric mirrors were explained in Example 1.6.5 in Ch. 1.

dielectric mirrors to obtain the required reflectance (~99%). The whole optical cavity in Figure 4.33 looks "vertical" if we keep the current flow the same as in a conventional laser diode cavity.

The active layer is generally very thin (< 0.1 μm) and is likely to be a multiple quantum well (MQW) for improved threshold current. The required semiconductor layers are grown by epitaxial growth on a suitable substrate which is transparent in the emission wavelength. For example, a 980 nm emitting VCSEL device has InGaAs as the active layer to provide the 980 nm emission, and a GaAs crystal is used as substrate which is transparent at 980 nm. The dielectric mirrors are then alternating layers of AlGaAs with different compositions and hence different bandgaps and refractive indices. The top dielectric mirror is etched after all the layers have been epitaxially grown on the GaAs substrate to arrive at the structure shown in Figure 4.33 (which is still highly simplified). In practice, the current flowing through the dielectric mirrors give rise to an undesirable voltage drop and methods are used to feed the current into the active region more directly, for example, by depositing "peripheral" contacts close to the active region. There are various sophisticated VCSEL structures and Figure 4.33 is only one simplified example.

The vertical cavity is generally circular in its cross section so that the emitted beam has a **circular cross-section**, which is an advantage. The height of the vertical cavity may be as small as several microns. Therefore the longitudinal mode separation is sufficiently large to allow only one longitudinal mode to operate. However, there may be one or more lateral (transverse) modes depending on the lateral size of the cavity. In practice there is only one single lateral mode (and hence one mode) in the output spectrum for cavity diameters less than ~8 μm. Various VCSELs in the market have several lateral modes but the spectral width is still only ~0.5 nm, substantially less than a conventional longitudinal multimode laser diode.

With cavity dimensions in the microns range, such a laser is referred to as a **microlaser**. One of the most significant advantages of microlasers is that they can be arrayed to construct a **matrix emitter** that is a broad area surface emitting laser source. Such laser arrays have important potential applications in optical interconnect and optical computing technologies. Further, such laser arrays can provide a higher optical power than that available from a single conventional laser diode. Powers reaching a few watts have been demonstrated using such matrix lasers.

An 850 nm VCSEL diode. *(Courtesy of Honeywell.)*

SEM (scanning electron microscope) of the first low-threshold VCSELs developed at Bell Laboratories in 1989. The largest device area is 5 μm in diameter *(Courtesy of Alex Scherer, CalTech)*

4.15 OPTICAL LASER AMPLIFIERS

A semiconductor laser structure can also be used as an optical amplifier that amplifies light waves passing through its active region as illustrated in Figure 4.34 (a). The wavelength of radiation to be amplified must fall within the optical gain bandwidth of the laser. Such a device would not be a laser oscillator, emitting lasing emission without an input, but an optical amplifier with input and output ports for light entry and exit. In the **traveling wave semiconductor laser** amplifier the ends of the optical cavity have antireflection (AR) coatings so that the optical cavity does not act as an efficient optical resonator, a condition for laser-oscillations. Light, for example, from an optical fiber, is

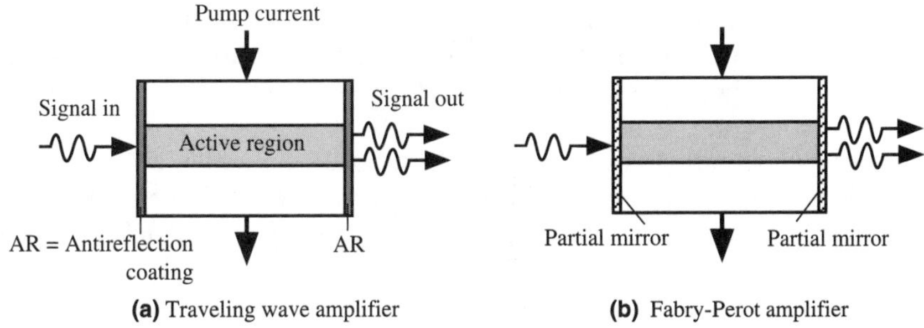

(a) Traveling wave amplifier **(b)** Fabry-Perot amplifier

FIGURE 4.34 Simplified schematic illustrations of two types of laser amplifiers.

coupled into the active region of the laser structure. As the radiation propagates through the active layer, optically guided by this layer, it becomes amplified by the induced stimulated emissions, and leaves the optical cavity with a higher intensity. Obviously the device must be pumped to achieve optical gain (population inversion) in the active layer. Random spontaneous emissions in the active layer feed "noise" into the signal and broaden the spectral width of the passing radiation. This can be overcome by using an optical filter at the output to allow the original light wavelength to pass through. Typically, such laser amplifiers are buried heterostructure devices and have optical gains of ~20 dB depending on the efficiency of the AR coating.

The **Fabry-Perot laser amplifier**, as shown in Figure 4.34 (b), is similar to the conventional laser oscillator, but is operated below the threshold current for lasing oscillations; the active region has an optical gain but not sufficient to sustain a self-lasing output. Light passing through such an active region will be amplified by stimulated emissions but, because of the presence of an optical resonator, there will be internal multiple reflections. These multiple reflections lead to the gain being highest at the resonant frequencies of the cavity within the optical gain bandwidth. Optical frequencies around the cavity resonant frequencies will experience higher gain than those away from resonant frequencies. Although the Fabry-Perot laser amplifier can have a higher gain than the traveling wave amplifier, it is less stable.

A 1550 nm semiconductor optical amplifier using an InGaAsP chip. *(Courtesy of Alcatel)*

4.16 HOLOGRAPHY

Holography is a technique of reproducing three dimensional optical images of an object by using a highly coherent radiation from a laser source. Although holography was invented by Denis Gabor in 1948 prior to the availability of highly coherent laser beams, its practical use and popularity increased soon after the commercial development of lasers.

The principle of holography is illustrated in Figure 4.35 (a) and (b). The object, a cat, is illuminated by a highly coherent beam as normally would be available from a laser[23] as in Figure 4.35 (a). The laser beam has both *spatial* and *temporal coherence.*

[23] A short pulse of laser light, such as from a ruby laser, will not harm the cat; this is how holograms of people are taken. The first successful laser holography was carried out by Juris Upatneiks and Emmett Leith in 1960.

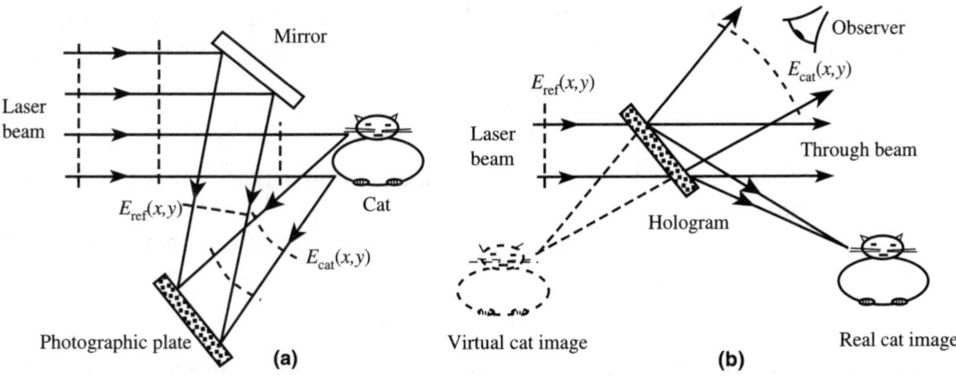

FIGURE 4.35 A highly simplified illustration of holography. (a) A laser beam is made to interfere with the diffracted beam from the subject to produce a hologram. (b) Shining the laser beam through the hologram generates a real and a virtual image.

Part of the coherent wave is reflected from a mirror to form a **reference beam** E_{ref} and travels towards a fine-grained photographic plate. The waves that are reflected from the cat, E_{cat}, will have both amplitude and phase variations that represent the cat's surface (topology). If we were looking at the cat, our eyes would register the wavefront of the reflected wave E_{cat}. Moving our head around we would capture different portions of the reflected wave and we would see the cat as a three dimensional object.

The reflected wave from the cat (E_{cat}) is made to interfere with the reference wave (E_{ref}) at the photographic plate and give rise to a complicated *interference pattern* that depends on the magnitude and phase variation in E_{cat}. The recorded interference pattern in the photographic film (after processing) is called a **hologram**. It contains all the information necessary to reconstruct the wavefront E_{cat} reflected from the cat and hence produce a three dimensional image.

To obtain the three dimensional image, we have to illuminate the hologram with the reference beam E_{ref} alone as in Figure 4.35 (b). Most of the beam goes right through but some of it becomes *diffracted* by the interference pattern in the hologram. One diffracted beam is an exact replica of the original wavefront E_{cat} from the cat. The observer sees this wavefront as if the waves were reflected from the original cat and registers a three dimensional image of the cat. This is the **virtual image**. We know that first order diffraction from a grating has to satisfy the Bragg condition, $d \sin\theta = m\lambda (m = \pm 1)$ where λ is the wavelength and d is the separation of the slits. We can qualitatively think of a diffracted beam from one locality in the hologram as being determined by the local separation d between interference fringes in this region. Since d changes in the hologram depending on the interference pattern produced by E_{cat}, the whole diffracted beam depends on E_{cat} and the diffracted beam wavefront is an exact scaled replica of E_{cat}. Just as normally there would be another diffracted beam on the other side of the zero-order (through) beam, there is a second image, called the **real image**, as in Figure 4.35 (b), which is of lower quality. (It may help to imagine what happens if the object consists of black and white stripes. We would then obtain periodic dark and bright interference fringes in the hologram. This periodic variation is just like a diffraction grating; the exact analysis is more complicated).

Holography can be explained by the following highly simplified analysis. Suppose that the photographic plate is in the xy plane in Figure 4.35. Assume that the reference wave can be represented by

$$E_{\text{ref}}(x, y) = U_r(x, y)e^{j\omega t} \tag{1}$$

where $U_r(x,y)$ is its amplitude, generally a complex number that includes magnitude and phase information.

The reflected wave from the cat will have a complex amplitude that contains the magnitude and phase information of the cat's surface, *i.e.*

$$E_{\text{cat}}(x, y) = U(x, y)e^{j\omega t} \tag{2}$$

where $U(x, y)$ is a complex number. Normally we would look at the cat and construct the image from the wavefront in Eq. (2). We thus have to reconstruct the wavefront in Eq. (2), *i.e.* obtain U(x, y).

When E_{ref} and E_{cat} interfere, the "brightness" of the photographic image depends on the intensity and hence on

$$I(x, y) = |E_{\text{ref}} + E_{\text{cat}}|^2 = |U_r + U|^2 = (U_r + U)(U_r^* + U^*)$$

where we have used the normal definition that the square of the magnitude of a complex number is product with its complex conjugate (indicated by an asterisk). Thus,

$$I(x, y) = UU^* + U_rU_r^* + U_r^*U + U_rU^* \tag{3}$$

This is the pattern on the photographic plate. The first and second terms are the intensities of the reflected and reference waves. It is the third and fourth terms that contain the information on the magnitude and phase of $U(x, y)$.

We now illuminate the hologram, $I(x, y)$, with the reference beam U_r so that the transmitted light wave has a complex magnitude $U_t(x, y)$ proportional to $U_rI(x, y)$,

$$U_t \propto U_rI(x, y) = U_r[UU^* + U_rU_r^* + U_r^*U + U_rU^*]$$

i.e

$$U_t \propto U_r(UU^* + U_rU_r^*) + (U_rU_r^*)U + U_r^2U^* \tag{4}$$

or

$$U_t \propto a + bU(x, y) + cU^*(x, y)$$

where a, b and c are appropriate constants. The first term $a = U_r(UU^* + U_rU_r^*)$ is the through beam. The second and third terms represent the diffracted beams.[24] Since $b = (U_rU_r^*)$ is a constant (intensity of the reference beam), the second term is a scaled version of U(x, y) and represents the original wavefront amplitude from the cat (includes the magnitude and phase information). We have effectively reconstructed the original wavefront and the observer will see a three dimensional view of the cat within the angle of original recording; $bU(x, y)$ is the **virtual image**. The third term $cU^*(x, y)$ is the **real image** and notice that its amplitude is the complex conjugate of $U(x, y)$; it is

[24] This is a simple factual statement as the mathematical treatment above has not proved this point.

called the **conjugate image**. It is apparent that holography is a method of **wavefront reconstruction**. It should be mentioned that the observer will always see the a positive image whether a positive or a negative image of the hologram is used. A negative hologram simply produces a 180° phase shift in the field of the transmitted wave and the eye cannot detect this phase shift.

Dennis Gabor (1900–1979), inventor of holography, is standing next to his holographic portrait. Professor Gabor was a Hungarian born British physicist who published his holography invention in *Nature* in 1948 while he was at Thomson-Houston Co. Ltd, at a time when coherent light from lasers was not yet available. He was subsequently a professor of applied electron physics at Imperial College, University of London. *(From M.D.E.C. Photo Lab, Courtesy AIP Emilio Segrè Visual Archives, AIP.)*

QUESTIONS AND PROBLEMS

4.1 The He-Ne Laser The He-Ne laser system energy levels can be quite complicated as shown in Figure 4.36. There are a number of lasing emissions in the laser output in the red (632.8 nm), green (543.5 nm), orange (612 nm), yellow (594.1 nm), and in the infrared regions at 1.52 μm and 3.39 μm, which give the He-Ne laser its versatility. The pumping mechanism for all these lasers operations is the same, energy transfer from excited He atoms to Ne atoms by atomic collisions in the gas discharge tube. There are two excited states for the He atom in the configuration $1s^1 2s^2$. The configuration with electron spins parallel has a lower energy than that with opposite electron spins as indicated in Figure 4.36. These two He states can excite Ne atoms to either the $2p^5 4s^1$ or $2p^5 5s^1$. There is then a population inversion between these levels and the $2p^5 3p^1$ and between $2p^5 5s^1$ and $2p^5 4p^1$ which leads to the above lasing transitions. Generally, we do not have a single discrete level for the energy of a many-electron atom of given n, l configuration. For example, the atomic configuration $2p^5 3p^1$ has 10 closely spaced energy levels resulting from various different values of m_l and m_s that can be assigned to the sixth excited electron $(3p^1)$ and the

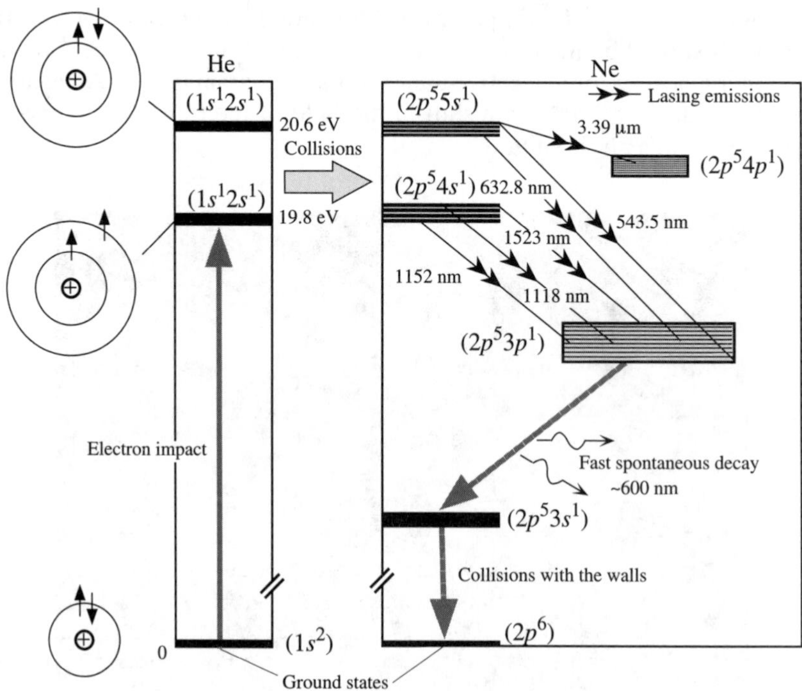

FIGURE 4.36 Various lasing transitions in the He-Ne laser.

remaining electrons $(2p^5)$ within quantum mechanical rules. Not all transitions to these energy levels are allowed as photon emission requires quantum number selection rules to be obeyed.

(a) Table 4.2 shows some typical commercial He-Ne laser characteristics (within 30-50%) for various wavelengths. Calculate the overall efficiencies of these lasers.

TABLE 4.2 Typical commercial He-Ne laser characteristics

Wavelength (nm)	543.5 Green	594.1 Yellow	612 Orange	632.8 Red	1523 Infrared
Optical output power (mW)	1.5	2	4	5	1
Typical current (mA)	6.5	6.5	6.5	6.5	6
Typical voltage (V)	2750	2070	2070	1910	3380

(b) The human eye is at least twice as sensitive to orange color light than for the red. Discuss typical applications where it is more desirable to use the orange laser.

(c) The 1523 nm emission has potential for use in optical communications by modulating the laser beam externally. By considering the spectral line width $(\Delta \upsilon \approx 1400 \text{ MHz})$, typical powers, stability, discuss the advantages and disadvantages of using a He-Ne laser over a semiconductor diode.

4.2 The He-Ne Laser A particular He-Ne laser operating at 632.8 nm has a tube that is 50 cm long. The operating temperature is 130°C

(a) Estimate the Doppler broadened linewidth ($\Delta\lambda$) in the output spectrum.

(b) What are the mode number m values that satisfy the resonant cavity condition? How many modes are therefore allowed?

(c) What is the separation $\Delta\upsilon_m$ in the frequencies of the modes? What is the mode separation $\Delta\lambda_m$ in wavelength.

(d) Show that if during operation, the temperature changes the length of the cavity by δL, the wavelength of a given mode changes by $\delta\lambda_m$,

$$\delta\lambda_m = \frac{\lambda_m}{L}\delta L$$

Given that typically a glass has a linear expansion coefficient $\alpha \approx 10^{-6}\,\text{K}^{-1}$, calculate the change $\delta\lambda_m$ in the output wavelength (due to one particular mode) as the tube warms up from 20 °C to 130 °C, and also per degree change in the operating temperature. Note that $\delta L/L = \alpha\delta T$, and $L' = L[1 + \alpha(T' - T)]$. Change in mode wavelength $\delta\lambda_m$ with the change δL in the cavity length L is called *mode sweeping.*

(e) How do the mode separations $\Delta\upsilon_m$ and $\Delta\lambda_m$ change as the tube warms up from 20 °C to 130 °C during operation?

(f) How can you increase the output intensity from the He-Ne laser?

4.3 The Ar ion laser The argon-ion laser can provide powerful CW visible coherent radiation of several watts. The laser operation is achieved as follows: The Ar atoms are ionized by electron collisions in a high current electrical discharge. Further multiple collisions with electrons excite the argon ion, Ar^+, to a group of $4p$ energy levels ~35 eV above the atomic ground state as shown in Figure 4.37. Thus a population inversion forms between the $4p$ levels and the $4s$ level which is about 33.5 eV above the Ar atom ground level. Consequently, the stimulated radiation from the $4p$ levels down to the $4s$ level contains a series of wavelengths ranging from 351.1 nm to 528.7 nm. Most of the power however is concentrated, approximately equally, in the 488 and 514.5 nm emissions. The Ar^+ ion at the lower laser level ($4s$) returns to its neutral atomic ground state via a radiative decay to the Ar^+ ion ground state, followed by recombination with an electron to form the neutral atom. The Ar atom is then ready for "pumping" again.

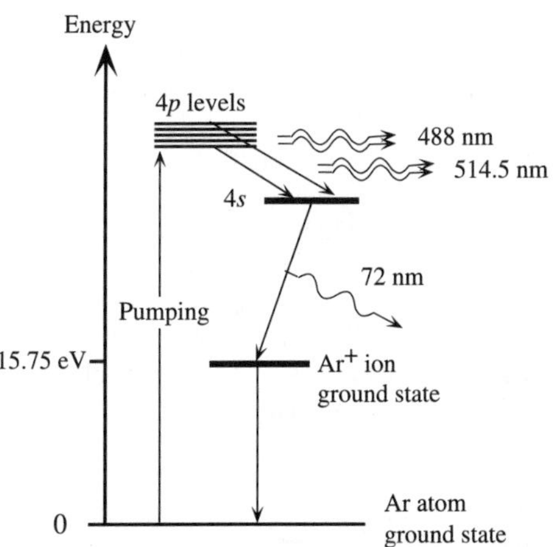

FIGURE 4.37 The Ar-ion laser energy diagram.

(a) Calculate the energy drop involved in the excited Ar^+ ion when it is stimulated to emit the radiation at 514.5 nm.

(b) The Doppler broadened linewidth of the 514.5 nm radiation is about 3500 MHz ($\Delta\upsilon$) and is between the half-intensity points.

(c) (i) Calculate the Doppler broadened width in the wavelength, $\Delta\lambda$.
(ii) Estimate the operation temperature of the argon ion gas; give the temperature in °C.

(d) In a particular argon-ion laser the discharge tube, made of Beryllia (Beryllium Oxide), is 30 cm long and has a bore of 3 mm in diameter. When the laser is operated with a current of 40 A at 200 V dc, the total output power in the emitted radiation is 3 W. What is the efficiency of the laser?

4.4 Einstein coefficients and critical photon concentration $\rho(h\upsilon)$ is the energy of the electromagnetic radiation per unit volume per unit frequency due to photons with energy $h\upsilon = E_2 - E_1$. Suppose that there are n_{ph} photons per unit volume. Each has an energy $h\upsilon$. The frequency range of emission is $\Delta\upsilon$. Then,

$$\rho(h\upsilon) = \frac{n_{ph}h\upsilon}{\Delta\upsilon}$$

Consider the Ar ion laser system. Given that the emission wavelength is at 488 nm and the linewidth in the output spectrum is about 5×10^9 Hz between half intensity points, *estimate* the photon concentration necessary to achieve more stimulated emission than spontaneous emission.

4.5 Photon concentration in a gas laser The Ar ion laser has a strong lasing emission at 488 nm. The laser tube is 1 m in length, and the bore diameter is 3 mm. The output power is 1 W. Assume that most of the output power is in the 488 nm emission. Assume that the tube end has a transmittance T of 0.1. Calculate the photon output flow (number of lasing photons emitted from the tube per unit time), photon flux (number of lasing photons emitted per unit area per unit time), and estimate the order of magnitude of the steady state photon concentration (at 488 nm) in the tube (assume that the gas refractive index is approximately 1).

4.6 Threshold gain and population inversion

(a) Consider a He-Ne gas laser operating at 632.8 nm. The tube length $L = 40$ cm, tube diameter is 1.5 mm and mirror reflectances are approximately 99.9% and 98%. The linewidth $\Delta\upsilon = 1.5$ GHz, the loss coefficient is $\gamma \approx 0.05\ \text{m}^{-1}$, spontaneous decay time constant $\tau_{sp} = 1/A_{21} \approx 300$ ns, $n \approx 1$. What is the threshold gain and population inversion?

(b) Consider a 488 nm Ar-ion gas laser. The tube length $L = 1$ m, tube mirror reflectances are approximately 99.9% and 95%. The linewidth $\Delta\upsilon = 3$ GHz, the loss coefficient is $\gamma \approx 0.1\ \text{m}^{-1}$, spontaneous decay time constant $\tau_{sp} = 1/A_{21} \approx 10$ ns, $n \approx 1$. What is the threshold population inversion?

(c) Consider a semiconductor laser operating at (λ_o) 870 nm with a GaAs laser cavity with cleaved facets. The cavity length is 50 μm. The refractive index (n) of GaAs is 3.6. The loss coefficient γ at normal temperatures is of the order of ~10 cm^{-1}. Estimate the required threshold gain. What is your conclusion?

[Note: γ depends on a number of factors including the injected carrier concentration and at best the above calculation is an estimate. We cannot (c) simply calculate the threshold population inversion, ΔN_{th}, from Eq. (12) in Section 4.7 which does not apply for a number of reasons; see Section 7.2 in P. Bhattacharya, *Semiconductor Optoelectronic Devices, Second Edition* (Prentice-Hall, New York, 1993).

4.7 Divergence of a laser beam Whenever the free propagation of a collimated beam is obstructed in one way or another, the beam becomes diffracted and "diverges out" from the obstruction. Consider diffraction by a circular aperture and a single slit as in Figure 1.23 and 1.25 in Chapter 1. The divergence angle 2θ that contains most of the light intensity is determined by the nature of diffraction. For example, for diffraction from a circular aperture of diameter D, Eq. (5) in §1.10, $\sin\theta = 1.22\lambda/D$.

(a) A He-Ne laser tube operating at 623.8 nm has a diameter $D = 1.5$ mm. Suppose that the laser beam emerging from the tube is Gaussian. What is the diameter of the beam at a distance of 20 m? What is this diameter if you assume a diffraction limited divergence?

(b) Consider a semiconductor laser with an active layer width of 2 μm and estimate the divergence angle 2θ based on diffraction effects. Take the refractive index to be typically 3.5. What is the divergence angle?

4.8 Fabry-Perot optical resonator

(a) Consider an idealized He-Ne laser optical cavity. Taking $L = 0.5$ m, $R = 0.99$, calculate the separation of the modes and the spectral width following Example 1.7.1.

(b) Consider a semiconductor Fabry-Perot optical cavity of length 200 micron with end-mirrors that have a reflectance of 0.8. If the semiconductor refractive index is 3.7, calculate the cavity mode nearest to the free space wavelength of 1300 nm. Calculate the separation of the modes and the spectral width following Example 1.7.1.

4.9 Population inversion in a GaAs laser diode Consider the energy diagram of a forward biased GaAs laser diode as in Figure 4.38. For simplicity we assume a symmetrical device ($n = p$) and we assume that population inversion has been just reached by A and B overlapping as illustrated in Figure 4.38 which results in $E_{Fn} - E_{Fp} = E_g$. Estimate the minimum carrier concentration $n = p$ for population inversion in GaAs at 300 K. The intrinsic carrier concentration in GaAs is of the order of 10^7 cm^{-3}. Assume for simplicity that

$$n = n_i \exp[(E_{Fn} - E_{Fi})/k_BT)] \quad \text{and} \quad p = n_i \exp[(E_{Fi} - E_{Fp})/k_BT)]$$

(Note: The analysis will only be an order of magnitude as the above equations do not hold in degenerate semiconductors. A better approach is to use the Joyce-Dixon equations as can be found in advanced textbooks)

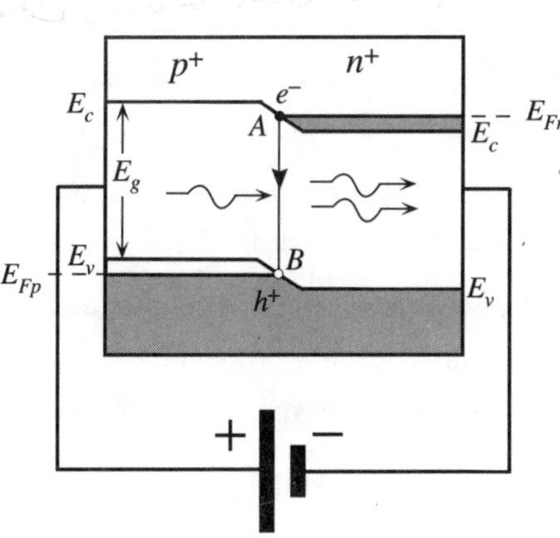

FIGURE 4.38 The energy band diagram of a degenerately doped *p-n* junction with with a sufficiently large forward bias to just cause population inversion where *A* and *B* overlap.

4.10 Threshold current and power output from a laser diode

(a) Consider the rate equations and their results in Section 4.10. It takes $\Delta t = nL/c$ second for photons to cross the laser cavity length L, where n is the refractive index. If N_{ph} is the coherent radiation photon concentration, then only half of the photons, $(1/2)N_{ph}$, in the cavity would be moving towards the output face of the crystal at any instant. Given that the active layer has a length L, width W and thickness d, show that the coherent optical output power and intensity are

$$P = \left[\frac{hc^2 N_{ph} dW}{2n\lambda}\right](1 - R) \quad \text{and} \quad I = \left[\frac{hc^2 N_{ph}}{2n\lambda}\right](1 - R)$$

where R is the reflectance of the semiconductor crystal face.

(b) If α is the attenuation coefficient for the coherent radiation within the semiconductor active layer due to various loss processes such as scattering and R is the reflectance of the crystal ends then the total attenuation coefficient α_t is,

$$\alpha_t = \alpha + \frac{1}{2L}\ln\left(\frac{1}{R^2}\right)$$

Consider a double heterostructure InGaAsP semiconductor laser operating at 1310 nm. The cavity length $L \approx 60$ μm, width $W \approx 10$ μm, and $d \approx 0.25$ μm. The refractive index $n \approx 3.5$. The loss coefficient $\alpha \approx 10$ cm^{-1}. Find α_t, τ_{ph}.

(c) For the above device, threshold current density $J_{th} \approx 500$ A cm^{-2} and $\tau_{sp} \approx 10$ ps. What is the threshold electron concentration? Calculate the lasing optical power and intensity when the current is 5 mA.

4.11 InGaAsP-InP Laser Consider an InGaAsP-InP laser diode which has an optical cavity of length 250 microns. The peak radiation is at 1550 nm and the refractive index of InGaAsP is 4. The optical gain bandwidth (as measured between half intensity points) will normally depend on the pumping current (diode current) but for this problem assume that it is 2 nm.

(a) What is the mode integer m of the peak radiation?

(b) What is the separation between the modes of the cavity?

(c) How many modes are there in the cavity?

(d) What is the reflection coefficient and reflectance at the ends of the optical cavity (faces of the InGaAsP crystal)?

(e) What determines the angular divergence of the laser beam emerging from the optical cavity?

4.12 Laser diode efficiency

(a) There are several laser diode efficiency definitions as follows:

The **external quantum efficiency**, η_{EQE}, of a laser diode is defined as

$$\eta_{EQE} = \frac{\text{Number of output photons from the diode (per unit second)}}{\text{Number of injected electrons into diode (per unit second)}}$$

The **external differential quantum efficiency**, η_{EDQE}, of a laser diode is defined as

$$\eta_{EDQE} = \frac{\text{Increase in number of output photons from diode (per unit second)}}{\text{Increase in number of injected electrons into diode (per unit second)}}$$

The **external power efficiency,** η_{EPE}, of the laser diode is defined by

$$\eta_{EPE} = \frac{\text{Optical ouput power}}{\text{Electical input power}}$$

If P_o is the emitted optical power, show that

$$\eta_{EQE} = \frac{eP_o}{E_g I}$$

$$\eta_{EDQE} = \left(\frac{e}{E_g}\right)\frac{dP_o}{dI}$$

$$\eta_{EPE} = \eta_{EQE}\left(\frac{E_g}{eV}\right)$$

(b) A commercial laser diode with an emission wavelength of 670 nm (red) has the following characteristics. The threshold current at 25°C is 76 mA. At $I = 80$ mA, the output optical power is 2 mW and the voltage across the diode is 2.3 V. If the diode current is increased to 82 mA, the optical output power increases to 3 mW. Calculate the external QE, external differential QE and the external power efficiency of the laser diode.

(c) Consider an InGaAsP laser diode operating at $\lambda = 1310$ nm for optical communications. The laser diode has an optical cavity of length 200 microns. The refractive index, $n = 3.5$. The threshold current at 25°C is 30 mA. At $I = 40$ mA, the output optical power is 3 mW and the voltage across the diode is 1.4 V. If the diode current is increased to 45 mA, the optical output power increases to 4 mW. Calculate external quantum efficiency (QE), external differential QE, external power efficiency of the laser diode.

4.13 Temperature dependence of laser characteristics The threshold current of a laser diode increases with temperature because more current is needed to achieve the necessary population inversion at higher temperatures. For a particular laser diode, the threshold current, I_{th}, is 76 mA at 25°C, 57.8 mA at 0°C and 100 mA at 50°C. Using these three points, show that I_{th} depends exponentially on the absolute temperature. What would be an empirical expression for it?

4.14 Single frequency lasers Consider a DFB laser operating at 1550 nm. Suppose that the refractive index $n = 3.4$ (InGaAsP). What should be the corrugation period Λ for a first order grating $q = 1$. What is Λ for a second order grating, $q = 2$. How many corrugations are needed for a first order grating if the cavity length is 20 μ? How many corrugations are there for $q = 2$? Which is easier to fabricate?

4.15 The SQW laser Consider a SQW (single quantum well) laser which has an ultrathin active InGaAs of bandgap 0.70 eV and thickness 10 nm between two layers of InAlAs which has a bandgap of 1.45 eV. Effective mass of conduction electrons in InGaAs is about $0.04m_e$ and that of the holes in the valence band is $0.44m_e$ where m_e is the mass of the electron in vacuum. Calculate the first and second electron energy levels above E_c and the first hole energy level below E_v in the QW. What is the lasing emission wavelength for this SQW laser? What is this wavelength if the transition were to occur in bulk InGaAs with the same bandgap?

4.16 A GaAs quantum well Effective mass of conduction electrons in GaAs is $0.07m_e$ where m_e is the electron mass in vacuum. Calculate the first three electron energy levels for a quantum well of thickness 8 nm. What is the hole energy below E_v if the effective mass of the hole is $0.47m_e$? What is the change in the emission wavelength with respect to bulk GaAs which has an energy bandgap of 1.42 eV.

4.17 Buried heterostructure laser diode Figure 4.20 shows a structure of a buried heterostructure laser diode based on GaAs and AlGaAs. Discuss how you would change the semiconducting materials to use the same structure for operation at 1.3 μm and at 1.55 μm?

4.18 Holography Consider holography as shown in Figure 4.35 (a) and (b). Suppose that the x and y coordinate system is placed on the photographic plate. We can write the reference beam E_{ref} and that reflected from the cat E_{cat} as,

$$E_{ref}(x, y) = U_r(x, y)e^{j\omega t} \quad \text{and} \quad E_{cat}(x, y) = U(x, y)e^{j\omega t}$$

where $U_r(x, y)$ is its amplitude of the reference beam and $U(x, y)$ is the complex amplitude of the wave from the cat at the plate. Further we can write U_r and $U(x, y)$ as,

$$U_r(x, y) = A_o \exp j(k_1 x + k_2 y)$$

where A_o is its amplitude and k_1 and k_2 are propagation constants along x and y.

The reflected wave from the cat will have a complex amplitude $U(x, y)$ that contains the magnitude and phase information which at the plate is,

$$U(x, y) = A \exp(j\phi); \quad A = A(x, y) \text{ and } \phi = \phi(x, y)$$

(a) Show that the intensity of the interference of E_{ref} and E_{cat} is

$$I(x, y) = |E_{\text{ref}} + E_{\text{cat}}|^2 = A^2 + A_o^2 + 2AA_o \cos[\phi(x, y) - k_1 x - k_2 y]$$

(b) Show that the transmitted wave U_t through the hologram, proportional to $U_r I(x, t)$ is

$$U_t(x, y) \propto A_o(A^2 + A_o^2)\exp[j(k_1 x + k_2 y)] + A_o^2 A \exp(j\phi) + A_o^2 A \exp[-j(\phi - 2k_1 x - 2k_2 y)]$$

Interpret each term in the above expression.

2,929,922

No. 2,929,922

THE UNITED STATES OF AMERICA

TO ALL TO WHOM THESE PRESENTS SHALL COME:

Whereas Arthur L. Schawlow, of Madison, New Jersey, and Charles H. Townes, of New York, N. Y., assignors to Bell Telephone Laboratories, Incorporated, of New York, N. Y., a corporation of New York,

presented to the Commissioner of Patents a petition praying for the grant of Letters Patent for an alleged new and useful invention the title and a description of which are contained in the specification of which a copy is hereunto annexed and made a part hereof, and complied with the various requirements of Law in such cases made and provided, and

Whereas upon due examination made the said Claimants are adjudged to be justly entitled to a patent under the Law.

Now therefore these Letters Patent are to grant unto the said Bell Telephone Laboratories, Incorporated, its successors or assigns for the term of SEVENTEEN years from the date of this grant the right to exclude others from making, using or selling the said invention throughout the United States.

In testimony whereof I have hereunto set my hand and caused the seal of the Patent Office to be affixed at the City of Washington this twenty-second day of March, in the year of our Lord one thousand nine hundred and sixty, and of the Independence of the United States of America the one hundred and eighty-fourth.

Attest:

Attesting Officer.

Robert C. Watson
Commissioner of Patents.

The patent for the invention of the laser by Charles H. Townes and Arthur L. Schawlow in 1960 (Courtesy of Bell Laboratories). The laser patent was later bitterly disputed for almost three decades in "the patent wars" by Gordon Gould, an American physicist, and his designated agents. Gordon Gould eventually received the US patent for optical pumping of the laser in 1977 since the original laser patent did not detail such a pumping procedure. In 1987 he also received a patent for the gas discharge laser, thereby winning his 30 year patent war. His original notebook even contained the word "laser". *(See "Winning the laser-patent war", Jeff Hecht, Laser Focus World, December 1994, pp. 49–51).*

CHAPTER 5

Photodetectors

"The detector is like the journalist who must determine what, where, when, which, and how? What is the identity of the particle? Exactly where is it when it is observed? When does the particle get to the detector? Which way is it going? How fast is it moving?"

—Sheldon L. Glashow[1]

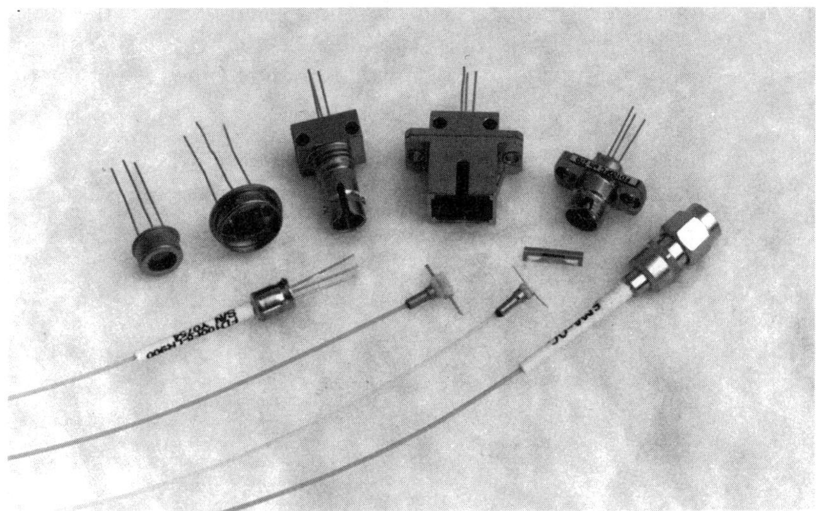

A selection of commercial InGaAs based photodetectors, including fiber-pigtailed photodiodes. *(Courtesy of Fermionics, California.)*

5.1 PRINCIPLE OF THE *pN* JUNCTION PHOTODIODE

Photodetectors convert a light signal to an electrical signal such as a voltage or current. In many photodetectors such as photoconductors and photodiodes this conversion is typically achieved by the creation of **free electron hole pairs** (EHPs) by the absorption of photons, that is, the creation of electrons in the conduction band (CB) and holes in the valence band (VB). In some devices such as **pyroelectric detectors** the energy conversion involves the generation of heat which increases the temperature of the device which changes its polarization and hence its relative permittivity. We will consider *pn*

[1] Sheldon L. Glashow, *Interactions* (Warner Books, New York, 1988), p. 101.

junction based photodiode type devices only as these devices are small and have high speed and good sensitivity for use in various optoelectronics applications, the most important of which is in optical communications.

Figure 5.1 (a) shows the simplified structure of a typical *pn* junction **photodiode** that has a p^+n type of junction, that is, the acceptor concentration N_a in the *p*-side is much greater than the donor concentration N_d in the *n*-side. The illuminated side has a window, defined by an annular electrode, to allow photons to enter the device. There is also an **antireflection coating**, typically Si_3N_4, to reduce light reflections. The p^+ side is generally very thin (less than a micron) and is usually formed by planar diffusion into an *n*-type epitaxial layer. Figure 5.1 (b) shows the net space charge distribution across the p^+n junction. These charges are in the **depletion region**, or in the **space charge layer** (SCL), and represent the exposed negatively charged acceptors in the p^+ side and exposed positively charged donors in the *n*-side. The depletion region extends almost entirely into the lightly doped *n*-side and, at most, it is a few microns.

The photodiode is normally reverse biased. The applied reverse bias V_r drops across the highly resistive depletion layer width W and makes the voltage across W equal to $V_o + V_r$ where V_o is the built-in voltage. The field is found by the integration of the net

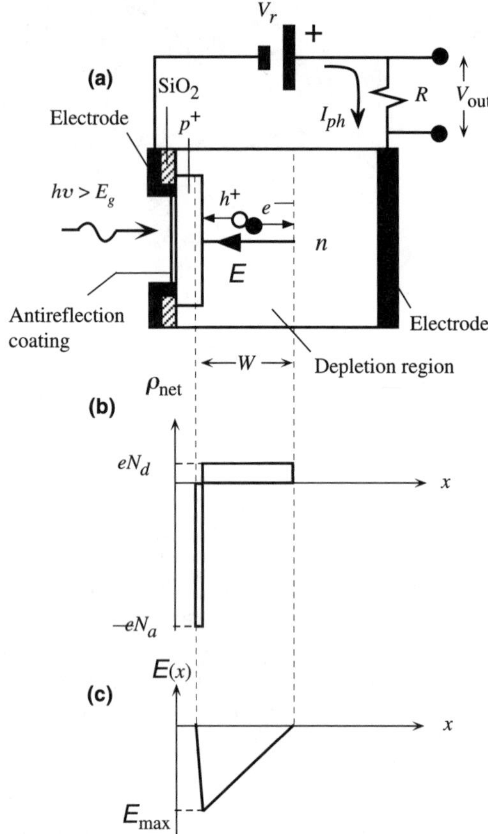

FIGURE 5.1 (a) A schematic diagram of a reverse biased *pn* junction photodiode. (b) Net space charge density across the diode in the depletion region. N_d and N_a are the donor and acceptor concentrations in the *n* and *p* sides. (c) The field in the depletion region.

space charge density ρ_{net} in Figure 5.1 (b) across W subject to a voltage difference of $V_o + V_r$. The field only exists in the depletion region and is not uniform. It varies across the depletion region as shown in Figure 5.1 (c) where it is maximum at the junction and penetrates into the n-side. The regions outside the depletion layer are the **neutral regions** in which there are majority carriers. It is sometimes convenient to treat these neutral regions simply as resistive extensions of electrodes to the depletion layer.

When a photon with an energy greater than the bandgap E_g is incident, it becomes absorbed to **photogenerate** a free EHP, that is an electron in the CB and a hole in the VB. Usually the energy of the photon is such that photogeneration takes place in the depletion layer. The field $\boldsymbol{E}$ in the depletion layer then separates the EHP and drifts them in opposite directions until they reach the neutral regions as depicted in Figure 5.1 (a). Drifting carriers generate a current, called **photocurrent** I_{ph}, in the external circuit that provides the electrical signal. The photocurrent lasts for the duration it takes for the electron and hole to cross the depletion layer (W) and reach the neutral regions. When the drifting hole reaches the neutral p^+-region it recombines with a electron entering the p^+-side from the negative electrode, that is from the battery. Similarly, when the drifting electron reaches the neutral n-side, an electron leaves the n-side into the electrode (battery). The photocurrent I_{ph} depends on the number of EHPs photogenerated and the drift velocities of the carriers while they are transiting the depletion layer. Since the field is not uniform and the absorption of photons occurs over a distance that depends on the wavelength, the time dependence of the photocurrent signal cannot be determined in a simple fashion.

It should be mentioned that the photocurrent in the external circuit is due to the flow of electrons only even though there are both electrons and holes drifting within the device. Suppose that there are N number of EHPs photogenerated. If we were to integrate the photocurrent I_{ph} to calculate how much charge has flowed we would find a charge Q that is due to the total number of photogenerated electrons (eN) and not due to both electrons and holes ($2eN$).

5.2 RAMO'S THEOREM AND EXTERNAL PHOTOCURRENT

Consider a semiconductor material with a negligible dark conductivity that is electroded and biased as shown in Figure 5.2 (a). The electrodes do not inject carriers but allow excess carriers in the sample to leave and become collected by the battery (they are termed noninjecting electrodes). The field $\boldsymbol{E}$ in the sample is uniform and it is V/L. We will later see that this situation is almost identical to the intrinsic region of a reverse biased *pin* photodiode. Suppose that a single photon is absorbed at a position $x = l$ from the left electrode and instantly creates an electron hole pair. The electron and the hole drift in opposite directions with respective drift velocities $v_e = \mu_e E$ and $v_h = \mu_h E$, where μ_e and μ_h are the electron and hole drift mobilities respectively. The **transit time** of a carrier is the time it takes for a carrier to drift from its generation point to the collecting electrode. The electron and hole transit times t_e and t_h respectively are marked on the t vs. x diagram in Figure 5.2 (b) where,

$$t_e = \frac{L - l}{v_e} \quad \text{and} \quad t_h = \frac{l}{v_h} \qquad (1)$$

Electron and hole transit times

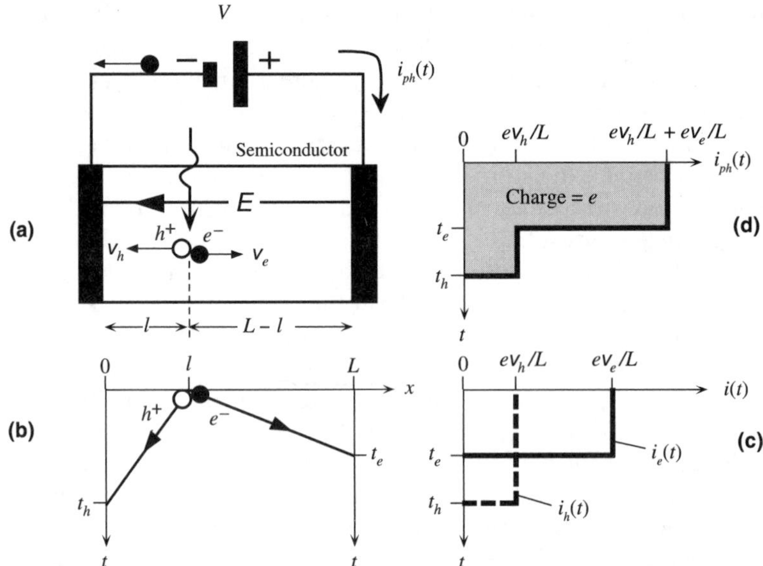

FIGURE 5.2 (a) An EHP is photogenerated at $x = l$. The electron and the hole drift in opposite directions with drift velocities v_h and v_e. (b) The electron arrives at time $t_e = (L - l)/v_e$ and the hole arrives at time $t_h = l/v_h$. (c) As the electron and hole drift, each generates an external photocurrent shown as $i_e(t)$ and $i_h(t)$. (d) The total photocurrent is the sum of hole and electron photocurrents each lasting a duration t_h and t_e respectively.

Consider first only the drifting electron. Suppose that the external photocurrent due to the motion of this electron is $i_e(t)$. The electron is acted on by the force $e\boldsymbol{E}$ of the electric field. When it moves a distance dx, work must be done by the external circuit. In time dt, the electron drifts a distance dx and does an amount of work $e\boldsymbol{E}dx$ which is provided by the battery in time dt as $Vi_e(t)\,dt$. Thus,

$$\text{Work done} = e\boldsymbol{E}dx = Vi_e(t)\,dt$$

Using $\boldsymbol{E} = V/L$ and $v_e = dx/dt$ we find the electron photocurrent

Electron photocurrent

$$i_e(t) = \frac{ev_e}{L}; \quad t < t_e \tag{2}$$

It is apparent that this current continues to flow as long as the electron is drifting (has a velocity v_e) in the sample. It lasts for a duration t_e at the end of which the electron reaches the battery. Thus, although the electron has been photogenerated instantaneously, the external photocurrent is *not* instantaneous and has a **time spread**. Figure 5.2 (c) shows the electron photocurrent $i_e(t)$.

We can apply similar arguments to the drifting hole as well which will generate a hole photocurrent $i_h(t)$ in the external circuit given, as in Figure 5.2 (c), by

Hole photocurrent

$$i_h(t) = \frac{ev_h}{L}; \quad t < t_h \tag{3}$$

The total external current will be the sum of $i_e(t)$ and $i_h(t)$ as shown in Figure 5.2 (d).

If we integrate the external current $i_{ph}(t)$ to evaluate the collected charge $Q_{\text{collected}}$ we would find,

$$Q_{\text{collected}} = \int_0^{t_e} i_e(t)dt + \int_0^{t_h} i_h(t)dt = e \tag{4}$$

This result can be verified by evaluating the area under the $i_{ph}(t)$ curve in Figure 5.2 (d). Thus, the collected charge is not $2e$ but just one electron as shown by the area in Figure 5.2 (d). Equations (2) to (4) constitute **Ramo's theorem.** In general, if a charge q is being drifted with a velocity $v_d(t)$ by a field between two biased electrodes separated by L, then this motion of q generates an external current given by

Ramo's theorem

$$i(t) = \frac{qv_d(t)}{L}; \quad t < t_{\text{transit}} \tag{5}$$

The total external current is the sum of all currents of the type in Eq. (5) from all drifting charges between the electrodes.

5.3 ABSORPTION COEFFICIENT AND PHOTODIODE MATERIALS

The photon absorption process for photogeneration, that is the creation of EHPs, requires the photon energy to be at least equal to the bandgap energy E_g of the semiconductor material to excite an electron from the valence band (VB) to the conduction band (CB). The **upper cut-off wavelength** (or the threshold wavelength) λ_g for photogenerative absorption is therefore determined by the bandgap energy E_g of the semiconductor so that $h(c/\lambda_g) = E_g$ or,

Cut-off wavelength and bandgap

$$\lambda_g(\mu\text{m}) = \frac{1.24}{E_g(\text{eV})} \tag{1}$$

For example, for Si, $E_g = 1.12$ eV and λ_g is 1.11 μm whereas for Ge $E_g = 0.66$ eV and the corresponding $\lambda_g = 1.87$ μm. It is clear that Si photodiodes cannot be used in optical communications at 1.3 and 1.55 μm whereas Ge photodiodes are commercially available for use at these wavelengths. Table 5.1 lists some typical bandgap energies and the corresponding cut-off wavelengths of various photodiode semiconductor materials.

Incident photons with wavelengths shorter than λ_g become absorbed as they travel in the semiconductor and the light intensity, which is proportional to the number of photons, decays exponentially with distance into the semiconductor. The light intensity I at a distance x from the semiconductor surface is given by

Absorption coefficient

$$I(x) = I_o \exp(-\alpha x) \tag{2}$$

where I_o is the intensity of the incident radiation and α is the **absorption coefficient** that depends on the photon energy or wavelength λ. Absorption coefficient α is a material property. Most of the photon absorption (63%) occurs over a distance $1/\alpha$ and $1/\alpha$ is

TABLE 5.1 Band gap energy E_g at 300 K, cut-off wavelength λ_g and type of bandgap (D = Direct and I = Indirect) for some photodetector materials.

Semiconductor	E_g (eV)	λ_g (μm)	Type
InP	1.35	0.91	D
$GaAs_{0.88}Sb_{0.12}$	1.15	1.08	D
Si	1.12	1.11	I
$In_{0.7}Ga_{0.3}As_{0.64}P_{0.36}$	0.89	1.4	D
$In_{0.53}Ga_{0.47}As$	0.75	1.65	D
Ge	0.66	1.87	I
InAs	0.35	3.5	D
InSb	0.18	7	D

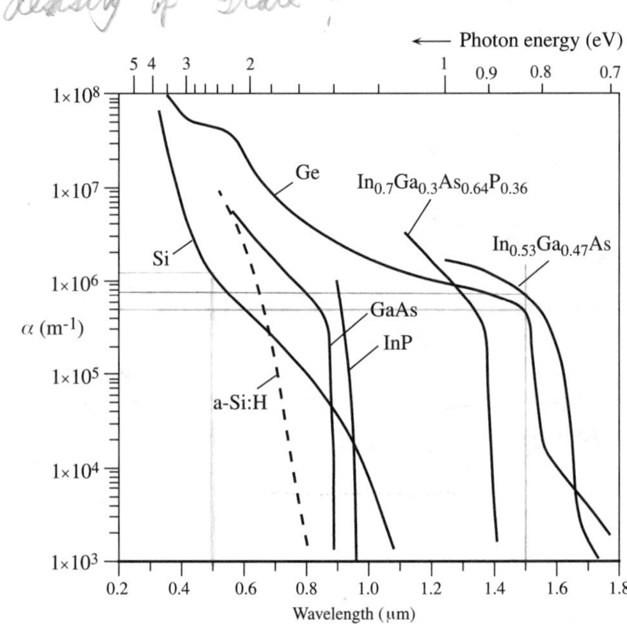

FIGURE 5.3 The absorption coefficient (α) vs. wavelength (λ) for various semiconductors (data selectively collected and combined from various sources including J. Gowar, Optical Communication Systems, Second Ed. (Prentice-Hall, New York, 1993), p. 444; H. Melchior, "Demodulation and Photodetection Techniques" in F.T. Arecchi and A.O. Schultz-Dubois, Eds, *Laser Handbook*, Vol. 1 (North-Holland, Amsterdam, 1972), pp. 725–835.

called the **penetration depth** δ. Figure 5.3 shows the α vs. λ characteristics of various semiconductors where it is apparent that the behavior of α with the wavelength λ depends on the semiconductor material.

In **direct bandgap** semiconductors such as III-V semiconductors (e.g. GaAs, InAs, InP, GaSb) and in many of their alloys (e.g. InGaAs, GaAsSb) the photon absorption process is a direct process that requires no assistance from lattice vibrations. The photon is absorbed and the electron is excited directly from the valence band to the conduction band without a change in its k-vector (or its crystal momentum $\hbar k$) inasmuch as the photon momentum is very small. The change in the electron momentum from the valence to the conduction band $\hbar k_{CB} - \hbar k_{VB}$ = photon momentum ≈ 0. This process corresponds to a vertical transition on the E-k diagram, that is electron energy (E) vs.

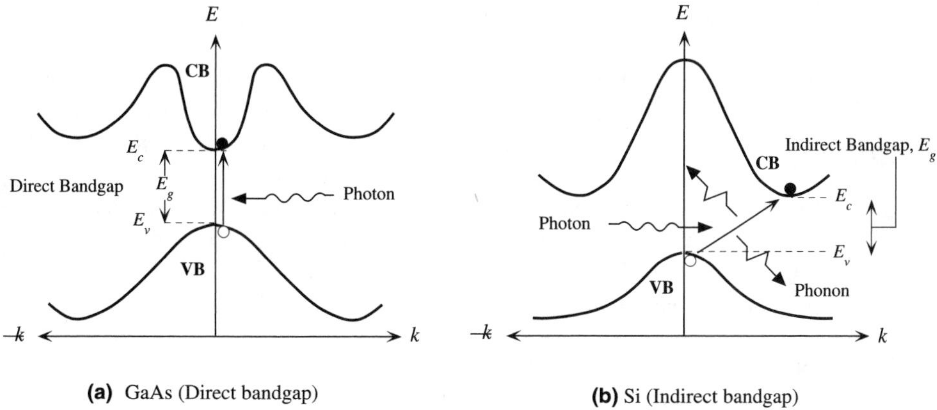

(a) GaAs (Direct bandgap)

(b) Si (Indirect bandgap)

FIGURE 5.4 (a) Photon absorption in a direct bandgap semiconductor. (b) Photon absorption in an indirect bandgap semiconductor (VB, valence band; CB, conduction band)

electron momentum ($\hbar k$) in the crystal in Figure 5.4 (a). The absorption coefficient of these semiconductors rises sharply with decreasing wavelength from λ_g as apparent for GaAs and InP in Figure 5.3.

In **indirect bandgap** semiconductors such as Si and Ge, the photon absorption for photon energies near E_g requires the absorption and emission of lattice vibrations, that is **phonons**,[2] during the absorption process as shown in Figure 5.4 (b). If K is the wavevector of a lattice wave (lattice vibrations travel in the crystal), then $\hbar K$ represents the momentum associated with such a lattice vibration, that is $\hbar K$ is a **phonon momentum**. When an electron in the valence band is excited to the conduction band there is a change in its momentum in the crystal and this change in the momentum cannot be supplied by the momentum of the incident photon which is very small. Thus, the momentum difference must be balanced by a phonon momentum:

$$\hbar k_{\mathrm{CB}} - \hbar k_{\mathrm{VB}} = \text{phonon momentum} = \hbar K.$$

The absorption process is said to be **indirect** as it depends on lattice vibrations which in turn depend on the temperature. Since the interaction of a photon with a valence electron needs a third body, a lattice vibration, the probability of photon absorption is not as high as in a direct transition. Furthermore, the cut-off wavelength is not as sharp as for direct bandgap semiconductors. During the absorption process, a phonon may be absorbed or emitted. If ϑ is the frequency of the lattice vibrations then the phonon energy is $h\vartheta$. The photon energy is hv where v is the photon frequency. Conservation of energy requires that

$$hv = E_g \pm h\vartheta$$

Thus, the onset of absorption does not exactly coincide with E_g, but typically it is very close to E_g inasmuch as $h\vartheta$ is small (<0.1 eV). The absorption coefficient initially

[2] As much as an electromagnetic radiation is quantized in terms of photons, lattice vibrations in the crystal are quantized in terms of phonons.

rises slowly with decreasing wavelength from about λ_g as apparent in Figure 5.3 for Ge and Si.

The choice of material for a photodiode must be such that the photon energies are greater than E_g. Further, at the wavelength of radiation, the absorption occurs over a depth covering the depletion layer so that the photogenerated EHPs can be separated by the field and collected at the electrodes. If the absorption coefficient is too large then absorption will occur very near the surface of the p^+ layer which is outside the depletion layer. First, the absence of a field means that the photogenerated electron can only make it to the depletion layer to cross to the n-side by diffusion. Secondly, photogeneration near the surface invariably leads to rapid recombination due to surface defects that act as recombination centers. On the other hand, if the absorption coefficient is too small, only a small portion of the photons will be absorbed in the depletion layer and only a limited number of EHPs can be photogenerated.

5.4 QUANTUM EFFICIENCY AND RESPONSIVITY

Not all the incident photons are absorbed to create *free* electron-hole pairs (EHPs) that can be collected and give rise to a photocurrent. The efficiency of the conversion process of received photons to free EHPs is measured by the **quantum efficiency (QE) η of the detector** defined as[3]

Quantum Efficiency, QE

$$\eta = \frac{\text{Number of free EHP generated and collected}}{\text{Number of incident photons}} \tag{1}$$

The measured photocurrent I_{ph} in the external circuit is due to the flow of electrons per second to the terminals of the photodiode. Number of electrons collected per second is I_{ph}/e. If P_o is the incident optical power then the number of photons arriving per second is $P_o/h\nu$. Then the QE η can also be defined by

Quantum Efficiency, QE

$$\eta = \frac{I_{ph}/e}{P_o/h\nu} \tag{2}$$

Not all of the absorbed photons may photogenerate free EHPs that can be collected. Some EHPs may disappear by recombination without contributing to the photocurrent or become immediately trapped. Further if the semiconductor length is comparable with the penetration depth $(1/\alpha)$ then not all the photons will be absorbed. The device QE is therefore always less than unity. It depends on the absorption coefficient α of the semiconductor at the wavelength of interest and on the structure of the device. QE can be increased by reducing the reflections at the semiconductor surface, increasing absorption within the depletion layer and preventing the recombination or trapping of carriers before they are collected. The QE defined in Eq. (1) is for the whole device. More specifically, it is known as the **external quantum efficiency**. Internal quantum efficiency is the number of free EHPs photogenerated per *absorbed photon* and is typically quite high for many devices. The QE

[3] "Free" implies carriers that can be collected, *i.e.* electrons in the conduction band and holes in the valence band.

definition in Eq. (1) incorporates the internal quantum efficiency as it applies to the whole device.

The **responsivity** R of a photodiode characterizes its performance in terms of the photocurrent generated (I_{ph}) per incident optical power (P_o) at a given wavelength,

$$R = \frac{\text{Photocurrent (A)}}{\text{Incident Optical Power (W)}} = \frac{I_{ph}}{P_o} \quad (3)$$

Responsivity

From the definition of QE, it is clear that,

$$R = \eta \frac{e}{h\upsilon} = \eta \frac{e\lambda}{hc} \quad (4)$$

Responsivity and QE

In Eq. (4) η depends on the wavelength. The responsivity therefore clearly depends on the wavelength. R is also called the **spectral responsivity** or **radiant sensitivity**. The R vs. λ characteristics represents the spectral response of the photodiode and is generally provided by the manufacturer. Ideally with a quantum efficiency of 100% $(\eta = 1)$, R should increase with λ up to λ_g as indicated in Figure 5.5. In practice, QE limits the responsivity to lie below the ideal photodiode line with upper and lower wavelength limits as shown for a typical Si photodiode in the same figure. The QE of a well designed Si photodiode in the wavelength range 700–900 nm can be close to 90–95%.

5.5 THE *pin* PHOTODIODE

The simple *pn* junction photodiode (Figure 5.1) has two major drawbacks. Its junction or depletion layer capacitance is not sufficiently small to allow photodetection at high modulation frequencies. This is an *RC* time constant limitation. Secondly, its depletion layer is at most a few microns. This means that at long wavelengths where the penetration depth is greater than the depletion layer width, the majority of photons are absorbed outside the depletion layer where there is no field to separate the EHPs and drift them. The QE is correspondingly low at these long wavelengths. These problems are substantially reduced in the ***pin* (*p*–intrinsic–*n*-type) photodiode**.[4]

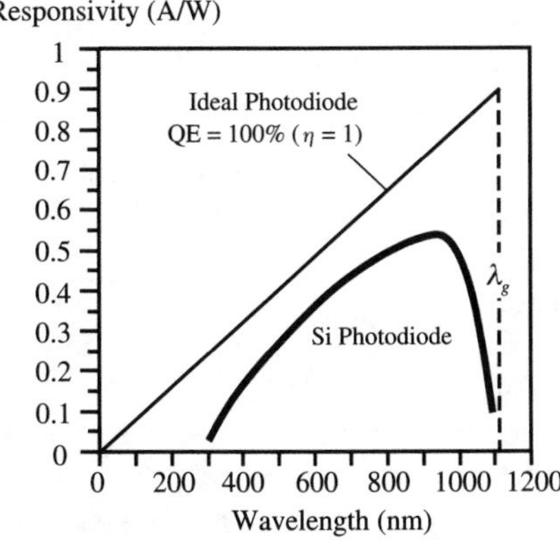

FIGURE 5.5 Responsivity (R) vs. wavelength (λ) for an ideal photodiode with QE = 100% $(\eta = 1)$ and for a typical commercial Si photodiode.

[4] The *pin* photodiode was invented by J. Nishizawa and his research group in Japan in 1950.

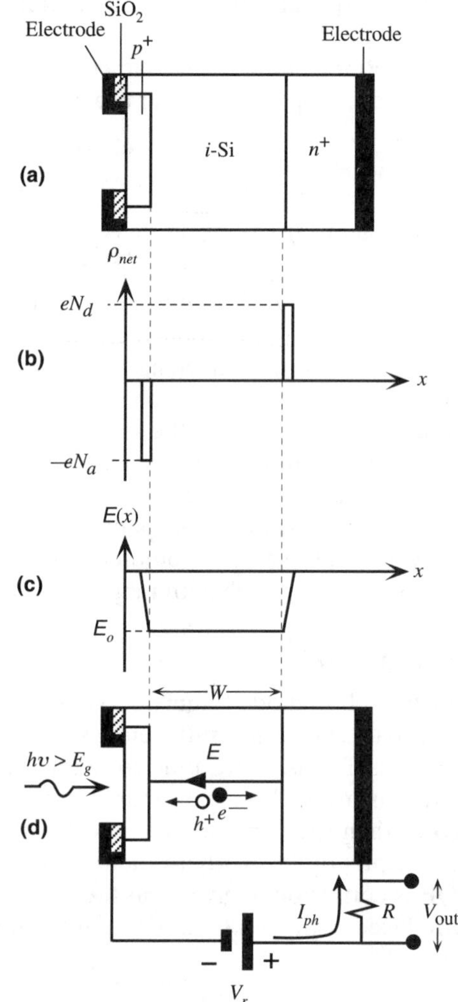

FIGURE 5.6 The schematic structure of an idealized *pin* photodiode. (b) The net space charge density across the photodiode. (c) The built-in field across the diode. (d) The *pin* photodiode in photodetection is reverse biased.

The *pin* refers to a semiconductor device that has the structure p^+–intrinsic–n^+ as schematically illustrated in the idealized structure in Figure 5.6 (a). The intrinsic layer has a much smaller doping than both p^+ and n^+ regions and it is much wider than these regions, typically 5–50 μm depending on the particular application. In the idealized ***pin* photodiode** we can take, for simplicity, the *i*-Si region to be truly intrinsic. When the structure is first formed, holes diffuse from the p^+-side and electrons from n^+-side into the *i*-Si layer where they recombine and disappear. This leaves behind a thin layer of exposed negatively charged acceptor ions in the p^+-side and a thin layer of exposed positively charged donor ions in the n^+-side as shown in Figure 5.6 (b). The two charges are separated by the *i*-Si layer of thickness W. There is a uniform built-in field $\boldsymbol{E}_o$ in *i*-Si layer from the exposed positive ions to exposed negative ions as illustrated in Figure 5.6 (c). In contrast, the built-in field in the depletion layer of a *pn* junction is not uniform. With no applied bias, equilibrium is maintained by the built-in field E_o which prevents further diffusion of majority carriers into the *i*-Si layer.

The separation of two very thin layers of negative and positive charges by a fixed distance, width W of the *i*-Si, is the same as that in a parallel plate capacitor. The **junction or depletion layer capacitance** of the *pin* diode is given by

$$C_{dep} = \frac{\varepsilon_o \varepsilon_r A}{W} \tag{1}$$

Junction capacitance of pin

where A is the cross sectional area and $\varepsilon_o \varepsilon_r$ is the permittivity of the semiconductor (Si). Further, since the width W of the *i*-Si layer is fixed by the structure, the junction capacitance does not depend on the applied voltage in contrast to that of the *pn* junction. C_{dep} is typically of the order of a picofarad in fast *pin* photodiodes so that with a 50 Ω resistor, the RC_{dep} time constant is about 50 ps.

When a reverse bias voltage V_r is applied across the *pin* device, it drops almost entirely across the width of *i*-Si layer. The depletion layer widths of the thin sheets of acceptor and donor charges in the p^+ and n^+ sides are negligible compared with W. The reverse bias V_r increases the built-in voltage to $V_o + V_r$ as shown in Figure 5.6 (d). The field E in the *i*-Si layer is still uniform and increases to

$$E = E_o + \frac{V_r}{W} \approx \frac{V_r}{W} \qquad (V_r \gg V_o) \tag{2}$$

Biased pin

The *pin* structure is designed so that photon absorption occurs over the *i*-Si layer. The photogenerated EHPs in the *i*-Si layer are then separated by the field E and drifted towards the n^+ and p^+ sides respectively as illustrated in Figure 5.6 (d). While the photogenerated carriers are drifting through the *i*-Si layer they give rise to an external photocurrent which is detected as a voltage across a small resistor R in Figure 5.6 (d). The **response time** of the *pin* photodiode is determined by the transit times of the photogenerated carriers across the width W of the *i*-Si layer. Increasing W allows more photons to be absorbed which increases the QE but it slows down the speed of response as carrier transit times become longer. For a charge carrier that is photogenerated at the edge on the *i*-Si layer, the transit time or drift time t_{drift} across the *i*-Si layer is

$$t_{\text{drift}} = \frac{W}{v_d} \tag{3}$$

Transit time

where v_d is its drift velocity. To reduce the drift time, that is increase the speed of response, we have to increase v_d and therefore increase the applied field E. At high fields v_d does not follow the expected $\mu_d E$ behavior, where μ_d is the drift mobility, but instead tends to saturate at v_{sat} which is of the order of 10^5 m s^{-1} at fields greater than 10^6 V m^{-1} in the case of Si. Figure 5.7 shows the variation of the drift velocity of electrons and holes with the field in Si. The $v_d = \mu_d E$ behavior is only observed at low fields. At high fields both the drift velocities saturate. For an *i*-Si layer of width 10 μm, with carriers drifting as saturation velocities, the drift time is about 0.1 ns which is longer than typical RC_{dep} time constants. The speed of *pin* photodiodes are invariably limited by the transit time of photogenerated carriers across the *i*-Si layer.

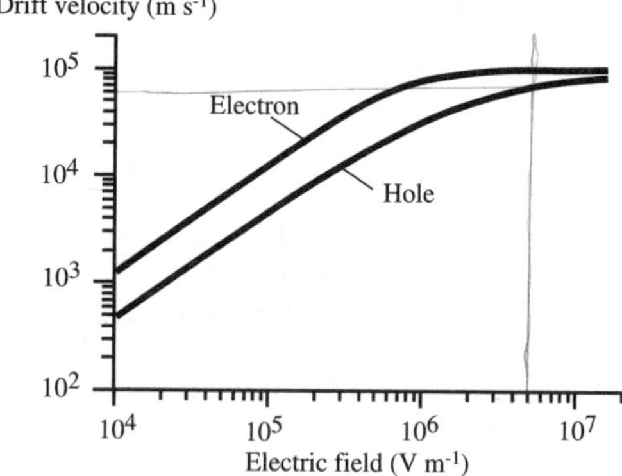

FIGURE 5.7 Drift velocity vs. electric field for holes and electrons in Si.

The *pin* photodiode structure shown in Figure 5.6 is, of course, idealized. In reality, the *i*-Si layer will have some small doping. For example, if the sandwiched layer is lightly *n*-type doped it is labeled as a ν-layer and the structure is $p^+\nu n^+$. The sandwiched ν-layer becomes a depletion layer with a small concentration of exposed positive donors. The field then is not entirely uniform across the photodiode. The field is maximum at the $p^+\nu$ junction and decreases slightly across ν-Si to reach the n^+side. As an approximation we can still consider the ν-Si layer as an *i*-Si layer.

EXAMPLE 5.5.1 Operation and speed of a *pin* photodiode

A Si *pin* photodiode has an *i*-Si layer of width 20 μm. The p^+ layer on the illumination side is very thin (0.1 μm). The *pin* is reverse biased by a voltage of 100 V and then illuminated with a very short optical pulse of wavelength 900 nm. What is the duration of the photocurrent if absorption occurs over the whole *i*-Si layer?

Solution The absorption coefficient at 900 nm is ~3 × 10⁴ m⁻¹ so that the absorption depth is ~33 μm as apparent in Figure 5.3. We can assume that absorption and hence photogeneration occurs over the entire width W of the *i*-Si layer. The field in the *i*-Si layer is

$$E \approx V_r/W = (100\text{ V})/(20 \times 10^{-6}\text{ m}) = 5 \times 10^6\text{ V m}^{-1}.$$

At this field the electron drift velocity v_e is very near its saturation at 10^5 m s^{-1}, whereas the hole drift velocity v_h, is about 7×10^4 m s^{-1} as shown in Figure 5.7. Holes are slightly slower than the electrons. The transit time t_h of holes across the *i*-Si layer is

$$t_h = W/v_h = (20 \times 10^{-6}\text{ m})/(7 \times 10^4\text{ m s}^{-1}) = 2.86 \times 10^{-10}\text{ s or } 0.3\text{ ns}.$$

This is the response time of the *pin* as determined by the transit time of the slowest carriers, holes, across the *i*-Si layer. To improve the response time the width of the *i*-Si layer has to be narrowed but this decreases the quantity of absorbed photons and hence reduces the responsivity. There is therefore a trade off between speed and responsivity.

EXAMPLE 5.5.2 Photocarrier Diffusion in a *pin* photodiode

A reverse biased *pin* photodiode is illuminated with a short wavelength photon that is absorbed very near the surface as shown in Figure 5.8. The photogenerated electron has to diffuse to the depletion region where it is swept into the *i*-layer and drifted across. What is the speed of response of this photodiode if the i-Si layer is 20 μm and the p^+ layer is 1 μm and the applied voltage is 120 V? The diffusion coefficient (D_e) of electrons in the heavily doped p^+ region is approximately 3×10^{-4} m² s⁻¹.

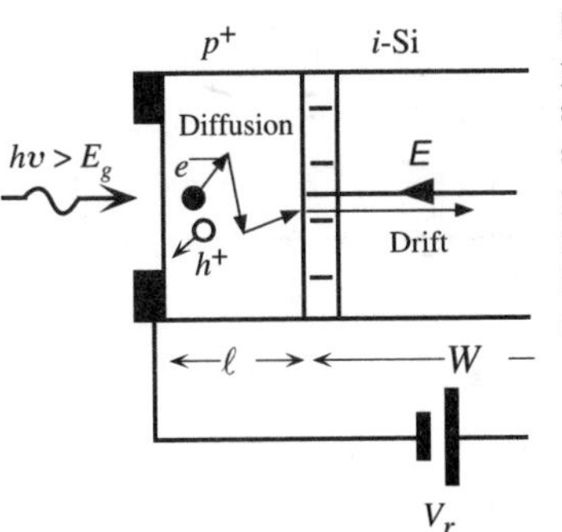

FIGURE 5.8 A reverse biased *pin* photodiode is illuminated with a short wavelength photon that is absorbed very near the surface. The photogenerated electron has to diffuse to the depletion region where it is swept into the i-layer and drifted across.

Solution There is no electric field in the p^+ side outside the depletion region as shown in Figure 5.8. The photogenerated electrons have to make it across to the n^+ side to give rise to a photocurrent. In the p^+ side, the electrons move by diffusion. In time t, an electron, on average, diffuses a distance ℓ given by[5]

$$\ell = [2D_e t]^{1/2}$$

The *diffusion time* t_{diff} is the time it takes for an electron to diffuse across the p^+ side (of length ℓ) to reach the depletion layer is

$$t_{\text{diff}} = \ell^2/(2D_e) = (1 \times 10^{-6}\,\text{m})^2/[2(3 \times 10^{-4}\,\text{m}^2\,\text{s}^{-1})] = 1.67 \times 10^{-9}\,\text{s or } 1.67\text{ ns.}$$

On the other hand, once the electron reaches the depletion region, it becomes drifted across the width W of the i-Si layer at the saturation drift velocity since the electric field here is $E = V_r/W = 120\,\text{V}/20\,\mu\text{m} = 6 \times 10^6\,\text{V m}^{-1}$ and at this field the electron drift velocity v_e saturates at 10^5 m s⁻¹. The *drift time* across the i-Si layer is

$$t_{\text{drift}} = W/v_e = (20 \times 10^{-6}\,\text{m})/(1 \times 10^5\,\text{m s}^{-1}) = 2.0 \times 10^{-10}\,\text{s or } 0.2\text{ ns.}$$

Thus, the response time of the *pin* to a pulse of short wavelength radiation that is absorbed near the surface is about $t_{\text{diff}} + t_{\text{drift}}$ or 1.87 ns.

EXAMPLE 5.5.3 Responsivity of a *pin* photodiode

A Si *pin* photodiode has an active light receiving area of diameter 0.4 mm. When radiation of wavelength 700 nm (red light) and intensity 0.1 mW cm⁻² is incident it generates a photocurrent of 56.6 nA. What is the responsivity and QE of the photodiode at 700 nm?

Solution The incident light intensity $I = 0.1$ mW cm⁻² means that the incident power for conversion is

[5] See, for example, *Principles of Electronic Materials and Devices*, Second Edition, S.O. Kasap (McGraw-Hill, 2001), Ch. 1 and Ch. 5 for the derivation of the root mean square diffusion distance.

$$P_o = AI = \pi(0.02 \text{ cm})^2(10^{-3} \text{ W cm}^{-2}) = 1.26 \times 10^{-7} \text{ W or } 0.126 \ \mu\text{W}.$$

The responsivity is

$$R = I_{ph}/P_o = (56.6 \times 10^{-9} \text{ A})/(1.26 \times 10^{-7} \text{ W}) = 0.45 \text{ A W}^{-1}$$

The QE can be found from

$$\eta = R\frac{hc}{e\lambda} = (0.45 \text{ A W}^{-1})\frac{(6.62 \times 10^{-34} \text{ J s})(3 \times 10^8 \text{ m s}^{-1})}{(1.6 \times 10^{-19} \text{ C})(700 \times 10^{-9} \text{ m})} = 0.80 = 80\%$$

5.6 AVALANCHE PHOTODIODE (APD)

Avalanche photodiodes (APDs), are widely used in optical communications due to their high speed and internal gain. A simplified schematic diagram of a Si **reach-through APD** is shown Figure 5.9 (a). The n^+ side is thin and it is the side that is illuminated through a window. There are three p-type layers of different doping levels next to the n^+ layer.

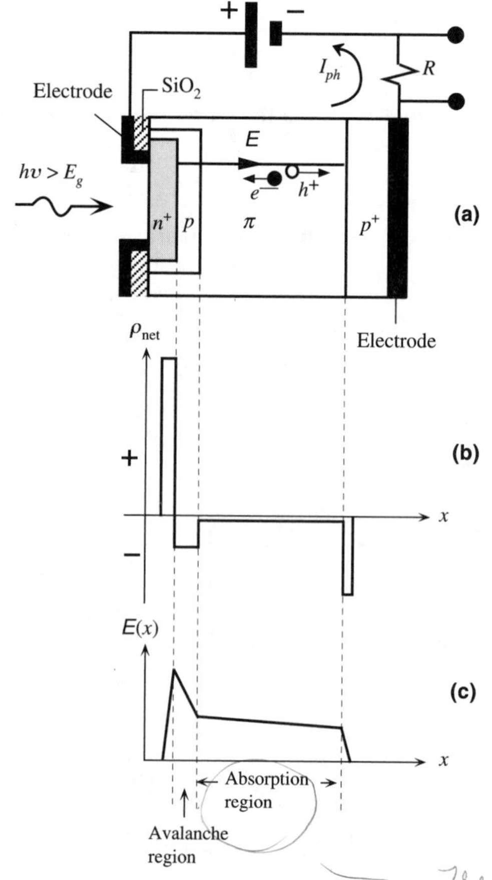

FIGURE 5.9 (a) A schematic illustration of the structure of an avalanche photodiode (APD) biased for avalanche gain. (b) The net space charge density across the photodiode. (c) The field across the diode and the identification of absorption and multiplication regions.

to suitably modify the field distribution across the diode. The first is a thin p-type layer and the second is a thick lightly p-type doped (almost intrinsic) π-layer and the third is a heavily doped p^+ layer. The diode is reverse biased to increase the fields in the depletion regions. The net space charge distribution across the diode due to exposed dopant ions is shown in Figure 5.9 (b). Under zero bias the depletion layer in the p-region does not normally extend across this layer to the π-layer. But when a sufficient reverse bias is applied the depletion region in the p-layer widens to *reach-through* to the π-layer (and hence the name *reach-through*). The field extends from the exposed positively charged donors in the thin depletion layer in n^+ side, all the way to the exposed negatively charged acceptors in the thin depletion layer in p^+-side.

The electric field is given by the integration of the net space charge density ρ_{net} across the diode subject to an applied voltage V_r across the device. The variation in the field across the diode is shown in Figure 5.9 (c). The field lines start at positive ions and end at negative ions which exist through the p, π and p^+ layers. This means that E is maximum at the n^+p junction, then decreases slowly through the p-layer. Through the π-layer it decreases only slightly as the net space charge density here is small. The field vanishes at the end of the narrow depletion layer in the p^+ side.

The absorption of photons and hence photogeneration takes place mainly in the long π-layer. The nearly uniform field here separates the electron hole pairs (EHPs) and drifts them at velocities near saturation towards the n^+ and p^+ sides respectively. When the drifting electrons reach the p-layer, they experience even greater fields and therefore acquire sufficient kinetic energy (greater than E_g) to **impact-ionize** some of the Si covalent bonds and release EHPs as illustrated in Figure 5.10. These generated EHPs themselves can also be accelerated by the high fields in this region to sufficiently large kinetic energies to further cause impact ionization and release more EHPs which leads to an **avalanche of impact ionization processes.** Thus from a single electron entering the p-layer one can generate a large number of EHPs all of which contribute to the observed photocurrent. The photodiode possesses an **internal gain mechanism** in that a single photon absorption leads to a large number of EHPs generated. The photocurrent

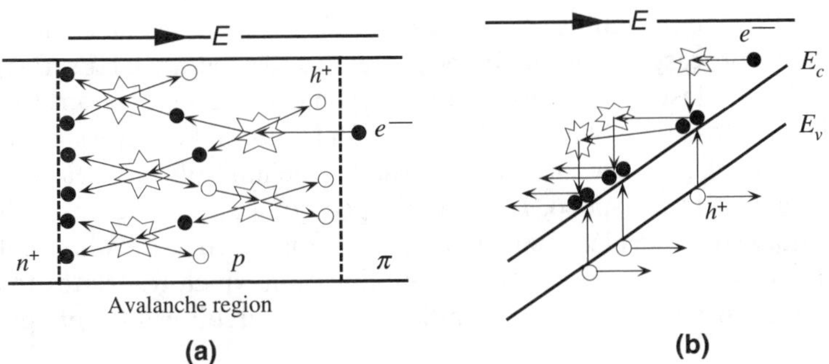

FIGURE 5.10 (a) A pictorial view of impact ionization processes releasing EHPs and the resulting avalanche multiplication. (b) Impact of an energetic conduction electron with crystal vibrations transfers the electron's kinetic energy to a valence electron and thereby excites it to the conduction band.

in the APD in the presence of avalanche multiplication therefore corresponds to an effective quantum efficiency in excess of unity.

The reason for keeping the photogeneration within the π-region and reasonably separate from the avalanche p-region in Figure 5.9 (a) is that avalanche multiplication is a statistical process and hence leads to carrier generation fluctuation which leads to **excess noise** in the avalanche multiplied photocurrent. This is minimized if impact ionization is restricted to the carrier with the highest impact ionization efficiency which in Si is the electron. Thus the structure in Figure 5.9 (a) allows the photogenerated electrons to drift and reach the avalanche region but not the photogenerated holes.

The multiplication of carriers in the avalanche region depends on the probability of impact ionization which depends strongly on the field in this region and hence on the reverse bias V_r. The overall or effective **avalanche multiplication factor** M of an APD is defined as[6]

$$M = \frac{\text{Multiplied photocurrent}}{\text{Primary unmultiplied photocurrent}} = \frac{I_{ph}}{I_{pho}} \tag{1}$$

where I_{ph} is the APD photocurrent that has been multiplied and I_{pho} is the **primary or unmultiplied photocurrent**, the photocurrent that is measured in the absence of multiplication, for example, under a small reverse bias V_r. The multiplication M is a strong function of the reverse bias and also the temperature. The multiplication M can empirically be expressed as

Multiplication

$$M = \frac{1}{1 - \left(\frac{V_r}{V_{br}}\right)^n} \tag{2}$$

where V_{br} is a parameter called the **avalanche breakdown voltage** and n is a characteristic index that provides the best fit to the experimental data (n depends on the temperature). Both V_{br} and n are strongly temperature dependent. For Si APDs M values can be as high as 100 but for many commercial Ge APDs, M are typically around 10.

The speed of the reach-through APD, shown in Figure 5.9 (a), depends on three factors. First is the time it takes for the photogenerated electron to cross the absorption region (the π-layer) to the multiplication region (the p-layer). Second is the time it takes for the avalanche process to build-up in the p-region and generate EHPs. The third is the time it takes for the last hole released in the avalanche process to transit through the π-region. The response time of an APD to an optical pulse is therefore somewhat longer than a corresponding *pin* structure but, in practice, the multiplicative gain often makes up for the reduction in the speed. The overall speed of a photodetector circuit also includes limitations from the electronic preamplifier connected to the photodetector. The APD requires less subsequent electronic amplification which translates to an overall speed that can be faster than a corresponding detector circuit using a *pin* photodiode.

[6] This definition is for an *average* avalanche multiplication for two reasons. First is that the avalanche process is a statistical process with a mean and a standard deviation. Secondly, the field is not entirely uniform in the avalanche region and impact ionization probability is extremely sensitive to the electric field which means multiplication is not uniform over the multiplication region.

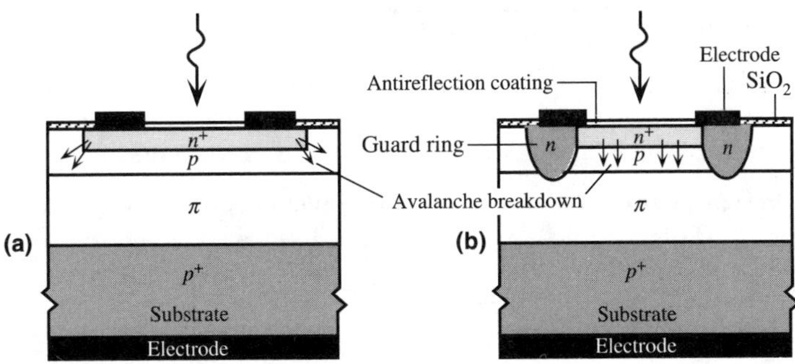

FIGURE 5.11 (a) A Si APD structure without a guard ring. (b) A schematic illustration of the structure of a more practical Si APD.

One of the drawbacks of the simple reach-through APD structure is that the field around the n^+p junction peripheral edge reaches avalanche breakdown before the n^+p regions under the illuminated area as illustrated in Figure 5.11 (a). Ideally the avalanche multiplication should occur uniformly in the illuminated region to encourage the avalanche multiplication of the primary photocurrent rather than the multiplication of the dark current (that is, thermally generated random electron hole pairs). In a practical Si APD, an n-type doped region acting as a **guard ring** surrounds the central n^+ region as shown Figure 5.11 (b) so that the breakdown voltage around the periphery is now higher and avalanche is confined more to illuminated region (n^+p junction). The n^+ and p-layers are very thin (<2 μm) to reduce any absorption in this region; the main absorption occurs in the thick π-region.

Table 5.2 summarizes some typical characteristics of *pn* junction, *pin* and APD photodetectors made from Si, Ge and InGaAs. The rise time t_r refers to the time it takes for the photocurrent to rise from 10% to 90% of its final steady state value from the instant that an optical step excitation is applied. It determines the speed of response of the photodiode. Typical parameters listed in the table are of course highly dependent on the particular device structure which is dictated by the designed application.

TABLE 5.2 Typical characteristics of some *pn* junction, pin and APD type photodetectors based on Si, Ge and InGaAs. t_r is the rise time of the photocurrent from 10% to 90% of its final value when an optical step excitation is applied with photodetector under normal operating conditions (under reverse bias). I_{dark} is typical dark current at normal operating conditions for photosensitive area less than 1 mm^2.

Photodiode	λ_{range} nm	λ_{peak} nm	R at λ_{peak} A/W	Gain	t_r (ns)	I_{dark}
Si *pn* junction	200–1100	600–900	0.5–0.6	<1	0.5	0.01–0.1 nA
Si *pin*	300–1100	800–900	0.5–0.6	<1	0.03–0.05	0.01–0.1 nA
Si APD	400–1100	830–900	40–130	10–100	0.1	1–10 nA
Ge *pn* junction	700–1800	1500–1600	0.4–0.7	<1	0.05	0.1–1 μA
Ge APD	700–1700	1500–1600	4–14	10–20	0.1	1–10 μA
InGaAs-InP *pin*	800–1700	1500–1600	0.7–0.9	<1	0.03–0.1	0.1–10 nA
InGaAs-InP APD	800–1700	1500–1600	7–18	10–20	0.07–0.1	10–100 nA

EXAMPLE 5.6.1 InGaAs APD Responsivity

An InGaAs APD has a quantum efficiency (QE, η) of 60% at 1.55 μm in the absence of multiplication ($M = 1$). It is biased to operate with a multiplication of 12. Calculate the photocurrent if the incident optical power is 20 nW. What is the responsivity when the multiplication is 12?

Solution The responsivity at $M = 1$ in terms of the quantum efficiency is

$$R = \eta\frac{e\lambda}{hc} = (0.6)\frac{(1.6 \times 10^{-19})(1550 \times 10^{-9})}{(6.626 \times 10^{-34})(3 \times 10^{8})} = 0.75 \text{ A W}^{-1}$$

If I_{pho} is the primary photocurrent (unmultiplied) and P_o is the incident optical power then by definition $R = I_{pho}/P_o$ so that

$$I_{pho} = RP_o = (0.75 \text{ A W}^{-1})(20 \times 10^{-9} \text{ W}) = 1.5 \times 10^{-8} \text{ A}.$$

The photodiode current I_{ph} in the APD will be I_{pho} multiplied by M,

$$I_{ph} = MI_{pho} = (12)(1.5 \times 10^{-8} \text{ A}) = 1.80 \times 10^{-7} \text{ A or } 180 \text{ nA}.$$

The responsivity at $M = 12$ is

$$R' = I_{ph}/P_o = MR = (12)/(0.75) = 9.0 \text{ A W}^{-1}$$

EXAMPLE 5.6.2 Silicon APD

A Si APD has a QE of 70% at 830 nm in the absence of multiplication, that is $M = 1$. The APD is biased to operate with a multiplication of 100. If the incident optical power is 10 nW what is the photocurrent?

Solution The unmultiplied responsivity is given by,

$$R = \eta\frac{e\lambda}{hc} = (0.70)\frac{(1.6 \times 10^{-19})(830 \times 10^{-9})}{(6.626 \times 10^{-34})(3 \times 10^{8})} = 0.47 \text{ A W}^{-1}$$

Then the unmultiplied primary photocurrent from the definition of R is

$$I_{pho} = RP_o = (0.47 \text{ A W}^{-1})(10 \times 10^{-9} \text{ W}) = 4.7 \text{ nA}.$$

The multiplied photocurrent is

$$I_{ph} = MI_{pho} = (100)(4.67 \text{ nA}) = 470 \text{ nA or } 0.47 \text{ μA}.$$

5.7 HETEROJUNCTION PHOTODIODES

A. Separate Absorption and Multiplication (SAM) APD

III-V compound based APDs have been developed for use at the communications wavelengths 1.3 μm and 1.55 μm. As in the reach-through Si APD, the **absorption** or **photogeneration region** is separated from the **avalanche** or **multiplication region** which allows the multiplication to be initiated by one type of carrier. Figure 5.12 is a simplified schematic diagram of the structure of an InGaAs-InP APD with a **separate absorption**

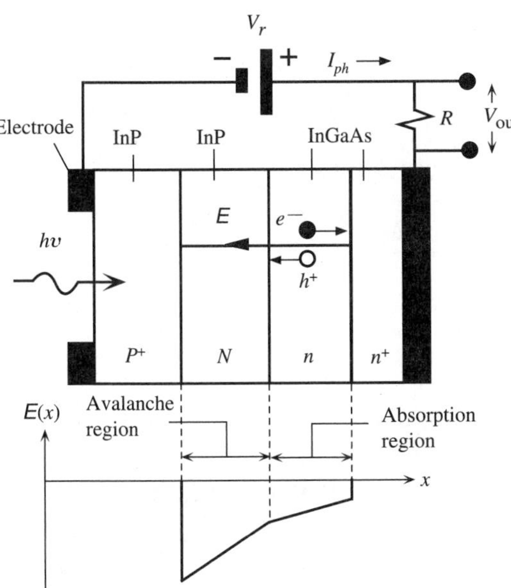

FIGURE 5.12 Simplified schematic diagram of a separate absorption and multiplication (SAM) APD using a heterostructure based on InGaAs-InP. P and N refer to p and n-type wider-bandgap semiconductors.

and multiplication (SAM). InP has a wider bandgap than InGaAs and the p and n type doping of InP is indicated by capital letters, P and N. The main depletion layer is between P^+-InP and N-InP layers and it is within the N-InP. This is where the field is greatest and therefore it is in this N-InP layer where avalanche multiplication takes place. With sufficient reverse bias the depletion layer in the n-InGaAs reaches through to the N-InP layer. The field in the depletion layer in n-InGaAs is not as great as that in N-InP. The variation of the field across the device is also shown Figure 5.12. Although the long wavelength photons are incident onto the InP side, they are not absorbed by InP since the photon energy is less than the bandgap energy of InP $(E_g = 1.35 \text{ eV})$. Photons pass through the InP layer and become absorbed in the n-InGaAs layers. The field in the n-InGaAs layer drifts the holes to the multiplication region where impact ionization multiplies the carriers.

There are a number of practical features that are not shown in the highly simplified diagram in Figure 5.12. Photogenerated holes drifting from n-InGaAs to N-InP become trapped at the interface because there is a sharp increase in the bandgap and a sharp change ΔE_v in E_v (valence band edge) between the two semiconductors and holes cannot easily surmount the potential energy barrier ΔE_v as depicted in Figure 5.13 (a). This problem is overcome by using thin layers of n-type InGaAsP with intermediate bandgaps to provide a graded transition from InGaAs to InP as illustrated in Figure 5.13 (b). Effectively ΔE_v has been broken up into two steps. The hole has sufficient energy to overcome the first step and enter the InGaAsP layer. It drifts and accelerates in the InGaAsP layer to gain sufficient energy to surmount the second step. These devices are called **separate absorption, grading and multiplication** (SAGM) APDs. Both the InP layers are grown epitaxially on an InP substrate. The substrate itself is not used directly to make the P-N junction to prevent crystal defects (for example, dislocations) in the substrate appearing in the

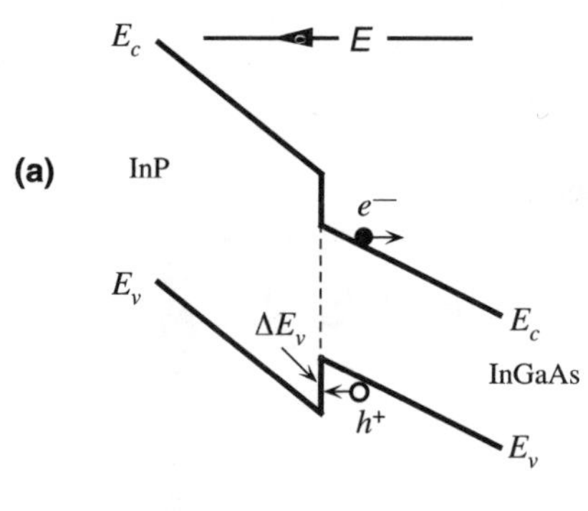

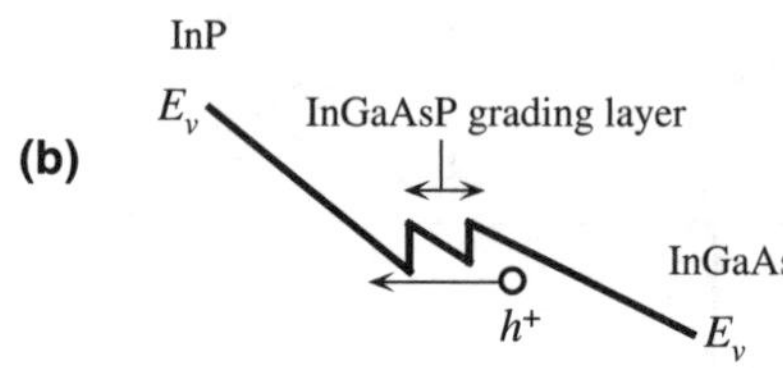

FIGURE 5.13 (a) Energy band diagram for a SAM heterojunction APD where there is a valence band step ΔE_v from InGaAs tro InP that slows hole entry into the InP layer. (b) An interposing grading layer (InGaAsP) with an intermediate bandgap breaks ΔE_v and makes it easier for the hole to pass to the InP layer.

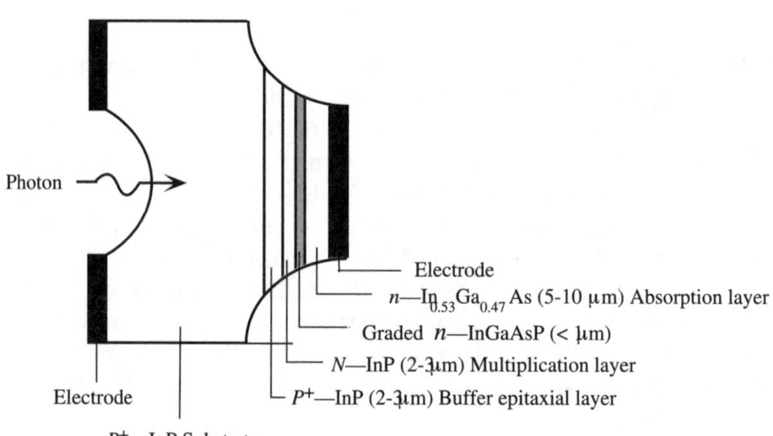

FIGURE 5.14 Simplified schematic diagram of a more practical mesa-etched SAGM layered APD.

multiplication region and hence deteriorating the device performance. The schematic diagram of a more practical SAGM APD is depicted in Figure 5.14.

B. Superlattice APDs

As mentioned previously, APDs exhibit excess noise in the photocurrent (above the expected shot noise) due to inherent statistical variations in the avalanche multiplication process. This excess avalanche noise is reduced to minimum when only one type of car-

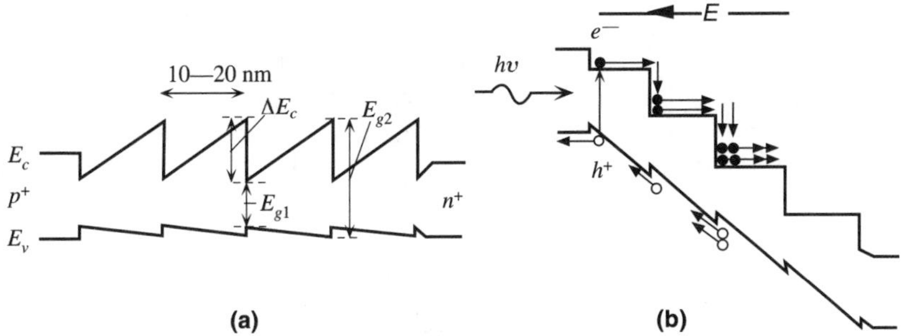

FIGURE 5.15 Energy band diagram of a staircase superlattice APD (a) No bias. (b) With an applied bias.

rier, for example the electron, is involved in impact ionizations. One method of achieving single carrier multiplication is by fabricating multilayer devices that have alternating layers of different bandgap semiconductors, as in **multiple quantum well** (MQW) devices discussed in Chapter 4. The multilayered structure consisting of many alternating layers of different bandgap semiconductors is called a **superlattice.** Figure 5.15 (a) shows the energy band diagram of a **staircase superlattice APD** in which the bandgap is graded within each layer. The bandgap in each layer changes from a minimum E_{g1} to a maximum E_{g2} which is more than twice E_{g1}. There is a step change ΔE_c in the conduction band edge between two neighboring graded layers that is greater than E_{g1}.

In very simple terms, as illustrated in Figure 5.15 (b), the photogenerated electron initially drifts in the graded layer conduction band. When the electron drifts into the neighboring layer, it now has a kinetic energy ΔE_c above E_c in this layer. It therefore enters the neighboring layer as a highly energetic electron and loses the excess energy ΔE_c by impact ionization. The process repeats itself from layer to layer leading to an avalanche multiplication of the photogenerated electron. Since the impact ionization is primarily achieved as a result of transition over ΔE_c, the device does not need the high fields typical of avalanche multiplication in bulk semiconductor; it can operate at lower fields. Further, the impact ionized holes experience only a small ΔE_v which is insufficient to lead to multiplication. Thus, effectively, only electrons are multiplied and the device is a **solid state photomultiplier.**

Such staircase superlattice APDs are difficult to fabricate and involve varying the composition of a quaternary semiconducting alloy (such as AlGaAsSb) to obtain the necessary bandgap grading. Superlattice structures that are simply alternating layers of low and high bandgap semiconductor layers, that is layers do not have a graded bandgap, are easier to fabricate and constitute **multiple quantum well** (MQW) detectors. Typically *molecular beam epitaxy* (MBE) is used to fabricate such multilayer structures.

5.8 PHOTOTRANSISTORS

The **phototransistor** is a bipolar junction transistor (BJT) that operates as a photodetectors with a photocurrent gain. The basic principle is illustrated in Figure 5.16. In an ideal device, only the depletion regions, or the space charge layers (SCL), contain an

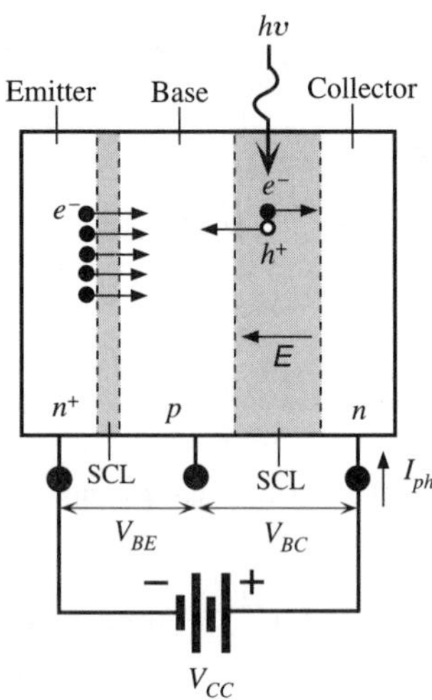

FIGURE 5.16 The principle of operation of the phototransistor. SCL is the space charge layer or the depletion region. The primary photocurrent acts as a base current and gives rise to a large photocurrent in the emitter-collector circuit.

electric field. The base terminal is normally open and there is a voltage applied between the collector and emitter terminals just as in the normal operation of a common emitter BJT. An incident photon is absorbed in the SCL between the base and collector to generate an electron hole pair (EHP). The field $\mathbf{E}$ in the SCL separates the electron hole pair and drifts them in opposite direction. This is the primary photocurrent and effectively constitutes a base current even though the base terminal is open circuit (current is flowing into the base from the collector). When the drifting electron reaches the collector, it becomes collected (and thereby neutralized) by the battery. On the other hand, when the hole enters the neutral base region, it can only be neutralized by injecting a large number of electrons into the base. It effectively "forces" a large number of electrons to be injected from the emitter. Typically the electron recombination time in the base is very long compared with the time it takes for electrons to diffuse across the base. This means that only a small fraction of electrons injected from the emitter can recombine with holes in the base. Thus, the emitter has to inject a large number of electrons to neutralize this extra hole in the base. These electrons (except one) diffuse across the base and reach the collector and thereby constitute an amplified photocurrent.

Alternatively, one can argue that the photogeneration of EHPs in the collector SCL decreases the resistance of this region which decreases the voltage V_{BC} across the base-collector junction. Consequently the base-emitter voltage V_{BE} must increase inasmuch as $V_{BE} + V_{BC} = V_{CC}$ (Figure 5.16). This increase in V_{BE} acts as if it were a forward bias across the base-emitter junction and injects electrons into the base due to the transistor action, that is the emitter current $I_E \propto \exp(eV_{BE}/kT)$.

Since the photon generated primary photocurrent I_{pho} is amplified as if it were a base current (I_B), the photocurrent flowing in the external circuit is

$$I_{ph} \approx \beta I_{pho}$$

where β is the current gain (or h_{FE}) of the transistor. The phototransistor construction is such that incident radiation is absorbed in the base-collector junction SCL.

It is possible to construct a **heterojunction phototransistor** that has different bandgap materials for the emitter, base and collector. For example, if the emitter in Figure 5.16 is InP $(E_g = 1.35 \text{ eV})$ and the base is an InGaAsP alloy (for example, $E_g \approx 0.85$ eV) then photons with energies less than 1.35 eV but more than 0.85 will pass through the emitter and become absorbed in the base. This means the device can be illuminated through the emitter.

5.9 PHOTOCONDUCTIVE DETECTORS AND PHOTOCONDUCTIVE GAIN

The **photoconductive detectors** have the simple structure schematically depicted in Figure 5.17 where two electrodes are attached to a semiconductor that has the desired absorption coefficient and quantum efficiency over the wavelengths of interest. Incident photons become absorbed in the semiconductor and photogenerate electron hole pairs (EHPs). The result is an increase in the conductivity of the semiconductor and hence an increase in the external current which constitutes the photocurrent I_{ph} as shown Figure 5.17.

The actual response of the detector depends whether the contacts to the semiconductor are ohmic or blocking (for example Schottky junctions that do not inject carriers) and on the nature of carrier recombination kinetics. We will consider a photoconductor with ohmic contacts (that is, the contacts do not limit the current flow as in the case of a Schottky junction contact). With ohmic contacts, the photoconductor exhibits **photoconductive gain**, that is *the external photocurrent is due to more than one electron flow per absorbed photon* as illustrated in Figure 5.18 as explained below.

An absorbed photon photogenerates an EHP which drift in opposite directions as shown in Figure 5.18 (a). The electron drifts much faster than the hole and therefore leaves the samples quickly. The sample however must be neutral which means another electrons must enter the sample from the negative electrode as in (b) (the electrode is ohmic). This new electron also drifts across quickly as in (b) and (c) to leave the sample while the hole is still drifting slowly in the sample. Thus another electron must enter the sample to maintain neutrality, and so on, until either the hole reaches the negative electrode or recombines with one of these electrons entering the sample. The external photocurrent therefore corresponds to the flow of many electrons per absorbed photon

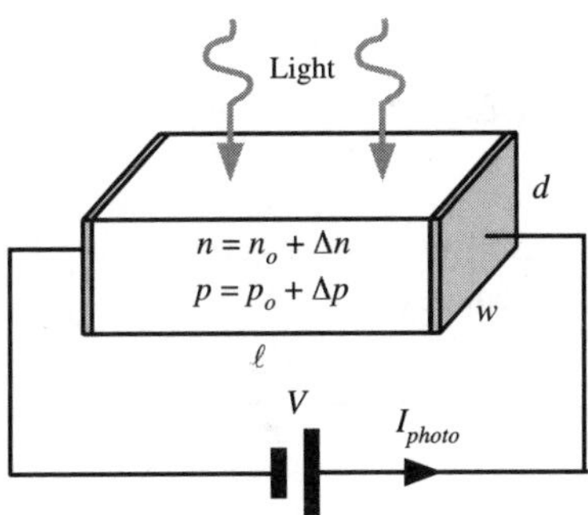

FIGURE 5.17 A semiconductor slab of length ℓ, width w and depth d is illuminated with light of wavelength λ.

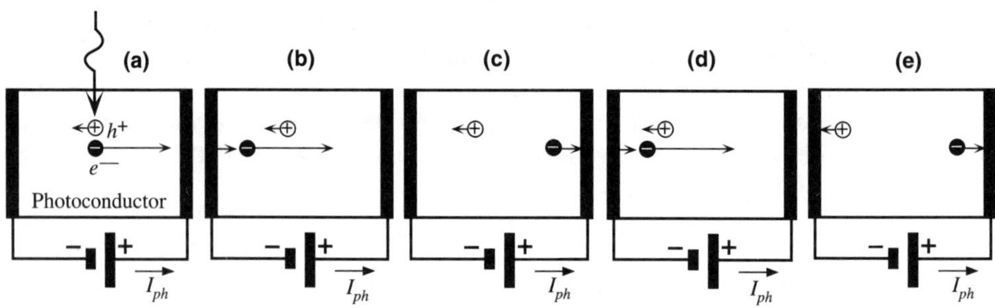

FIGURE 5.18 A photoconductor with ohmic contacts (contacts not limiting carrier entry) can exhibit gain. As the slow hole drifts through the photoconductors, many fast electrons enter and drift through the photoconductor because, at any instant, the photoconductor must be neutral. Electrons drift faster which means as one leaves, another must enter.

which represents a **gain**. The gain depends on the drift time of the carriers and their recombination lifetime.

Suppose that the photoconductor is suddenly illuminated by a step light. If Γ_{ph} is the number of photons arriving per unit area per unit second (the photon flux), then $\Gamma_{ph} = I/h\nu$ where I is the light intensity (energy flowing per unit area per second) and $h\nu$ is the energy per photon. Thus, the number of EHPs generated per unit volume per second, photogeneration rate per unit volume g_{ph}, is given by

$$g_{ph} = \frac{\eta A \Gamma_{ph}}{Ad} = \frac{\eta\left(\frac{I}{h\nu}\right)}{d} = \frac{\eta I \lambda}{hcd} \tag{1}$$

where A is the illuminated surface area.

Suppose that at any instant the electron concentration is n (includes photogenerated electrons) and the thermal equilibrium concentration (in the dark) is n_o. Then $\Delta n = n - n_o$, is the *excess electron concentration*. For photogeneration, obviously $\Delta n = \Delta p$.

The rate of increase in the excess electron concentration
= Rate of photogeneration of excess electrons
−Rate of recombination of excess electrons

If τ is the mean recombination time of excess electrons, then

$$\frac{d\Delta n}{dt} = g_{ph} - \frac{\Delta n}{\tau} \tag{2}$$

It is clear from Eq. (2), that Δn increases exponentially from the instant the light is turned on until a steady state is reached when

$$\frac{d\Delta n}{dt} = g_{ph} - \frac{\Delta n}{\tau} = 0$$

so that

$$\Delta n = \tau g_{ph} = \frac{\tau \eta I \lambda}{hcd}. \tag{3}$$

The conductivity of a semiconductor is given by $\sigma = e\mu_e n + e\mu_h p$, so that the change in the conductivity, called **photoconductivity**, is

$$\Delta\sigma = e\mu_e \Delta n + e\mu_h \Delta p = e\Delta n(\mu_e + \mu_h)$$

since electrons and holes are generated in pairs, $\Delta n = \Delta p$. Thus, substituting for Δn in the $\Delta\sigma$ expression above we get,

$$\Delta\sigma = \frac{e\eta I \lambda \tau(\mu_e + \mu_h)}{hcd}. \tag{4}$$

The photocurrent density is simply

$$J_{ph} = \Delta\sigma \frac{V}{\ell} = \Delta\sigma E \tag{5}$$

The number of electrons flowing in the external circuit can be found from the photocurrent because

$$\text{Rate of electron flow} = \frac{I_{ph}}{e} = \frac{wdJ_{ph}}{e} = \frac{\eta I w \lambda \tau(\mu_e + \mu_h)E}{hc}. \tag{6}$$

However, the rate of electron (that is EHP) photogeneration is

$$\text{Rate of electron generation} = (\text{Volume})g_{ph} = (wd\ell)g_{ph} = w\ell\frac{\eta I \lambda}{hc}$$

The photoconductive gain is then simply

$$G = \frac{\text{Rate of electron flow in external circuit}}{\text{Rate of electron generation by light absorption}} = \frac{\tau(\mu_e + \mu_h)E}{\ell}. \tag{7}$$

Equation (7) can be simplified further by noting that the drift velocities of the electrons and holes in the photoconductor are $\mu_e E$ and $\mu_h E$ respectively so that their transit times (times to cross the semiconductor) are

$$t_e = \ell/(\mu_e E) \qquad \text{and} \qquad t_h = \ell/(\mu_h E)$$

Using these transit times in Eq. (6) we find,

$$G = \frac{\tau}{t_e} + \frac{\tau}{t_h} = \frac{\tau}{t_e}\left(1 + \frac{\mu_h}{\mu_e}\right) \tag{8}$$

Photoconductive gain

The photoconductive gain can be quite high if τ/t_e is kept large which requires a long recombination time and a short transit time. The transit time can be made shorter by applying a greater field but this will also lead to an increase in the dark current and more noise. The speed of response of the device is limited by the recombination time of the injected carriers. A long τ means a slow device.

5.10 NOISE IN PHOTODETECTORS

A. The *pn* Junction and the *pin* Photodiodes

The lowest signal that a photodetectors can detect is determined by the extent of random fluctuations in the current through the detector and the voltage across it as a result of various statistical processes in the device. When a *pn* junction is reverse biased there is still a dark current I_d present which is mainly due to thermal generation of electron-hole pairs (EHPs) in the depletion layer and within diffusion lengths to the depletion layer. If the dark current were absolutely constant with no fluctuations then any change in the diode current, however small (even a tiny fraction of I_d), due to an optical signal could be easily detected by blocking or removing I_d. The dark current however exhibits **shot noise** or fluctuations about I_d, as shown in Figure 5.19. This shot noise is due to the fact that electrical conduction is by *discrete charges* which means that there is a statistical distribution in the transit times of the carriers across the photodiode. Carriers are collected as discrete amounts of charge (e) that arrive at random times and not continuously. It is unlike a continuous flow of water through a pipe but rather like rolling ball bearings down a pipe at random times. There will be fluctuations in the outflow rate of ball bearings at the collection end.

The *root mean square* (rms) value of the fluctuations in the dark current represents the **shot noise** current $i_{n\text{-dark}}$

Dark current shot noise

$$i_{n\text{-dark}} = \left[2eI_dB\right]^{1/2} \tag{1}$$

where B is the frequency bandwidth of the photodetector. The photocurrent signal must be greater than this shot noise in the dark current.

The photodetection process involves the interaction of discrete photons with valence electrons. The discrete nature of photons means that there is an unavoidable random fluctuation in the rate of arrival of photons even if we did our best to keep the rate constant. The quantum nature of the photon therefore gives rise to a statistical randomness in the EHP photogeneration process. This type of fluctuation is called **quantum noise** (or **photon noise**) and it is equivalent to shot noise as far as its effects are concerned. Thus the photocurrent will always exhibit fluctuations about its mean value

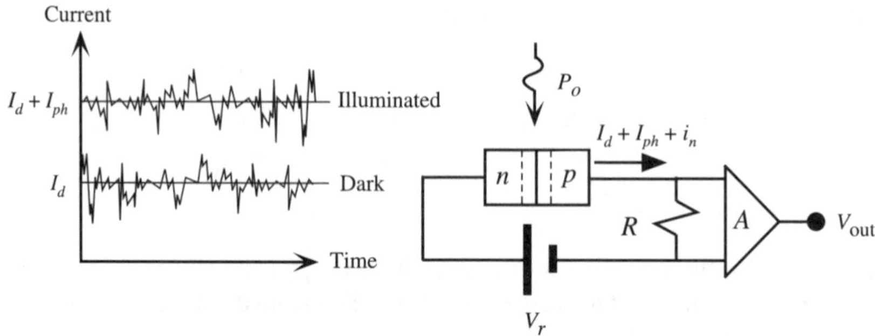

FIGURE 5.19 In *pn* junction and *pin* devices the main source of noise is shot noise due to the dark current and photocurrent.

due to quantum noise. If I_{ph} is the mean photocurrent, the fluctuations about this mean has an rms value that is called **shot noise current** due to **quantum noise**,

$$i_{n\text{-quantum}} = \left[2eI_{ph}B\right]^{1/2} \quad \textbf{(2)}$$

Quantum noise

Generally the dark current shot noise and quantum noise are the main sources of noise in *pn* junction and *pin* type photodiodes. The total shot noise generated by the photodetector is not a simple sum of Eqs. (1) and (2) because the two processes are due to independent random fluctuations. We need to sum the power in each or add the mean squares of the shot noise currents, *i.e.*

$$i_n^2 = i_{n\text{-dark}}^2 + i_{n\text{-quantum}}^2$$

so that the rms total shot noise current is,

$$i_n = \left[2e\left(I_d + I_{ph}\right)B\right]^{1/2} \quad \textbf{(3)}$$

Shot noise in pn junction

In Figure 5.19, the photodetector current, $I_d + I_{ph} + i_n$, flows through a load resistor R which acts as a sampling resistor for measuring the current. The voltage across R is amplified. In considering the noise of the receiver we also have to include the thermal noise in the resistor and the noise in the input stage of the amplifier. Thermal noise is random voltage fluctuations across any conductor due to random motions of the conduction electrons. In receiver design we are often interested in the **signal to noise ratio**, SNR or S/N, which is defined as the ratio of signal power to noise power,

$$\text{SNR} = \frac{\text{Signal Power}}{\text{Noise Power}} \quad \textbf{(4)}$$

For the photodetector alone, SNR is simply the ratio of I_{ph}^2 to i_n^2. SNR for the receiver must include the noise power generated in the sampling resistor R (thermal noise) and in the input elements of the amplifier (for example, resistors and transistors).

The **noise equivalent power** (NEP) is an important property of a photodetector that is frequently quoted. NEP is the optical signal power required to generate a photocurrent signal $\left(I_{ph}\right)$ that is equal to the total noise current $\left(i_n\right)$ in the photodetector at a given wavelength and within a bandwidth of 1 Hz. It is apparent that NEP represents the required optical power to achieve a SNR of 1 within a bandwidth of 1 Hz. The **detectivity** D is the reciprocal of NEP, $D = 1/\text{NEP}$.

If R is the responsivity and P_o is the monochromatic incident optical power then the generated photocurrent is,

$$I_{ph} = RP_o \quad \textbf{(5)}$$

Suppose that the photogenerated current I_{ph} is equal to the noise current i_n in Eq. (3), when the incident optical power P_o is P_1. Then

$$RP_1 = \left[2e\left(I_d + I_{ph}\right)B\right]^{1/2}$$

From this we find the optical power per square root of bandwidth as

$$\frac{P_1}{B^{1/2}} = \frac{1}{R}\left[2e\left(I_d + I_{ph}\right)\right]^{1/2}$$

The quantity $P_1/B^{1/2}$ represents the optical power necessary, per square root of frequency bandwidth, that generates a photocurrent equal to the noise current which is the quantitative definition of NEP, that is,

Noise equivalent power

$$\text{NEP} = \frac{P_1}{B^{1/2}} = \frac{1}{R}\left[2e(I_d + I_{ph})\right]^{1/2} \quad \textbf{(6)}$$

It is clear that if we put $B = 1$ Hz, we obtain numerically NEP $= P_1$, that value of P_o which makes I_{ph} equal to the total noise current. From Eq. (6), the units for NEP are W Hz$^{-1/2}$.

EXAMPLE 5.10.1 NEP of a Si *pin* Photodiode

A Si *pin* photodiode has a quoted NEP of 1×10^{-13} W Hz$^{-1/2}$. What is the optical signal power it needs for a signal to noise ratio (SNR) of 1 if the bandwidth of operation is 1 GHz?

Solution By definition NEP is that optical power per square root of bandwidth which generates a photocurrent equal to the noise current in the detector.

$$\text{NEP} = P_1/B^{1/2}.$$

Thus,

$$P_1 = \text{NEP}B^{1/2} = (10^{-13}\text{ W Hz}^{-1/2})(10^9\text{ Hz})^{1/2} = 3.16 \times 10^{-9}\text{ W or } 3.16\text{ nW}.$$

EXAMPLE 5.10.2 Noise of an ideal photodetector

Consider an ideal photodiode with $\eta = 1$ (QE $= 100\%$) and no dark current, $I_d = 0$. Show that the minimum optical power required for a signal to noise ratio (SNR) of 1 is

$$P_1 = \frac{2hc}{\lambda}B \quad \textbf{(7)}$$

Calculate the minimum optical power for an SNR $= 1$ for an ideal photodetector operating at 1300 nm with a bandwidth of 1 GHz. What is the corresponding photocurrent?

Solution We need the incident optical power P_1 that makes the photocurrent I_{ph} equal to the noise current i_n, so that SNR $= 1$. The photocurrent (signal) is equal to the noise current when,

$$I_{ph} = \left[2e(I_d + I_{ph})B\right]^{1/2} = \left[2eI_{ph}B\right]^{1/2}$$

since $I_d = 0$.

Thus,

$$I_{ph} = 2eB$$

From Eqs. (3) and (4) in §5.4, the photocurrent I_{ph} and the incident optical power P_1 are related by

$$I_{ph} = \frac{\eta e\lambda P_1}{hc} = 2eB$$

Thus,

$$P_1 = \frac{2hc}{\eta\lambda} B$$

For an ideal photodetector, $\eta = 1$ which leads to Eq. (7). We note that for a bandwidth of 1 Hz, NEP is numerically equal to P_1 or NEP $= 2hc/\lambda$.

For an ideal photodetector operating at 1.3 μm and at 1 GHz,

$$P_1 = 2hcB/\eta\lambda = 2(6.626 \times 10^{-34})(3 \times 10^8)(10^9)/(1)(1.3 \times 10^{-6})$$
$$= 3.1 \times 10^{-10} \text{ W or } 0.31 \text{ nW}.$$

This is the minimum signal for a SNR = 1. *The noise current is due to quantum noise.* The corresponding photocurrent is,

$$I_{ph} = 2eB = 2(1.6 \times 10^{-19})(10^9) = 3.2 \times 10^{-10} \text{ A or } 0.32 \text{ nA}.$$

Alternatively we can calculate I_{ph} from $I_{ph} = \eta e\lambda P_1/hc$ with $\eta = 1$.

EXAMPLE 5.10.3 SNR of a Receiver

Consider an InGaAs *pin* photodiode used in a receiver circuit as in Figure 5.19 with a load resistor of 1 kΩ. The photodiode has a dark current of 5 nA. The bandwidth of the amplifier is 500 MHz. Assuming that the amplifier is noiseless, calculate the SNR when the incident optical power generates a mean photocurrent of 15 nA (corresponding to an incident optical power of about 20 nW).

Solution The noise generated comes from the photodetector as shot noise and from R as thermal noise. The mean thermal noise power in the resistor R is $4k_BTB$. If I_{ph} is the photocurrent and i_n is the shot noise in the photodetector then

$$\text{SNR} = \frac{\text{Signal Power}}{\text{Noise Power}} = \frac{I_{ph}^2 R}{i_n^2 R + 4k_BTB} = \frac{I_{ph}^2}{\left[2e(I_d + I_{ph})B\right] + \dfrac{4k_BTB}{R}}$$

The term $4k_BTB/R$ in the denominator represents the mean square of the thermal noise current in the resistor. We can evaluate the magnitude of each noise current by substituting, $I_{ph} = 15$ nA, $I_d = 5$ nA, $B = 500$ MHz, $R = 1000\ \Omega$, $T = 300$ K.

$$\text{Shot noise current from the detector} = \left[2e(I_d + I_{ph})B\right]^{1/2} = 1.79 \text{ nA}$$

and

$$\text{Thermal noise current from } R = \left[\frac{4k_BTB}{R}\right]^{1/2} = 5.26 \text{ nA}$$

Thus the noise contribution from R is greater than the photodetector. The SNR is

$$\text{SNR} = \frac{(15 \times 10^{-9})^2}{(1.79 \times 10^{-9})^2 + (5.26 \times 10^{-9})^2} = 7.26$$

Generally SNR is quoted in decibels. In that case we need 10 log(SNR), or 10 log(7.26) i.e. 8.6 dB.

B. Avalanche Noise in the APD

In the avalanche photodiode, both photogenerated and thermally generated carriers entering the avalanche zone are multiplied. The shot noise associated with these carriers are also multiplied. If I_{do} and I_{pho} are the unmultiplied ($M = 1$) dark current and photocurrent (primary photocurrent) respectively in the APD then the total shot noise current (as an rms value) in the APD should be

$$i_{n\text{-APD}} = M\left[2e\left(I_{do} + I_{pho}\right)B\right]^{1/2} = \left[2e\left(I_{do} + I_{pho}\right)M^2B\right]^{1/2} \quad \textbf{(9)}$$

The APDs exhibit excess avalanche noise that is above this multiplied shot noise of the photocurrent and dark current. This excess noise is due to the randomness of the impact ionization process in the multiplication region. Some carriers travel far and some short distances within this zone before they cause impact ionization. Furthermore, the impact ionization does not occur uniformly over the multiplication region but is more frequent in the highest field zone. Thus the multiplication M fluctuates about a mean value. The result of the statistics of impact ionization is an excess noise contribution, called **avalanche noise,** to the multiplied shot noise. The noise current in an APD is given by,

APD noise current

$$i_{n\text{-APD}} = \left[2e\left(I_{do} + I_{pho}\right)M^2FB\right]^{1/2} \quad \textbf{(10)}$$

where F is called the **excess noise factor** and is a function of M and the impact ionization probabilities (called coefficients). Generally, F is approximated by the relationship $F \approx M^x$ where x is an index that depends on the semiconductor, the APD structure and the type of carrier that initiates the avalanche (electron or hole). For Si APDs, x is $0.3 - 0.5$ whereas for Ge and III-V (such as InGaAs) alloys it is $0.7 - 1$.

EXAMPLE 5.10.4 Noise in an APD

Consider an InGaAs APD with $x \approx 0.7$ which is biased to operate at $M = 10$. The unmultiplied dark current is 10 nA and bandwidth is 700 MHz.

a. What is the APD noise current per square root of bandwidth?

b. What is the APD noise current for a bandwidth of 700 MHz?

c. If the responsivity (at $M = 1$) is 0.8 what is the minimum optical power for an SNR of 10?

Solution In the absence of any photocurrent, the noise in the APD comes from the dark current, If the unmultipled dark current is I_{do} then the noise current (rms) is,

$$i_{n\text{-dark}} = \left[2eI_{do}M^{2+x}B\right]^{1/2} \quad \textbf{(11)}$$

Thus,

$$\frac{i_{n\text{-dark}}}{\sqrt{B}} = \sqrt{2eI_{do}M^{2+x}} = \sqrt{2(1.6 \times 10^{-19})(10 \times 10^{-9})(10)^{2+0.7}}$$

$$= 1.27 \times 10^{-12}\ \text{A}/\sqrt{\text{Hz}} \text{ or } 1.27\ \text{pA}/\sqrt{\text{Hz}}\,.$$

In a bandwidth B of 700 MHz, the noise current is

$$i_{n\text{-dark}} = \left(700 \times 10^6\ \text{Hz}\right)^{1/2}\left(1.27\ \text{pA}/\sqrt{\text{Hz}}\right) = 3.35 \times 10^{-8}\ \text{A or } 0.335\ \text{nA}.$$

The SNR with a primary photocurrent I_{pho} in the APD is

$$\text{SNR} = \frac{\text{Signal Power}}{\text{Noise Power}} = \frac{M^2 I_{pho}^2}{\left[2e(I_{do} + I_{pho})M^{2+x}B\right]} \tag{12}$$

Rearranging to obtain I_{pho} we get,

$$(M^2)I_{pho}^2 - \left[2eM^{2+x}B(\text{SNR})\right]I_{pho} - \left[2eM^{2+x}B(\text{SNR})I_{do}\right] = 0$$

This is a quadratic equation in I_{pho} with defined coefficients since M, x, B, I_{do} and SNR are given. Solving this quadratic with an SNR $= 10$ for I_{pho} we find,

$$I_{pho} = 1.75 \times 10^{-8}\ \text{A or } 17.5\ \text{nA}.$$

By the definition of responsivity, $R = I_{pho}/P_o$, we find,

$$P_o = I_{pho}/R = (0.175 \times 10^{-9}\ \text{A})/(0.8\ \text{A/W}) = 2.19 \times 10^{-8}\ \text{W or } 21.9\ \text{nW}$$

QUESTIONS AND PROBLEMS

5.1 Band gap and photodetection

(a) Determine the maximum value of the energy gap (bandgap) which a semiconductor, used as a photoconductor, can have if it is to be sensitive to yellow light (600 nm).

(b) A photodetector whose area is 5×10^{-2} cm^2 is irradiated with yellow light whose intensity is 2 mW cm^{-2}. Assuming that each photon generates one electron-hole pair (EHP), calculate the number of EHPs generated per second.

(c) From the known energy gap of the semiconductor GaAs ($E_g = 1.42$ eV), calculate the primary wavelength of photons emitted from this crystal as a result of electron-hole recombination. Is this wavelength in the visible?

(d) Will a silicon photodetector be sensitive to the radiation from a GaAs laser? Why?

5.2 Absorption coefficient

(a) If d is the thickness of a photodetector material, I_o is the intensity of the incoming radiation, show that the number of photons absorbed per unit volume of sample is

$$n_{ph} = \frac{I_o[1 - \exp(-\alpha d)]}{dh\nu}$$

(b) What is the thickness of a Ge and $In_{0.53}Ga_{0.47}As$ crystal layer that is needed for absorbing 90% of the incident radiation at 1.5 μm?

(c) Suppose that each absorbed photon liberates one electron (or electron hole pair) in a unity quantum efficiency photodetector and that the photogenerated electrons are immediately collected. Thus, the rate of charge collection is limited by rate of photon generation. What is the external photocurrent density for the photodetectors in (b) if the incident radiation is 100 μW mm^{-2}?

5.3 Ge Photodiode Consider a commercial Ge *pn* junction photodiode which has the responsivity shown in Figure 5.20. Its photosensitive area is 0.008 mm^2. It is used under a reverse bias of 10 V when the dark current is 0.3 μA and the junction capacitance is 4 pF. The rise time of the photodiode is 0.5 ns.

(a) Calculate its quantum efficiency at 850, 1300 and 1550 nm.

(b) What is the intensity of light at 1.55 μm that gives a photocurrent equal to the dark current.

(c) What would be the effect of lowering the temperature on the responsivity curve?

(d) Given that the dark current is in the range of microamperes, what would be the advantage in reducing the temperature?

(e) Suppose that the photodiode is used with a 100 Ω resistance to sample the photocurrent. What limits the speed of response?

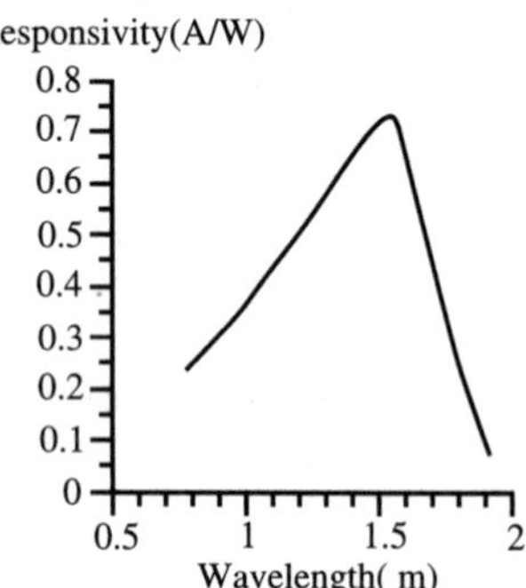

FIGURE 5.20 The responsivity of a commercial Ge *pn* junction photodiode.

5.4 Si *pin* Photodiodes Consider two commercial Si *pin* photodiodes, type A and type B, both classified as fast *pin* photodiodes. They have the responsivity shown Figure 5.21. Differences in the responsivity are due to the *pin* device structure. The photosensitive area is 0.125 mm^2 (0.4 mm in diameter).

(a) Calculate the photocurrent from each when they are illuminated with blue light of wavelength 450 nm and light intensity 1 μW cm^{-2}. What is the QE of each device?

(b) Calculate the photocurrent from each when they are illuminated with red light of wavelength 700 nm and light intensity 1 μW cm^{-2}. What is the QE of each device?

(c) Calculate the photocurrent from each when they are illuminated with infrared light of wavelength 1000 nm and light intensity 1 μW cm^{-2}. What is the QE of each device?

(d) What is your conclusion?

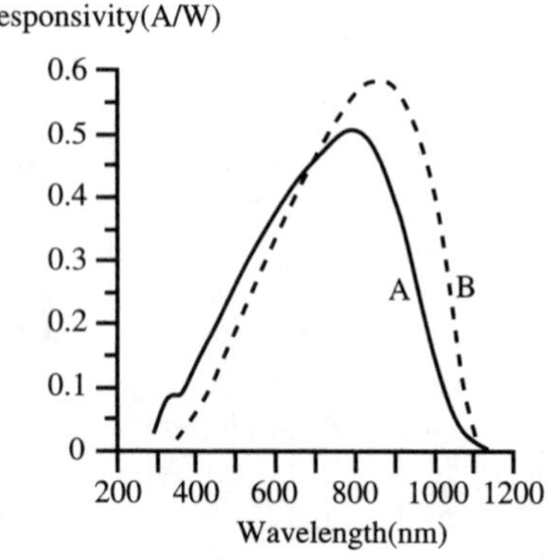

FIGURE 5.21 The responsivity of two commercial Si *pin* photodiodes.

5.5 InGaAs *pin* Photodiodes Consider a commercial InGaAs *pin* photodiode whose responsivity is shown in Figure 5.22. Its dark current is 5 nA.

(a) What optical power at a wavelength of 1.55 μm would give a photocurrent that is twice the dark current? What is the QE of the photodetector at 1.55 μm?

(b) What would be the photocurrent if the incident power was at 1.3 μm? What is the QE at 1.3 μm operation?

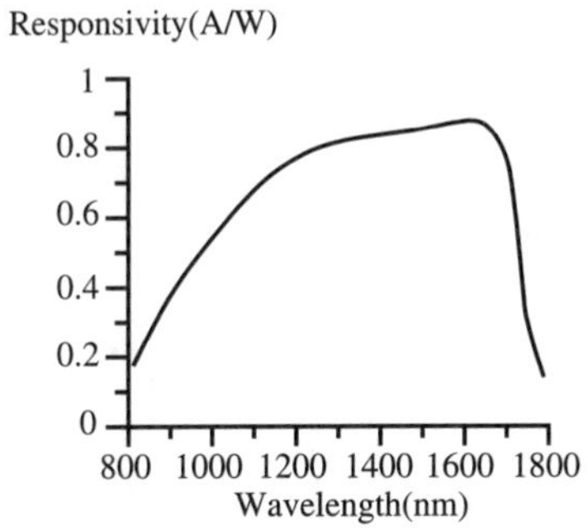

FIGURE 5.22 The responsivity of an InGaAs *pin* photodiode.

5.6 Maximum QE Show that maximum QE occurs when

$$\frac{dR}{d\lambda} = \frac{R}{\lambda}$$

that is, when the tangent at λ passes through the origin ($R = 0, \lambda = 0$). Hence determine the wavelengths where the QE is maximum for the InGAs *pin*, two Si *pin* and Ge photodiodes in Figure 5.22, Figure 5.21 and Figure 5.20 respectively.

5.7 Si *pin* photodiode speed Consider Si *pin* photodiodes that has a p^+ layer of thickness 0.75 μm, *i*-Si layer of width 10 μm. It is reverse biased with a voltage of 20 V.

(a) What is the speed of response due to bulk absorption? What wavelengths would lead to this type of speed of response?

(b) What is the speed of response due to absorption near the surface? What wavelengths would lead to this type of speed of response?

5.8 Transient photocurrents in a *pin* photodiode Consider a reverse biased Si *pin* photodiode as shown in Figure 5.23. It is appropriately reverse biased so that the field in the depletion region (*i*-Si layer) $E = V_r/W$ is the saturation field. Thus, photogenerated electrons and holes in this layer drift at saturation velocities v_{de} and v_{dh}. Assume that the field E is uniform and that the thickness of the p^+ is negligible. A very short light pulse (infinitesimally short) photogenerates EHPs in the depletion layer as shown in Figure 5.23 which results in an exponentially decaying EHP concentrations across W. Figure 5.23 shows the photogenerated electron concentration at time $t = 0$ and also at a later time t when the electrons have drifted a distance $\Delta x = v_{de}\Delta t$. Those that reach the back electrode B become collected. The electron distribution shifts at a constant velocity until the initial electrons at A reach B which represents the longest transit time $\tau_e = W/v_{de}$. Similar argument apply to holes but they drift in the opposite direction and their transit time $\tau_h = W/v_{dh}$ where v_{dh} is their saturation velocity. The photocurrent density at any instant is

$$j_{ph} = j_e(t) + j_h(t) = eN_e v_{de} + eN_h v_{dh}$$

where N_e and N_h are the overall electron and hole concentration in the sample at time t. Assume for convenience that the cross sectional area $A = 1$ (derivations below are not affected as we are interested in the photocurrent densities).

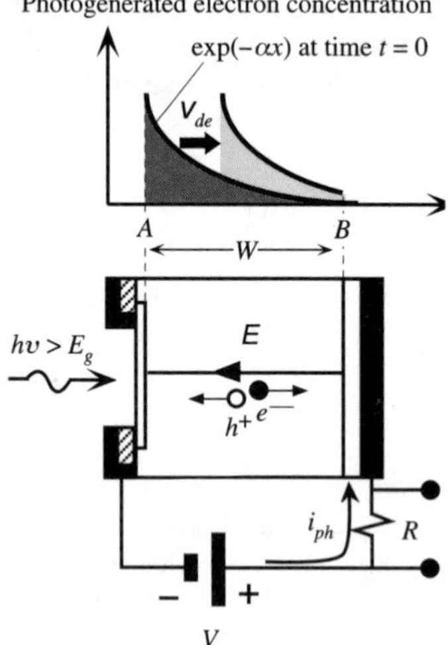

FIGURE 5.23 An infinitesimally short light pulse is absorbed throughout the depletion layer and creates an EHP concentration that decays exponentially.

(a) Sketch the hole distribution at a time t where $\tau_h > t > 0$ and τ_h = hole drift time $= W/v_{dh}$.

(b) The electron concentration distribution $n(x)$ at time t corresponds to that at $t = 0$ shifted by $v_{de}t$. Thus the total electrons in W is proportional to integrating this distribution $n(x)$ from A at $x = v_{de}t$ to B at $x = W$.

Given $n(x) = n_o \exp(-\alpha x)$ at $t = 0$, where n_o is the electron concentration at $x = 0$ at $t = 0$ we have

$$\text{Total number of electrons at time } t = \int_{v_{de}t}^{W} n_o \exp[-\alpha(x - v_{de}t)]\,dx$$

and

$$N_e(t) = \frac{\text{Total number electrons at time } t}{\text{Volume}}$$

Then

$$N_e(t) = \frac{1}{W}\int_{v_{de}t}^{W} n_o \exp[-\alpha(x - v_{de}t)]\,dx$$

$$= \frac{n_o}{W\alpha}\left\{1 - \exp\left[-\alpha W\left(1 - \frac{t}{\tau_e}\right)\right]\right\}$$

where $N_e(0)$ is the initial overall electron concentration at time $t = 0$, that is,

$$N_e(0) = \frac{1}{W}\int_{0}^{W} n_o \exp(-\alpha x)\,dx = \frac{n_o}{W\alpha}[1 - \exp(-\alpha W)]$$

We note that n_o depends on the intensity I of the light pulse so that $n_o \propto I$. Show that for holes,

$$N_h(t) = \frac{n_o \exp(-\alpha W)}{W\alpha}\left\{\exp\left[\alpha W\left(1 - \frac{t}{\tau_h}\right)\right] - 1\right\}$$

(c) Given $W = 40$ μm, $\alpha = 5 \times 10^4$ m^{-1}, $v_{de} = 10^5$ m s^{-1}, $v_{dh} = 0.8 \times 10^5$ m s^{-1}, $n_o = 10^{13}$ cm^{-3}, calculate the electron and hole transit times, sketch the photocurrent densities $j_e(t)$ and $j_h(t)$ and hence $j_{ph}(t)$ as a function of time, and calculate the initial photocurrent. What is your conclusion?

5.9 Fiber attenuation and InGaAs *pin* Photodiode Consider the commercial InGaAs *pin* photodiode whose responsivity is shown in Figure 5.22. This is used in a receiver circuit that needs a minimum of 5 nA photocurrent for a discernible output signal (acceptable signal to noise ratio for the customer). Suppose that the InGaAs *pin* PD is used at 1.3 μm operation with a single mode fiber whose attenuation is 0.35 dB km^{-1}. If the laser diode emitter can launch at most 2 mW of power into the fiber, what is the maximum distance for the communication without a repeater?

5.10 Photoconductive detector An n-type Si photoconductor has a length $L = 100$ μm and a hole lifetime of 1μs. The applied bias to the photoconductor is 10 V.

(a) What are the transit times, t_e and t_h, of an electron and a hole across L? What is the photoconductive gain?

(b) It should be apparent that as electrons are much faster than holes, a photogenerated electron leaves the photoconductor very quickly. This leaves behind a drifting hole and therefore a positive charge in the semiconductor. Secondary (*i.e.* additional electrons) then flow into the photoconductor to maintain neutrality in the sample and the current continues to flow. These events will continue until the hole has disappeared by recombination, which takes on average a time τ. Thus more charges flow through the contact per unit time than charges actually photogenerated per unit time. What will happen if the contacts are not ohmic, *i.e.* they are not injecting ?

(c) What can you say about the product $\Delta\sigma$ and the speed of response which is proportional to $1/\tau$.

5.11 NEP and Ge and InGaAs photodiodes

(a) Show that the noise equivalent power of a photodiode PD is given by

$$NEP = \frac{P_1}{B^{1/2}} = \frac{hc}{\eta e\lambda}\left[2e\left(I_d + I_{ph}\right)\right]^{1/2}$$

How would you improve the *NEP* of a photodiode? What is *NEP* for an ideal PD operating at $\lambda = 1.55$ μm?

(b) Given the dark current I_d of a PD, show that for SNR = 1, the photocurrent is

$$I_{ph} = eB\left[1 + \left(1 + \frac{2I_d}{eB}\right)^{1/2}\right]$$

What is the corresponding optical power P_1?

(c) Consider a fast Ge *pn* junction PD that has a photosensitive area of diameter 0.3 mm. It is reverse biased for photodetection and has a dark current of 0.5 μA. Its peak responsivity is 0.7 A/V at 1.55 μm (see Figure 5.20). The bandwidth of the photodetector and the amplifier circuit together is 100 MHz. Calculate its *NEP* at the peak wavelength and find the minimum optical power and hence minimum light intensity that gives an SNR of 1. How would you improve the minimum detectable optical power?

(d) Table 5.3 shows the characteristics of typical Ge *pn* junction and InGaAs *pin* photodiodes in terms of the responsivity and the dark current. Fill in the remainder of the columns in the table assuming that there is an ideal, noiseless, preamplifier to detect the photocurrent from the photodiode. Assume a working bandwidth, B, of 1 MHz. What is your conclusion?

TABLE 5.3 Ge *pn* junction and InGaAs *pin* PDs. Photosensitive area has a diameter of 1 mm.

Photodiode	R at 1.55 μm ($A W^{-1}$)	I_d (nA)	I_{ph} for SNR = 1 at B = 1 MHz (nA)	Optical power for SNR = 1 at B = 1 MHz (nW)	NEP $W Hz^{-1/2}$	Comment
Ge at 25 °C	0.8	400				
Ge at −20 °C	0.8	5				Thermoelectric cooling
InGaAs *pin*	0.95	3				

5.12 The APD and excess avalanche noise APDs exhibit excess avalanche noise which contributes to the shot noise of the diode current. The total noise current in the APD is given by

$$i_{n\text{-APD}} = \left[2e\left(I_{do} + I_{pho}\right)M^2FB\right]^{1/2} \tag{1}$$

where F is the excess noise factor which depends in a complicated way not only on M but also on the ionization probabilities of the carriers in the device. It is normally taken simply to be M^x where x in an index that depends on the semiconductor material and device structure.

(a) Table 5.4 provides measurements of F vs. M on a Ge APD using photogeneration at 1.55 μm. Find x in $F = M^x$. How good is the fit?

(b) The above Ge APD has an unmultiplied dark current of 0.5 μA and an unmultiplied responsivity of 0.8 A. V^{-1} at its peak response at 1.55 μm and is biased to operate at $M = 6$ in a receiver circuit with a bandwidth of 500 MHz. What is the minimum photocurrent that will give an SNR = 1? If the photosensitive area is 0.3 mm in diameter what are the corresponding minimum optical power and light intensity?

(c) What should be the photocurrent and incident optical power for SNR = 10?

TABLE 5.4 Data for excess avalanche noise as F vs. M for a Ge APD (from D. Scansen and S.O. Kasap, *Cnd. J. Physics.* **70**, 1070–1075, 1992)

M	1	3	5	7	9
F	1.1	2.8	4.4	5.5	7.5

5.13 Photodetector materials, devices and their limitations

(a) What limits the operation of a photodetector when the absorption depth ($\delta = 1/\alpha$) at short wavelengths becomes so narrow that EHPs are generated almost at the crystal surface?

(b) Quantum efficiency defined in terms of incident photons applies to the whole device and includes the effects arising from reflections from the semiconductor surface. What is the percentage of photons lost due to reflections at a Si crystal surface if the refractive index of Si is 3.5. How can you improve the transmitted number of photons into the semiconductor crystal?

(c) In some applications such as measuring the light intensity in the visible range from a source that also emits extensively in the infrared (such as an incandescent light source), it is necessary to use an infrared heat filter. Why?

(d) Consider a heterojunction APD such as that shown in Figure 5.12. For InP, $E_g = 1.35$ eV and for InGaAs $E_g = 0.75$ eV. Obviously 1.5 μm photons will not be absorbed in InP. What is the effect of mismatch in the refractive index n between the two semiconductors? Is it important if $\Delta n \approx 0.2$, and $n \approx 3.5$.

(e) What determines the speed of operation of the phototransistor in Figure 5.16? Consider how long it takes for the photoinjected hole into the p-type base to become neutralized by recombination.

"Photoconductors have appeared to offer generous possibilities for high-sensitivity devices by virtue of the known fact that a weak incident stream of photons can give rise to a large stream of electrons passing through the photoconductor. This photoconductive gain (the ratio of the current of electrons to the current of photons) has been observed to reach values as high as 10^6 in some sensitive photoconductors such as cadmium sulfide." —Albert Rose

Vision, Human and Electronic (Plenum Press, New York, 1973, p. 143)

C H A P T E R 6

Photovoltaic Devices

"From an energy point of view, for about 99.99% of the existence of humanity on this planet, the technological level of our species was only one step above that of animals. It has only been within the past few hundred years that humans have had more than 1 horsepower available to them."

—Michio Kaku[1]

Honda's two seated *Dream* car is powered by photovoltaics. The Honda Dream was first to finish 3,010 km in four days in the 1996 World Solar Challenge. *(Courtesy of Photovoltaics Special Research Centre, University of New South Wales, Sydney, Australia)*

6.1 SOLAR ENERGY SPECTRUM

Photovoltaic devices or **solar cells** convert the incident solar radiation energy into electrical energy. Incident photons are absorbed to photogenerate charge carriers that pass through an external load to do electrical work. Photovoltaic device applications cover a wider range from small consumer electronics, such as a solar cell calculator using less than a few milliwatts, to photovoltaic power generation by a central power plant (generating a few

---

[1] Michio Kaku (Henry Semat Professor of Theoretical Physics at the City College of New York) *Hyperspace* (Anchor Books, Doubleday, New York, USA, 1994), p. 275.

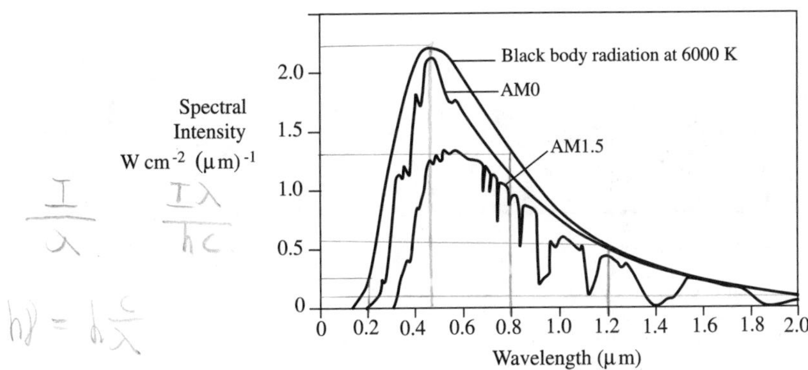

FIGURE 6.1 The spectrum of the solar energy represented as spectral intensity (I_λ) vs wavelength above Earth's atmosphere (AM0 radiation) and at the earth's surface (AM1.5 radiation). Black body radiation at 6000 K is shown for comparison (After H.J. Möller, *Semiconductors for Solar Cells*, Artech House Press, Boston, 1993, p. 10).

megawatts). There are several megawatt photovoltaic power generation plants and tens of thousands of small 1 kW scale photovoltaic generation systems currently in use.

The intensity of radiation emitted from the sun has a spectrum that resembles a *black body* radiation at a temperature of about 6,000 K (or at about 5,700 °C). Figure 6.1 shows the intensity spectrum of the sun under two different conditions corresponding to solar radiation above the earth's atmosphere and at the earth's surface. This spectrum is modified by the effects of the solar atmosphere, Fraunhofer absorption (absorption by hydrogen) and temperature variations on the surface of the sun facing us. Light intensity[2] variation with wavelength is typically represented by intensity per unit wavelength, called **spectral intensity** I_λ, so that $I_\lambda\,\delta\lambda$ is the intensity in a small interval $\delta\lambda$. Integration of I_λ over the whole spectrum gives the integrated or **total intensity**, I.

The integrated intensity above Earth's atmosphere gives the total power flow through a unit area perpendicular to the direction of the sun. This quantity is called the **solar constant** or **air-mass zero** (AM0) radiation and it is approximately constant at a value of 1.353 kW m^{-2}.

The actual intensity spectrum on Earth's surface depends on the absorption and scattering effects of the atmosphere and hence on the atmospheric composition and the radiation path length through the atmosphere. These atmospheric effects depend on the wavelength. Clouds increase the absorption and scattering of sun light and hence substantially reduce the incident intensity. On a clear sunny day, the light intensity arriving on Earth's surface is roughly 70% of the intensity above the atmosphere. Absorption and scattering effects increase with the sun beam's path through the atmosphere. The

[2] Intensity definition used here represents flow of energy per unit time per unit area (Wm^{-2}). In some texts this quantity is defined as *irradiance* and the term intensity is reserved for energy flow per unit solid angle.

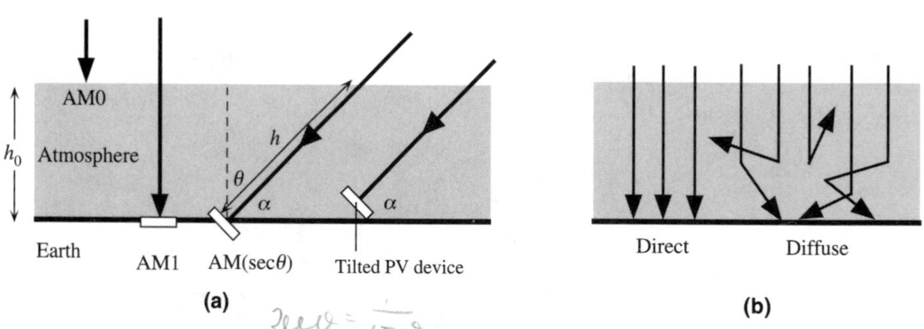

FIGURE 6.2 (a) Illustration of the effect of the angle of incidence θ on the ray path length and the definitions of AM0, AM1 and AM(sec θ). The angle α between the sun beam and the horizon is the solar latitude. (b) Scattering reduces the intensity and gives rise to a diffused radiation.

shortest path through the atmosphere is when the sun is directly above that location and the received spectrum is called air mass one (AM1) as illustrated in Figure 6.2 (a). All other angles of incidence ($\theta \neq 90°$ in Figure 6.2 (a)) increase the optical path through the atmosphere, and hence the atmospheric losses. **Air mass** m (AMm) is defined as the ratio of the actual radiation path h to the shortest path h_0, that is $m = h/h_0$, as illustrated in Figure 6.2 (a). Since $h = h_0 \sec\theta$, AMm is AMsecθ. The spectral distribution for AM1.5 is shown in Figure 6.1. This spectrum refers to incident energy on a unit area *normal* to sun rays (which have to travel the atmospheric length h as shown in Figure 6.2 (a)).

It is apparent that the spectrum in Figure 6.1 has several sharp absorption peaks at certain wavelengths which are due to those wavelengths being absorbed by various molecules in the atmosphere, such as ozone (at high altitudes), air and water vapor molecules. In addition, atmospheric molecules and dust particles scatter the sun light. Scattering not only reduces the intensity in the direction towards Earth but also gives rise to the sun's rays arriving at random angles as shown in Figure 6.2 (b). Consequently, the terrestrial light has a **diffuse** component in addition to the **direct** component. The diffuse component increases with cloudiness and sun's position, and has a spectrum shifted toward the blue light. The scattering of light increases with decreasing wavelength so that shorter wavelengths in the original sun beam experience more scattering than longer wavelengths. On a clear day, the diffuse component can be roughly 20% of the total radiation, and significantly higher on cloudy days.

According to Figure 6.2 (a), the amount of incident radiation depends on the position of the sun which changes cyclically over the course of the day and over the year. A photovoltaic device flat on Earth's surface will receive even less solar energy by a factor $\cos\theta$. However, the photovoltaic device can be tilted to directly face the sun to maximize the collection efficiency as shown in Figure 6.2 (a).

EXAMPLE 6.1.1 Solar energy conversion

Suppose that a particular family house in a sunny geographic location over a year consumes a daily average electrical power of 500 W. If the annual average solar intensity incident per day is about 6 kW h m^{-2}, and a photovoltaic device that converts solar energy to electrical energy has an efficiency of 15%, what is the required device area?

Solution Since we know the average light intensity incident,

$$\text{Total energy available for 1 day} = \text{Incident solar energy in 1 day per unit area} \times \text{Area} \times \text{Efficiency}$$

which must equal to the average energy consumed per house in 1 day. Thus,

$$\text{Area} = \frac{\text{Energy per house}}{\text{Incident solar energy per unit area} \times \text{Efficiency}}$$

$$= \frac{500\ \text{W} \times 60\,\dfrac{\text{s}}{\text{min}}\,60\,\dfrac{\text{min}}{\text{hr}}\,24\ \text{hrs}}{(6 \times 10^3\ \text{W}\cdot\text{hr}\cdot\text{m}^{-2}\ \text{day}^{-1})\left(60\,\dfrac{\text{s}}{\text{min}}\,60\,\dfrac{\text{min}}{\text{hr}}\right) \times 0.15}$$

$$= 13.3\ \text{m}^2 \text{ or a panel } 3.6\ \text{m} \times 3.6\ \text{m}.$$

The problem is that this area is based on averages over the year. Such a panel cannot supply the peak power when a number of electrical appliances are running at the same time (consuming several kilowatts). An energy storage device can be used to store the surplus energy generated during low-power consumption periods but this adds to the cost and complexity of the system.

6.2 PHOTOVOLTAIC DEVICE PRINCIPLES

A simplified schematic diagram of a typical solar cell is shown Figure 6.3. Consider a *pn* junction with a very narrow and more heavily doped *n*-region. The illumination is through the thin *n*-side. The depletion region (W) or the space charge layer (SCL) extends primarily into the *p*-side. There is a **built-in field** E_o in this depletion layer. The

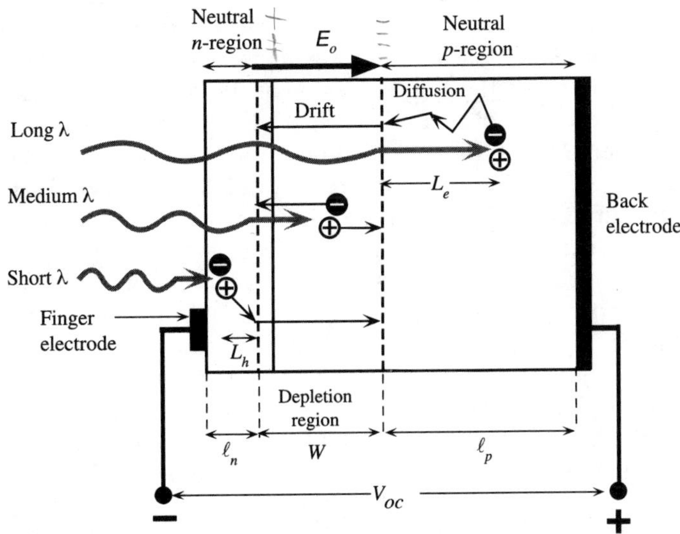

FIGURE 6.3 The principle of operation of the solar cell (exaggerated features to highlight principles).

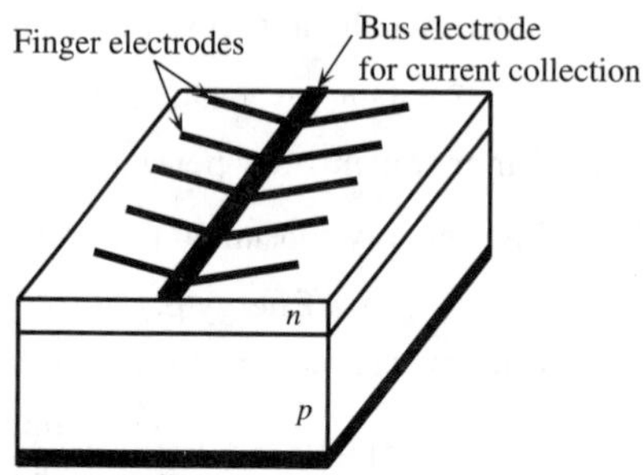

FIGURE 6.4 Finger electrodes on the surface of a solar cell reduce the series resistance.

electrodes attached to the n-side must allow illumination to enter the device and at the same time result in a small series resistance. They are deposited on to n-side to form an array of **finger electrodes** on the surface as depicted in Figure 6.4. A thin **antireflection coating** on the surface (not shown in the figure) reduces reflections and allows more light to enter the device.

As the n-side is very narrow most of the photons are absorbed within the depletion region (W) and within the neutral p-side (ℓ_p) and photogenerate electron-hole pairs (EHPs) in these regions. EHPs photogenerated in the depletion region are immediately separated by the built-in field $\boldsymbol{E}_o$ which drifts them apart. The electron drifts and reaches the neutral n^+ side whereupon it makes this region negative by an amount of charge $-e$. Similarly the hole drifts and reaches the neutral p-side and thereby makes this side positive. Consequently an **open circuit voltage** develops between the terminals of the device with the p-side positive with respect to the n-side. If an external load is connected then the excess electron in the n-side can travel around the external circuit, do work, and reach the p-side to recombine with the excess hole there. It is important to realize that without the internal field $\boldsymbol{E}_o$ it is not possible to drift apart the photogenerated EHPs and accumulate excess electrons on the n-side and excess holes on the p-side.

The EHPs photogenerated by long wavelength photons that are absorbed in the neutral p-side can only diffuse in this region as there is no electric field. If the recombination lifetime of the electron is τ_e, it diffuses a mean distance L_e given by $L_e = \sqrt{(2D_e\tau_e)}$ where D_e is its diffusion coefficient in the p-side. Those electrons within a distance L_e to the depletion region can readily diffuse and reach this region whereupon they become drifted by $\boldsymbol{E}_o$ to the n-side as shown in Figure 6.3. Consequently only those EHPs photogenerated within the **minority carrier diffusion length** L_e to the depletion layer can contribute to the photovoltaic effect. Again the importance of the built-in field $\boldsymbol{E}_o$ is apparent. Once an electron diffuses to the depletion region it is swept over to the n-side by $\boldsymbol{E}_o$ to give an additional negative charge there. Holes left behind in the p-side contribute a net positive charge to this region. Those photogenerated EHPs further away from depletion region than L_e are lost by recombination. It is therefore important to have the minority carrier diffusion length L_e as long as possible. This is the reason for choosing this side of a Si pn junction to be p-type which makes electrons to be the minority carriers; the electron diffusion length in Si is longer than the hole diffusion length. The same ideas also apply to EHPs photogenerated by short-wavelength photons absorbed in the n-side. Those holes photogenerated within a diffusion length L_h can reach the depletion layer and become swept across to the p-side. The photogener-

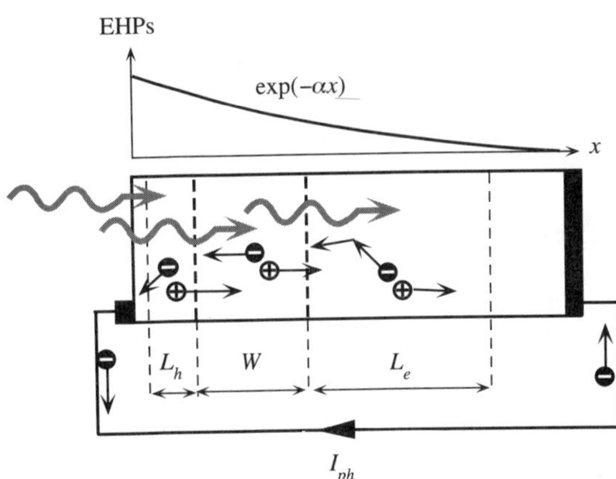

FIGURE 6.5 Photogenerated carriers within the volume $L_h + W + L_e$ give rise to a photocurrent I_{ph}. The variation in the photogenerated EHP concentration with distance is also shown where α is the absorption coefficient at the wavelength of interest.

ation of EHPs that contribute to the photovoltaic effect therefore occurs in a volume covering $L_h + W + L_e$. If the terminals of the device are shorted, as in Figure 6.5, then the excess electron in the n-side can flow through the external circuit to neutralize the excess hole in the p-side. This current due to the flow of the photogenerated carriers is called the **photocurrent**.

It is important to realize that the picture in Figure 6.3 is not complete. Under a steady state operation, there can be no net current through an open circuit solar cell. This means the photocurrent inside the device due to the flow of photogenerated carriers must be exactly balanced by a flow of carriers in the opposite direction. The latter carriers are minority carriers that become injected by the appearance of the photovoltaic voltage across the pn junction as in a normal diode. This is not shown in Figure 6.3.

EHPs photogenerated by energetic photons absorbed in the n-side near the surface region or outside the diffusion length L_h to the depletion layer are lost by recombination as the lifetime in the n side is generally very short (due to heavy doping). The n side is therefore made very thin, typically less than 0.2 μm or less. Indeed, the length ℓ_n of the n-side may be shorter than the hole diffusion length L_h. The EHP photogenerated very near the surface of the n-side however disappear by recombination due to various surface defects acting as recombination centers as discussed below.

At long wavelengths, around 1–1.2 μm, the absorption coefficient α of Si is small and the *absorption depth* $(1/\alpha)$ is typically greater than 100 μm. To capture these long wavelength photons we therefore need a thick p-side and at the same time a long minority carrier diffusion length L_e. Typically the p-side is 200–500 μm and L_e tends to be shorter than this.

Crystalline silicon has a bandgap of 1.1 eV which corresponds to a threshold wavelength of 1.1 μm. As can be see from Figure 6.1, the incident energy in the wavelength region greater than 1.1 μm is then wasted; this is not a negligible amount (about ~25%). The worst part of the efficiency limitation however comes from the high energy photons becoming absorbed near the crystal surface and being lost by recombination in the surface region. Crystal surfaces and interfaces contain a high concentration of **recombination centers** which facilitate the recombination of photogenerated EHP near the surface.

Losses due to EHP recombinations ne
combined effects bring the efficiency c
coating is not perfect which reduces
0.8-0.9. When we also include the limit
below), the upper limit to a photovolt
24–26% at room temperature.

EXAMPLE 6.2.1 The photocurrent

Consider a particular photovoltaic device th
togeneration occurs over the device thickn
G_{ph}, number of EHPs photogenerated p
where G_o is the *photogeneration rate* at th
that the device is shorted as in Figure 6.5 t
the external circuit (only electrons flow in the external circuit). Suppose that L_h is greater than the n-layer thickness ℓ_n so that all the EHPs generated within the volume $(\ell_n + W + L_e)$ contribute to the photocurrent. Further, assume that EHP recombination near the crystal surface is negligible. Show that the **photocurrent** I_{ph} is then

Photocurrent

$$I_{ph} = \frac{eG_oA}{\alpha}\{1 - \exp[-\alpha(\ell_n + W + L_e)]\} \quad \textbf{(1)}$$

where A is the device surface area under illumination (not blocked by finger electrodes).

Solution The EHP photogeneration rate from the illuminated crystal surface follows

$$G_{ph} = G_o \exp(-\alpha x)$$

Total number of EHP generated per unit time in a small volume $A\delta x$ is $G_{ph}(A\delta x)$. Thus:

$$\text{Total number EHP generated per unit time in } \ell_n + W + L_e = A\int_{x=0}^{\ell_n+W+L_e} G_o \exp(-\alpha x)\,dx$$

or

$$\frac{dN_{\text{EHP}}}{dt} = \frac{G_oA}{\alpha}\{1 - \exp[-\alpha(\ell_n + W + L_e)]\}$$

Since the photogenerated electrons flow through the external circuit, the photocurrent I_{ph} is then $e(dN_{\text{EHP}}/dt)$

$$I_{ph} = \frac{eG_oA}{\alpha}\{1 - \exp[-\alpha(\ell_n + W + L_e)]\}$$

For long wavelengths, α will be small. Expanding the exponential we find,

$$I_{ph} = eG_oA(\ell_n + W + L_e) \quad \textbf{(2)}$$

which applies under nearly uniform photogeneration conditions.

Taking a crystalline Si device that has $A = 5$ cm $\times$ 5 cm, $\ell_n = 0.5$ μm, $W = 2$ μm, $L_e = 50$ μm, small α such as $\alpha = 2000$ m^{-1} (absorption depth $\delta = 1/\alpha = 500$ μm) for Si at $\lambda \approx 1.1$ μm, and using $G_o = 1 \times 10^{18}$ cm^{-3} s^{-1} in Eq. (1), we find $I_{ph} \approx 20$ mA whereas Eq. (2) gives 21 mA. On the other hand for strong absorption at $\lambda \approx 0.83$ μm, $\alpha = 10^5$ m^{-1} (absorption depth $\delta = 1/\alpha = 10\ \mu\text{m}$), Eq. (1) gives $I_{ph} \approx 40$ mA. The current is doubled simply because more photons are now absorbed in the volume $(\ell_h + W + L_e)$. Further increases in α with decreasing wavelength will eventually (when $\lambda < 450$ nm) constrict the photogeneration to the surface region where surface defects will facilitate EHP recombination and thereby diminish the photocurrent.

6.3 *pn* JUNCTION PHOTOVOLTAIC *I-V* CHARACTERISTICS

Consider an ideal *pn* junction photovoltaic device connected to a resistive load R as shown in Figure 6.6 (a). Note that I and V in the figure define the convention for the direction of positive current and positive voltage. If the load is a short circuit, then the only current in the circuit is that generated by the incident light as shown in Figure 6.6 (b). This is called the **photocurrent,** I_{ph}, which depends on the number of EHPs photogenerated within the volume enclosing the depletion region (W) and the diffusion lengths to the depletion region (Figure 6.3). The greater is the light intensity, the higher is the photogeneration rate and larger is I_{ph}. If I is the light intensity then the **short circuit current** is

Short circuit current in light

$$I_{sc} = -I_{ph} = -KI \qquad (1)$$

where K is a constant that depends on the particular device. The photocurrent does not depend on the voltage across the *pn* junction because there is always some internal field to drift the photogenerated EHP. We exclude the secondary effect of the voltage modulating the width of the depletion region. The photocurrent, I_{ph}, therefore flows even when there is not a voltage across the device.

If R is not a short circuit then a positive voltage V appears across the *pn* junction as a result of the current passing through it as shown in Figure 6.6 (c). This voltage reduces the built in potential V_o of the *pn* junction and hence leads to minority carrier injection and diffusion just as it would in a normal diode. Thus, in addition to I_{ph} there is also a forward diode current I_d in the circuit as shown in Figure 6.6 (c) which arises from

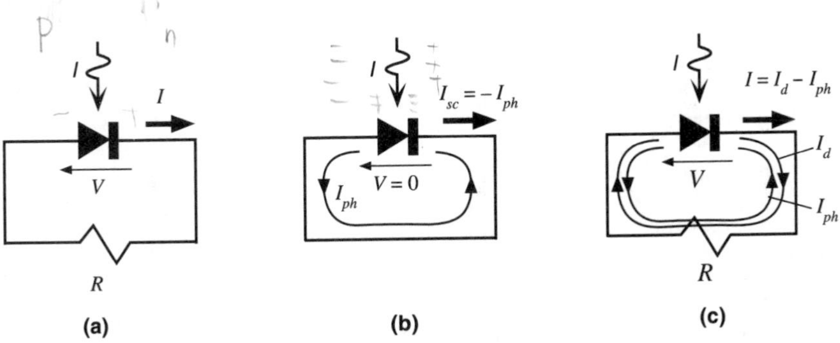

FIGURE 6.6 (a) The solar cell connected to an external load R and the convention for the definitions of positive voltage and positive current. (b) The solar cell in short circuit. The current is the photocurrent, I_{ph}. (c) The solar cell driving an external load R. There is a voltage V and current I in the circuit.

the voltage developed across R. Since I_d is due to the normal pn junction behavior it is given by the **diode characteristics**,

$$I_d = I_o\left[\exp\left(\frac{eV}{nk_BT}\right) - 1\right]$$

where I_o is the "reverse saturation current" and n is the ideality factor that depends on the semiconductor material and fabrication characteristics ($n = 1 - 2$). In an open circuit, the net current is zero. This means that the photocurrent I_{ph} develops just enough photovoltaic voltage V_{oc} to generate a diode current $I_d = I_{ph}$.

Thus the **total current** through the solar cell, as shown in Figure 6.6 (c) is

Solar cell I-V

$$I = -I_{ph} + I_o\left[\exp\left(\frac{eV}{nk_BT}\right) - 1\right] \quad \textbf{(2)}$$

The overall I-V characteristics of a typical Si solar cell is shown in Figure 6.7. It can be seen that it corresponds to the normal dark characteristics being shifted down by the photocurrent, I_{ph}, which depends on the light intensity, *I*. The open circuit output voltage, V_{oc}, of the solar cell is given by the point where the I–V curve cuts the V-axis ($I = 0$). It is apparent that although V_{oc} depends on the light intensity, its value typically lies in the range 0.4–0.6 V.

Equation (2) gives only the I–V characteristics of the solar cell. When the solar cell is connected to a load as in Figure 6.6 (a), the load has the same voltage as the solar cell and carries the same current. But the current I through R is now in the opposite direction to the conventional that current flows from high to low potential. Thus, as shown in Figure 6.8 (a),

The load line

$$I = -\frac{V}{R} \quad \textbf{(3)}$$

The actual current I' and voltage V' in the circuit must satisfy both the I–V characteristics of the solar cell, Eq. (2), and that of the load, Eq. (3). We can find I' and V' by solving eqs. (2) and (3) simultaneously but this is not a trivial analytical procedure. A graphical solution using the solar cell characteristics however is straightforward.

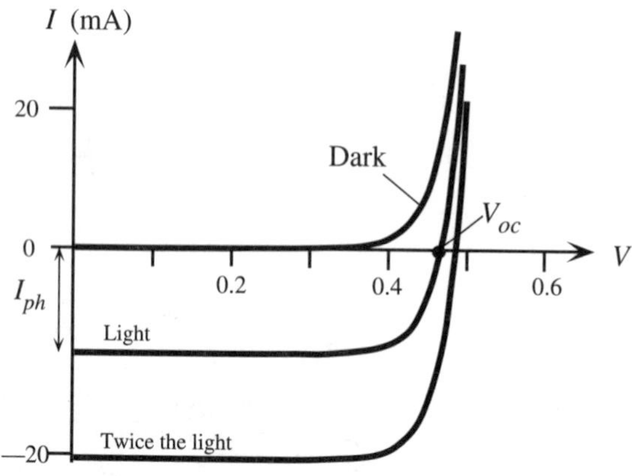

FIGURE 6.7 Typical I-V characteristics of a Si solar cell. The short circuit current is I_{ph} and the open circuit voltage is V_{oc}. The I-V curves for positive current requires an external bias voltage. Photovoltaic operation is always in the negative current region.

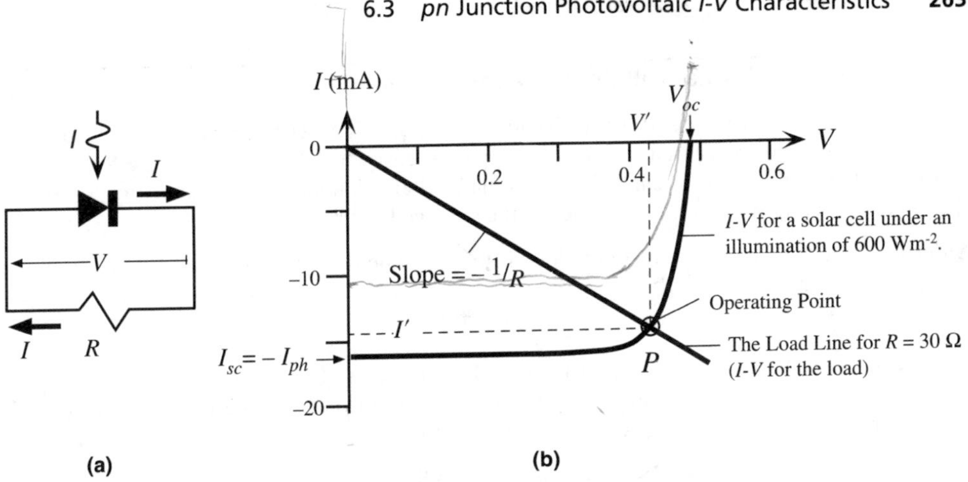

FIGURE 6.8 **(a)** When a solar cell drives a load R, R has the same voltage as the solar cell but the current through it is in the opposite direction to the convention that current flows from high to low potential. **(b)** The current I' and voltage V' in the circuit of (a) can be found from a load line construction. Point P is the operating point (I', V'). The load line is for $R = 30\ \Omega$.

The current I' and voltage V' in the solar cell circuit are most easily found by using a load line construction. The I–V characteristics of the load in Eq. (3) is a straight line with a negative slope $-1/R$. This is called the **load line** and is shown in Figure 6.8 (b) along with the I–V characteristics of the solar cell under a given intensity of illumination. The load line cuts the solar cell characteristic at P. At P, the load and the solar cell have the same current and voltage I' and V'. Point P therefore satisfies both eqs. (2) and (3) and thus represents the **operating point of the circuit**. The current I' and voltage V' in the circuit are hence given by point P.

The **power delivered** to the load is $P_{out} = I'V'$, which is the area of the rectangle bound by I- and V-axes and the dashed lines shown in Figure 6.8 (b). Maximum power is delivered to the load when this rectangular area is maximized (by changing R or the intensity of illumination), when $I' = I_m$ and $V' = V_m$. Since the maximum possible current is I_{sc} and the maximum possible voltage is V_{oc}, $I_{sc}V_{oc}$, represents the desirable goal in power delivery for a given solar cell. It is therefore useful to compare the maximum power output, I_mV_m, with $I_{sc}V_{oc}$. The **fill factor** FF, which is a figure of merit for the solar cell, is defined as

$$\text{FF} = \frac{I_mV_m}{I_{sc}V_{oc}} \qquad (4)$$

Definition of fill factor

FF is a measure of the closeness of the solar cell I–V curve to the rectangular shape (the ideal shape). It is clearly advantageous to have FF as close to unity as possible but the exponential *pn* junction properties prevent this. Typically FF values are in the range 70–85% and depend on the device material and structure.

EXAMPLE 6.3.1 A solar cell driving a resistive load

Consider a solar cell driving a 30 Ω resistive load as in Figure 6.8 (a). Suppose that the cell has an area of 1 cm × 1 cm and is illuminated with light of intensity 600 W m^{-2} and has the I–V characteristics in Figure 6.8 (b). What are the current and voltage in the circuit? What is the power delivered to the load? What is the efficiency of the solar cell in this circuit?

Solution The I–V characteristic of the load is the load line described by Eq. (3),

$$I = -\frac{V}{30\ \Omega}$$

The line is drawn in Figure 6.8 (b) with a slope $1/(30\ \Omega)$. It cuts the I–V characteristics of the solar cell at $I' = 14.2$ mA and $V' = 0.425$ V which are the current and voltage in the photovoltaic circuit of Figure 6.8 (a). The power delivered to the load is

$$P_{out} = I'V' = (14.2 \times 10^{-3})(0.425\ \text{V}) = 0.006035\ \text{W}, \quad \text{or} \quad 6.035\ \text{mW}.$$

This is *not* necessarily the maximum power available from the solar cell. The input sun-light power is

$$P_{in} = (\text{Light Intensity})(\text{Surface Area}) = (600\ \text{W m}^{-2})(0.01\ \text{m})^2 = 0.060\ \text{W}$$

The efficiency is

$$\eta = 100\frac{P_{out}}{P_{in}} = 100\frac{0.006035}{0.060} = 10.06\ \%$$

This will increase if the load is adjusted to extract the maximum power from the solar cell but the increase will be small as the rectangular area $I'V'$ in Figure 6.8 (b) is already close to the maximum.

EXAMPLE 6.3.2 Open circuit voltage and illumination

A solar cell under an illumination of 600 W m^{-2} has a short circuit current I_{sc} of 16.1 mA and an open circuit output voltage V_{oc}, of 0.485 V. What are the short circuit current and open circuit voltages when the light intensity is doubled?

Solution The general I–V characteristics under illumination is given by Eq. (2). Setting $I = 0$ for open circuit we have

$$I = -I_{ph} + I_o[\exp(eV_{oc}/nk_BT) - 1] = 0$$

Open circuit output voltage

Assuming that $V_{oc} \gg nk_BT/e$, rearranging the above equation we can find V_{oc},

$$V_{oc} = \frac{nk_BT}{e}\ln\left(\frac{I_{ph}}{I_o}\right) \tag{5}$$

In Eq. (5), the photocurrent, I_{ph}, depends on the light intensity I via, $I_{ph} = KI$. At a given temperature, then the change in V_{oc} is

$$V_{oc2} - V_{oc1} = \frac{nk_BT}{e}\ln\left(\frac{I_{ph2}}{I_{ph1}}\right) = \frac{nk_BT}{e}\ln\left(\frac{I_2}{I_1}\right)$$

The short circuit current is the photocurrent so that at double the intensity this is

$$I_{sc2} = I_{sc1}\left(\frac{I_2}{I_1}\right) = (16.1\ \text{mA})(2) = 32.2\ \text{mA}$$

Assuming $n = 1$, the new open circuit voltage is

$$V_{oc2} = V_{oc1} + \frac{nk_BT}{e}\ln\left(\frac{I_2}{I_1}\right) = 0.485 + 1(0.0259)\ln(2) = 0.503 \text{ V}$$

This is a 3.7% increase compared with the 100 % increase in illumination and the short circuit current. Ideally do we want V_{oc} to be always the same?

6.4 SERIES RESISTANCE AND EQUIVALENT CIRCUIT

Practical devices can deviate substantially from the ideal *pn* junction solar cell behavior depicted in Figure 6.7 due to a number of reasons. Consider an illuminated *pn* junction driving a load resistance R_L and assume that photogeneration takes place in the depletion region. As illustrated in Figure 6.9, the photogenerated electron has to traverse a surface semiconductor region to reach the nearest finger electrode. All these electron paths in the *n*-layer surface region to finger electrodes introduce an **effective series resistance** R_s into the photovoltaic circuit as indicated in Figure 6.9. If the finger electrodes are thin, then the resistance of the electrodes themselves will further increase R_s. There is also a series resistance due to the neutral *p*-region but this is generally small compared with the resistance of the electron paths to the finger electrodes.

Figure 6.10 a shows the equivalent circuit of an ideal *pn* junction solar cell. The photogeneration process is represented by a constant current generator I_{ph} where the current is proportional to the light intensity. The flow of photogenerated carriers across the junction gives rise to a photovoltaic voltage difference V across the junction and this voltage leads to the normal diode current $I_d = I_o[\exp(eV/nk_BT) - 1]$. This diode current I_d is represented by an ideal *pn* junction diode in the circuit as shown in Figure 6.10 (a). As apparent, I_{ph} and I_d are in opposite directions (I_{ph} is "up" and I_d is "down") so that in open circuit the photovoltaic voltage is such that I_{ph} and I_d have the same magnitude and cancel each other.

Figure 6.10 (b) shows the equivalent circuit of a more practical solar cell. The **series resistance** R_s in Figure 6.10 (b) gives rise to a voltage drop and therefore prevents

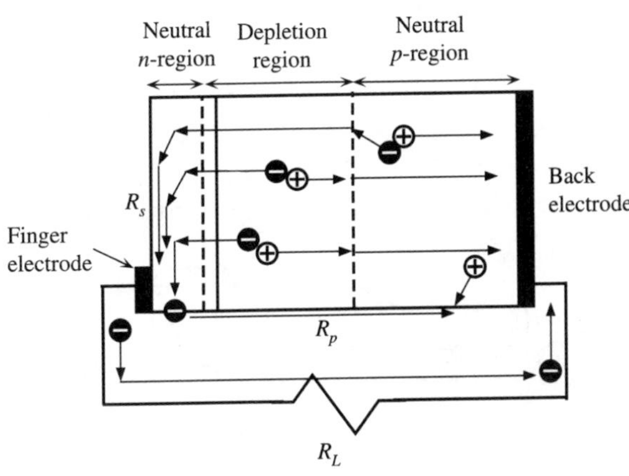

FIGURE 6.9 Series and shunt resistances and various fates of photogenerated EHPs.

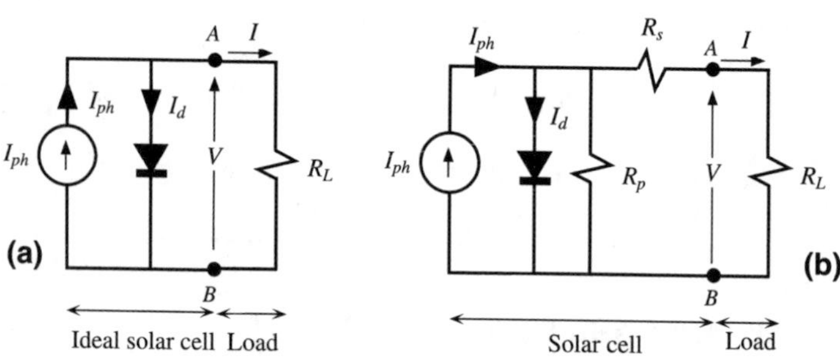

FIGURE 6.10 The equivalent circuit of a solar cell **(a)** Ideal *pn* junction solar cell **(b)** Parallel and series resistances

the full photovoltaic voltage from developing at the output between A and B. A fraction (usually small) of the photogenerated carriers can also flow through the crystal surfaces (edges of the device) or through *grain boundaries in polycrystalline devices* instead of flowing though the external load R_L. These effects that prevent photogenerated carriers from flowing in the external circuit can be represented by an effective internal **shunt** or **parallel resistance** R_p that diverts the photocurrent away from the load R_L. Typically R_p is less important than R_s in overall device behavior, unless the device is highly polycrystalline and the current component flowing through grain boundaries is not negligible.

The series resistance R_s can significantly deteriorate the solar cell performance as illustrated in Figure 6.11 where $R_s = 0$ is the best solar cell case. It is apparent that the available maximum output power decreases with the series resistance which therefore reduces the cell efficiency. Notice also that when R_s is sufficiently large, it limits the short circuit current. Similarly, low shunt resistance values, due to extensive defects in the material , also reduce the efficiency. The difference is that although R_s does not affect the open circuit voltage V_{oc}, low R_p leads to a reduced V_{oc}.

FIGURE 6.11 The series resistance broadens the *I-V* curve and reduces the maximum available power and hence the overall efficiency of the solar cell. The example is a Si solar cell with $n \approx 1.5$ and $I_o \approx 3 \times 10^{-6}$ mA. Illumination is such that the photocurrent $I_{ph} = 10$ mA.

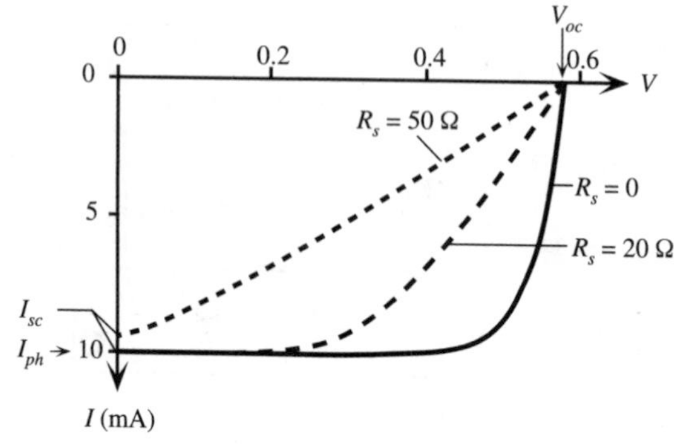

EXAMPLE 6.4.1 Solar cells in parallel

Consider two identical solar cells with the properties $I_o = 25 \times 10^{-6}$ mA, $n = 1.5$, $R_s = 20\ \Omega$, subjected to the same illumination so that $I_{ph} = 10$ mA. Explain the characteristics of two solar cells connected in parallel. Find the maximum power that can be delivered by one cell and two cells in series and also find the corresponding voltage and current at the maximum power point (assume $R_p = \infty$).

Solution Consider one individual solar cell as in Figure 6.10. The voltage V_d across the diode is $V - R_sI$ so that the external current I is,

$$I = -I_{ph} + I_o \exp\left(\frac{eV_d}{nk_BT}\right) - I_o = -I_{ph} + I_o \exp\left[\frac{e(V - IR_s)}{nk_BT}\right] - I_o \qquad \textbf{(1)}$$

Equation (1) gives the I–V characteristic of 1 cell and is plotted in Figure 6.12.[3] The output power P is simply IV which is also plotted in Figure 6.12. The power is maximum at 2.2 mW when the current is about 8 mA and the voltage is 0.27 V. The load must be 34 Ω.

Figure 6.13 shows the equivalent circuit of the two solar cells in parallel running a load R_L. I and V now refer to the whole system of two devices in parallel. Each device is now delivering a

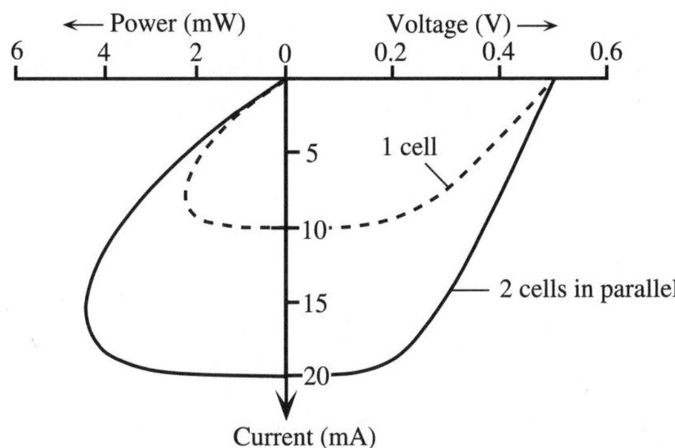

FIGURE 6.12 Current vs. Voltage and Power vs. Current characteristics of one cell and two cells in parallel. The two parallel devices have $R_s/2$ and $2I_{ph}$.

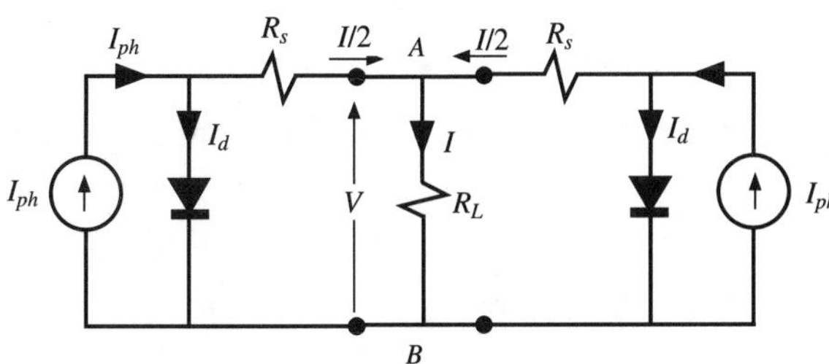

FIGURE 6.13 Two identical solar cells in parallel under the same illumination and driving a load R_L.

[3] This is most conveniently plotted by selecting I values and calculating V from Eq. (1): $V = (nk_BT/e)\ln[(I + I_{ph} + I_o)/I_o] + R_sI$.

current $I/2$. The diode voltage for one cell is $V_d - R_s(I/2)$. Thus

$$\tfrac{1}{2}I = -I_{ph} + I_o \exp\left(\frac{V - \frac{1}{2}IR_s}{nk_BT}\right) - I_o$$

or

$$I = -2I_{ph} + 2I_o \exp\left(\frac{V - \frac{1}{2}IR_s}{nk_BT}\right) - 2I_o \qquad \textbf{(2)}$$

Comparing this with Eq. (1), we see that the parallel combination has halved the series resistance, doubled the photocurrent, and also doubled the diode reverse saturation current I_o. All these are in line with intuitive expectation as the device area has now been effectively doubled. Figure 6.12 shows the *I–V* and *P* vs. *I* characteristics of the combined device. The power is maximum at about 4.4 mW when the $I \approx 16$ mA, V ≈ 0.27 V. The corresponding load is 17 Ω. It is clear that the parallel combination increases the available current and allows a lower resistance load to be driven.

If we were to use the two solar cells in series, then V_{oc} would be doubled at 1 V, I_{sc} would be the same as I_{ph} at 10 mA, and the maximum power would be at 4.4 mW available at about 8 mA and 0.55 V. This output power would need a load of about 34 Ω. These simple ideas however do not work when the cells are not identical. The connections of such mismatched cells can lead to much poorer performance than idealized predictions based on parallel and series connections of matched devices.

6.5 TEMPERATURE EFFECTS

The output voltage and the efficiency of a solar cell increases with decreasing temperature; solar cells operate best at lower temperatures. Consider the open circuit voltage V_{oc} of the device in Figure 6.8 b. As the total cell current is zero, the photocurrent I_{ph} generated by light must be balanced by I_d which is generated by the photovoltaic voltage V_{oc}, that is, $I_d = I_o \exp(eV_{oc}/nk_BT)$. If n_i is the intrinsic concentration then I_o is proportional to n_i^2 which means that I_o decreases rapidly with decreasing temperature. Consequently a greater voltage is developed to generate the necessary I_d that balances I_{ph}.

The output voltage, V_{oc}, when $V_{oc} \gg nk_BT/e$ is given by

$$V_{oc} = \frac{nk_BT}{e}\ln\left(\frac{I_{ph}}{I_o}\right) \qquad \textbf{(1)}$$

In Eq. (1), I_o is the reverse saturation current which is *strongly temperature dependent* because it depends on n_i^2 where n_i is the intrinsic concentration. Further, since $I_{ph} = KI$ where K is a constant and I is the light intensity, we can write Eq. (1) as

$$V_{oc} = \frac{nk_BT}{e}\ln\left(\frac{KI}{I_o}\right) \quad \text{or} \quad \frac{eV_{oc}}{nk_BT} = \ln\left(\frac{KI}{I_o}\right)$$

Assuming $n = 1$, then at two different temperatures T_1 and T_2 but at the same illumination level, by subtraction,

$$\frac{eV_{oc2}}{k_BT_2} - \frac{eV_{oc1}}{k_BT_1} = \ln\left(\frac{KI}{I_{o2}}\right) - \ln\left(\frac{KI}{I_{o1}}\right) = \ln\left(\frac{I_{o1}}{I_{o2}}\right) \approx \ln\left(\frac{n_{i1}^2}{n_{i2}^2}\right)$$

where the subscripts 1 and 2 refer to the temperatures T_1 or T_2 respectively.

We can substitute $n_i^2 = N_cN_v\exp(-E_g/k_BT)$ and neglect the temperature dependences of N_c and N_v compared with the exponential part to obtain,

$$\frac{eV_{oc2}}{k_BT_2} - \frac{eV_{oc1}}{k_BT_1} = \frac{E_g}{k_B}\left(\frac{1}{T_2} - \frac{1}{T_1}\right)$$

Rearranging for V_{oc2} in terms of other parameters we find,

$$V_{oc2} = V_{oc1}\left(\frac{T_2}{T_1}\right) + \frac{E_g}{e}\left(1 - \frac{T_2}{T_1}\right) \quad (1)$$

Open circuit-voltage vs. temperature

For example, a Si solar cell that has $V_{oc1} = 0.55$ V at 20 °C ($T_1 = 293$ K) will have V_{oc2} at 60 °C ($T_2 = 333$ K) given by

$$V_{oc2} = (0.55\text{ V})\left(\frac{333}{293}\right) + (1.1\text{ V})\left(1 - \frac{333}{293}\right) = 0.475\text{ V}$$

If we assume to first order that the absorption characteristics are unaltered (E_g, diffusion lengths etc. remaining roughly the same), so that I_{ph} remains the same, the efficiency decreases at least by this factor.

6.6 SOLAR CELLS MATERIALS, DEVICES AND EFFICIENCIES

The efficiency of a solar cell is one of its most important characteristics because it allows the device to be assessed economically in comparison to other energy conversion devices. The solar cell efficiency invariably refers to the fraction of incident light energy converted to electrical energy. For a given solar spectrum, this conversion efficiency depends on the semiconductor material properties and the device structure. In addition, it is controlled by the effect of ambient conditions such as the temperature and high radiation damage by high energy particles (for space applications). In addition, there can be significant changes in the sun's spectrum from one location to another which can alter the cell efficiency. In locations where there is a substantial diffuse component to the spectrum, a device using a higher bandgap semiconductor may be more efficient. Using solar concentrators to focus the light onto a solar cell can substantially increase the overall efficiency. Efficiency itself is meaningless if the cost of an efficient cell is prohibitive for an application. It is therefore also essential to know the cost per unit electrical power generation that is \$/W. The latter is difficult to asses because mass production reduces overall costs and other forms of energy costs does not include environmental effects such as pollution.

Most solar cells are silicon based because silicon based semiconductor fabrication is now a mature technology that enables cost effective devices to be manufactured. Typical Si based solar cell efficiencies range from about 18% for polycrystalline to 22–24% in high efficiency single crystal devices that have special structures to absorb as many of the incident photons as possible. Figure 6.14 illustrates how various factors typically reduce the efficiency of a Si solar cell. Some 25% of the solar energy is wasted because of photons not having sufficient energy to photogenerate EHPs. At the end of the spectrum, high energy photons are absorbed near the crystal surface and these photogenerated EHPs disappear by recombination. This portion depends on the surface passivation conditions and can vary from one particular device design to another. The cell has to absorb as many of the useful photons as possible and this photon collection efficiency factor depends on the particular device structure. Maximum electrical output power scales with V_{oc} and FF so that the final overall efficiency is ~20% for a high efficiency Si solar cell.

Solar cells fabricated by making a *pn* junction in the same crystal are called *homojunctions*. Best Si homojunction solar cell efficiencies are about 24% for expensive

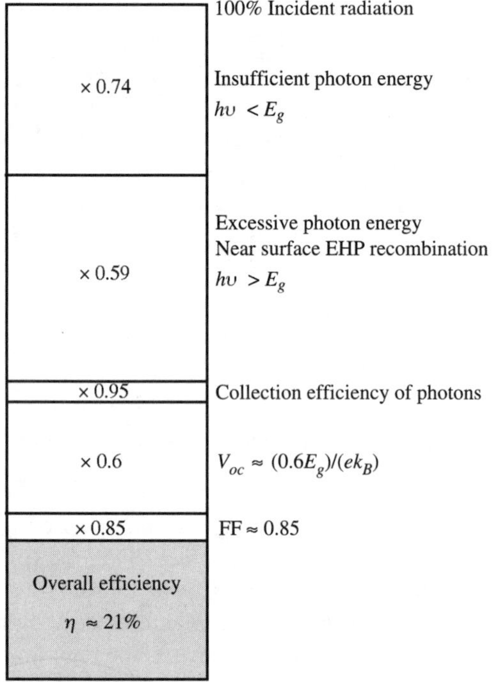

FIGURE 6.14 Accounting for various losses of energy in a high efficiency Si solar cell (Adapted from C. Hu and R. M. White, *Solar Cells* (McGraw-Hill Inc, New York, 1983, Figure 3.17, p. 61).

single crystal PERL (Passivated Emitter Rear Locally-diffused) cells. The PERL and similar cells have a textured surface that is an array of "inverted pyramids" etched into the surface to capture the incoming light as depicted in Figure 6.15. Normal reflections from a flat crystal surface lead to a loss of light whereas reflections inside the pyramid allow a second or even a third chance for absorption. Further, after refraction, photons would be entering the semiconductor at oblique angles which means that they will be absorbed in the useful photogeneration volume, that is within L_e of the depletion layer as shown in the Figure 6.15.

Table 6.1 summarizes some of typical characteristics of various solar cells. GaAs and Si solar cell have comparable efficiencies though theoretically GaAs with a higher bandgap is supposed to have a better efficiency. The largest factors reducing the efficiency

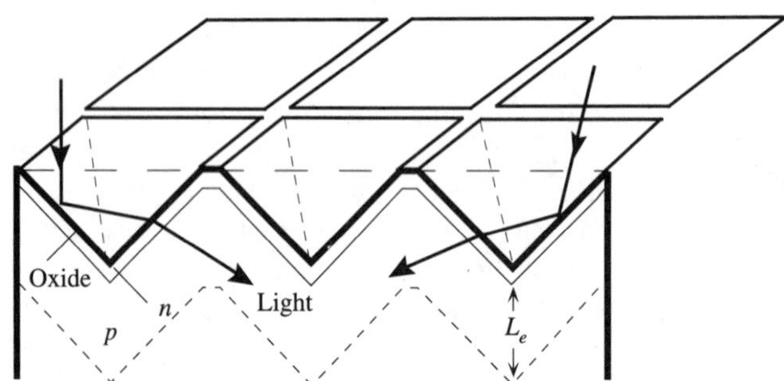

FIGURE 6.15 Inverted pyramid textured surface substantially reduces reflection losses and increases absorption probability in the device.

TABLE 6.1 Room temperature typical values typically under AM1.5 illumination 1000 W m^{-2} (data from various sources)

Semiconductor	E_g(eV)	V_{oc}	Max J_{sc} (mA cm^{-2})	FF	η (%)	Comment
Si, single crystal	1.1	0.5–0.69	42	0.7–0.8	16–24	
Si, polycrystalline	1.1	0.5	38	0.7–0.8	12–19	
Amorphous Si:Ge:H film					8–10	Amorphous film with tandem structure. Convenient large area fabrication
GaAs, single crystal	1.42	1.03	27.6	0.85	24–25	
AlGaAs/GaAs Tandem		1.03	27.9	0.864	24.8	Different bandgap materials in tandem increases absorption efficiency
GaInP/GaAs Tandem	-	2.5	14	0.86	25–30	Different bandgap materials in tandem increases absorption efficiency
CdTe Thin film	1.5	0.84	26	0.73	15–16	polycrystalline films
InP, single crystal	1.34	0.88	29	0.85	21–22	
$CuInSe_2$	1.0				12–13	

of a Si solar cell, as apparent in Figure 6.14, are the unabsorbed photons with $h\upsilon < E_g$ and short wavelength photons absorbed near the surface. Both these factors are improved if we use a tandem cell structure or heterojunctions as discussed below.

There are a number of III-V semiconductor alloys that can be prepared with different bandgaps but with the same lattice constant. *Heterojunctions* (junctions between different materials) from these semiconductors have negligible interface defects. AlGaAs has a wider bandgap than GaAs and would allow most solar photons to pass through. If we use a thin AlGaAs layer on a GaAs *pn* junction, as shown Figure 6.16, then this layer passivates the surface defects normally present in a homojunction GaAs cell. The AlGaAs window layer therefore overcomes the surface recombination limitation and improves the cell efficiency (such cells have efficiencies of about 24%).
Heterojunctions between different bandgap III-V semiconductors that are lattice matched offer the potential of developing high-efficiency solar cells. The simplest single heterojunction example, shown in Figure 6.17, consists of a *pn* junction using a wider bandgap *n*-AlGaAs with *p*-GaAs. Energetic photons ($h\upsilon > 2$ eV) are absorbed in AlGaAs where as those with energies less ($1.4 < h\upsilon < 2$ eV) are absorbed in the GaAs layer. In more sophisticated cells, the bandgap of AlGaAs is graded slowly from the surface by varying the composition of the AlGaAs layer.

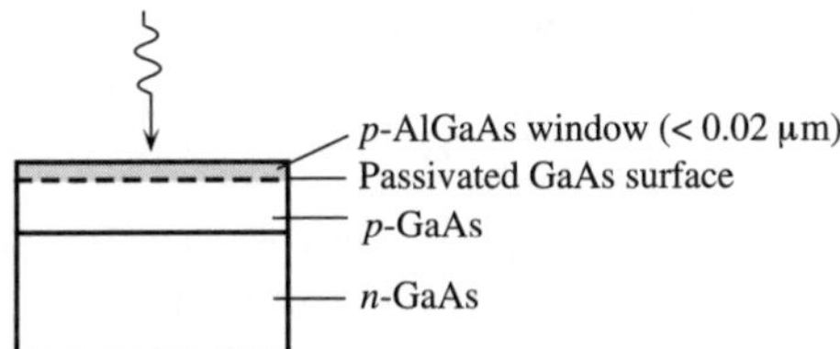

FIGURE 6.16 AlGaAs window layer on GaAs passivates the surface states and thereby increases the low wavelength photogeneration efficiency.

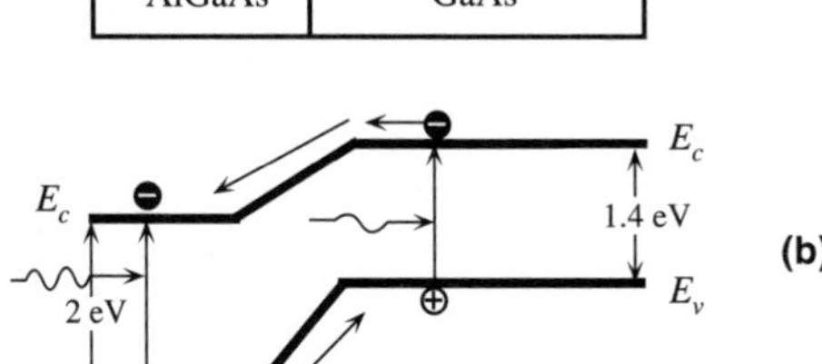

FIGURE 6.17 A heterojunction solar cell between two different bandgap semiconductors (GaAs and AlGaAs).

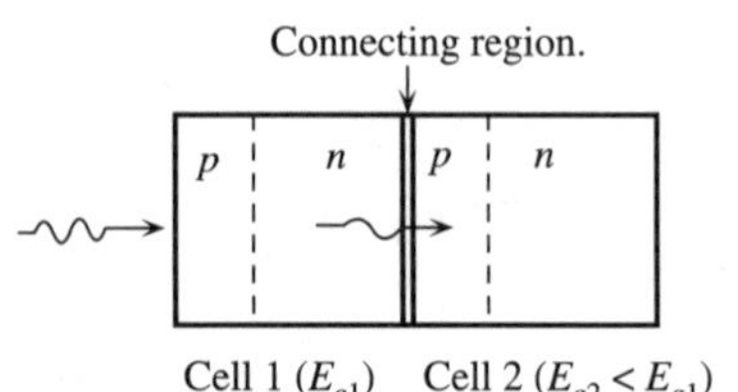

FIGURE 6.18 A tandem cell. Cell 1 has a wider handgap and absorbs energetic photons with $h\upsilon > E_{g1}$. Cell 2 absorbs photons that pass cell 1 and have $h\upsilon > E_{g2}$.

Tandem or cascaded cells use two or more cells in tandem or in cascade to increase the absorbed photons from the incident light as illustrated in Figure 6.18. The first cell is made from a wider bandgap material and only absorbs photons with $h\upsilon > E_{g1}$. The second cell absorbs photons that pass the first cell and have $h\upsilon > E_{g2}$. The whole structure can be grown within a single crystal by using lattice matched crystalline layers leading to a monolithic tandem cell. If, in addition, light concentrators are also used, the efficiency can be further increased. For example, a GaAs-GaSb tandem cell operating under a 100-sun condition, that is 100 times that of ordinary sunlight, have exhibited an efficiency of about 34%. Tandem cells have been used in thin film a-Si:H (amorphous hydrogenated amorphous Si) *pin* solar cells to obtain efficiencies up to about 12%. These tandem cells have a-Si:H and a-Si:Ge:H cells and are easily fabricated in large areas.

QUESTIONS AND PROBLEMS

6.1 Solar energy spectrum Given Figure 6.1, how would you obtain $I_{h\upsilon}$ vs. $h\upsilon$ where $I_{h\upsilon}$ is the spectral intensity per unit photon energy and $h\upsilon$ is the photon energy? Sketch the energy spectrum by taking 5 points on the I_λ vs. λ curve.

6.2 Photocurrent The absorption coefficient of crystalline Si is shown in Figure 5.3. Consider a crystalline Si device that has $A = 4\text{ cm} \times 4\text{ cm}$, $\ell_n = 0.5\ \mu\text{m}$, $W = 1.5\ \mu\text{m}$, $L_e = 70\ \mu\text{m}$ and consider a photogeneration rate $G_o = 1 \times 10^{18}\text{ cm}^{-3}\text{ s}^{-1}$ at $x = 0$. What is the photocurrent when $\lambda \approx 1.1\ \mu\text{m}$? What is the photocurrent when $\lambda \approx 500$ nm? Show that

$$G_o = \frac{I_o \alpha}{h\upsilon}$$

where I_o is the transmitted light intensity into the device at $x = 0$. Hence estimate the incident light intensity required in each case assuming a transmittance of 1.

6.3 Solar cell driving a load

(a) A Si solar cell of area 4 cm^2 is connected to drive a load R as in Figure 6.8 (a). It has the I–V characteristics in Figure 6.8 (b) under an illumination of 600 W m^{-2}. Suppose that the load is 20 Ω and it is used under a light intensity of 1 kW m^{-2}. What are the current and voltage in the circuit? What is the power delivered to the load? What is the efficiency of the solar cell in this circuit?

(b) What should the load be to obtain maximum power transfer from the solar cell to the load at 1 kW m^{-2} illumination. What is this load at 600 W m^{-2}?

(c) Consider using a number of such a solar cells to drive a calculator that needs a minimum of 3 V and draws 3.0 mA at 3–4 V. It is to be used indoors at a light intensity of about 400 W m^{-2}. How many solar cells would you need and how would you connect them? At what light intensity would the calculator stop working?

6.4 Open circuit voltage A solar cell under an illumination of 100 W m^{-2} has a short circuit current I_{sc} of 50 mA and an open circuit output voltage V_{oc}, of 0.55 V. What are the short circuit current and open circuit voltages when the light intensity is halved?

6.5 Shunt resistance Consider the equivalent circuit of a solar cell as shown in Figure 6.10

(a) Show that

$$I = -I_{ph} + I_d + \frac{V}{R_p} = -I_{ph} + I_o \exp\left(\frac{eV}{nk_BT}\right) - I_o + \frac{V}{R_p}$$

(b) Plot I vs. V for a polycrystalline Si solar cell that has $n = 2$ and $I_o = 3 \times 10^{-4}$ mA, for an illumination such that $I_{ph} = 5$ mA. Use $R_p = \infty$, 1000 Ω and then R_p 100 Ω. What is your conclusion?

6.6 Series connected solar cells Consider two identical solar cells with the properties $I_o = 25 \times 10^{-6}$ mA, $n = 1.5$, $R_s = 20$ Ω, subjected to the same illumination so that $I_{ph} = 10$ mA. Plot the individual I–V characteristics and the I–V characteristics of the two cells in series. Find the maximum power that can be delivered by one cell and two cells in series. Find the corresponding voltage and current at the maximum power point.

6.7 Series connected solar cells Consider two odd solar cells. Cell 1 has $I_{o1} = 25 \times 10^{-6}$ mA, $n_1 = 1.5, R_{s1} = 10$ Ω and cell 2 has $I_{o1} = 1 \times 10^{-7}$ mA, $n_1 = 1, R_{s2} = 50$ Ω. The illumination is such that $I_{ph1} = 10$ mA and $I_{ph2} = 15$ mA. Plot the individual I–V characteristics and the I–V characteristics of the two cells in series. Find the maximum power that can be delivered by each cell and two cells in series. Find the corresponding voltages and currents at the maximum power point. What is your conclusions?

6.8 Solar cells Show that the open circuit output voltage from a *pn* junction solar cell is approximately

$$V_{oc} \approx \frac{nk_BT}{e} \ln\left(\frac{BI}{n_i^2}\right)$$

where I is the light intensity, n_i is the intrinsic concentration and B is a constant that is only weakly temperature dependent (*e.g.*, $B \propto T^{\gamma}$ and $\gamma < 1$) compared with n_i.

6.9 Solar cell efficiency The fill factor FF of a solar cell is given by the empirical expression

$$\text{FF} \approx \frac{v_{oc} - \ln(v_{oc} + 0.72)}{v_{oc} + 2}$$

where $v_{oc} = V_{oc}/(nk_BT/e)$ is the normalized open circuit voltage (normalized with respect to the thermal voltage k_BT/e). The maximum power output from a solar cell is

$$P = \text{FF} I_{sc} V_{oc}$$

Taking $V_{oc} = 0.58$ V and $I_{sc} = I_{ph} = 35$ mA cm^{-2}, calculate the power available per unit area of solar cell at room temperature 20 °C, at −40 °C and at 40 °C?

6.10 Output voltage when it is cold

(a) The intensity of sun light arriving at a point on Earth, where the solar latitude is α can be approximated by the **Meinel and Meinel equation**

$$I = 1.353(0.7)^{(\mathrm{cosec}\,\alpha)^{0.678}} \quad \mathrm{kW\ m^{-2}} \tag{1}$$

where $\mathrm{cosec}\,\alpha = 1/\sin\alpha$. The solar latitude α is the angle between the sun rays and the horizon. Around September 23rd and March 22nd, the sun rays arrive parallel to the plane of the equator. What is the maximum power available for a photovoltaic device panel of area 1 m^2 if its efficiency of conversion is 10%.

(b) Manufacturer' characterization tests on a particular Si *pn* junction solar cell at 27 °C specifies an open circuit output voltage of 0.45 V and a short circuit current of 400 mA when illuminated directly with light of intensity 1 kW m^{-2}. The fill factor for the solar cell is 0.73. This solar cell is to be used in a portable equipment application near Eskimo Point (NWT) at a latitude (ϕ) of 63°. Calculate the open circuit output voltage and the maximum available power when the solar cell is used at noon on 23rd September when the temperature is around −10 °C. What is the maximum current this solar cell can supply to an electronic equipment? What is your conclusion?

6.11 Maximum Power and Fill Factor Given the solar cell current equation, and that the power extracted from the solar cell is $-IV$, show that maximum power occurs when $V = V_m$ and $I = I_m$ *when,*

$$\frac{V_m}{nV_T}\exp\left(\frac{V_m}{nV_T}\right) \approx \frac{I_{ph}}{I_o} \quad \text{and} \quad I_m = -I_{ph}\left(1 - \frac{nV_T}{V_m}\right) \tag{3}$$

where $V_T = k_BT/e$ is the "thermal voltage". Given n, $I_{sc} \approx I_{ph}$ and V_{oc}, suggest ways to estimate I_m and V_m and hence the FF.

6.12 Energyband diagram

(a) Explain the photovoltaic action by using an energy band diagram with and without illumination.

(b) Sketch the energy band diagram of an n-type semiconductor that has a decreasing E_g from left to right. What happens to an electron and a hole that are photogenerated?

Silicon solar cells on a house roof. *(Courtesy of Mobil.)*

C H A P T E R 7

Polarization and Modulation of Light

"As there are two different refractions, I conceived also that there are two different emanations of the waves of light..."

—Christiaan Huygens

Sailors visiting Iceland during the 17th century brought back to Europe calcite crystals (Iceland spar) which had the unusual property of showing double images when objects were viewed through it. Erasmus Bartholinus described this property as the effect of double refraction and later Christiaan Huygens (1629–1695), a Dutch physicist, explained this double refraction in terms of ordinary and extraordinary waves. Christiaan Huygens made many contributions to optics and wrote prolifically on the subject. *(Courtesy of AIP Emilio Segrè Visual Archives.)*

7.1 POLARIZATION

A. State of Polarization

A propagating electromagnetic (EM) wave has its electric and magnetic field at right angles to the direction of propagation. If we place a z-axis along the direction of propagation, then the electric field can be in any direction in the plane perpendicular to the

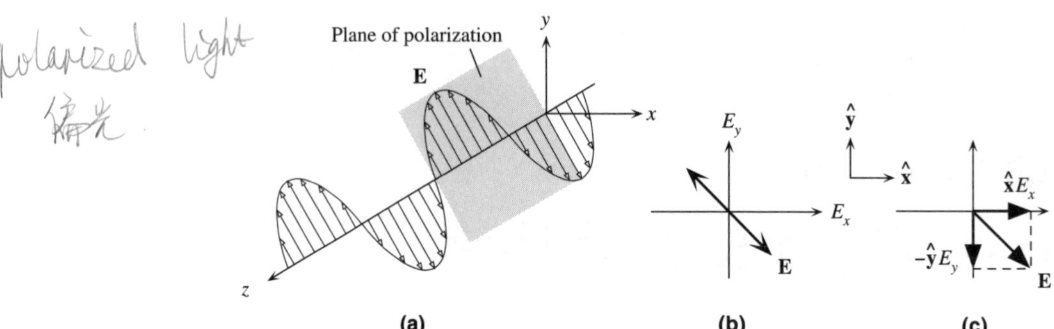

FIGURE 7.1 (a) A linearly polarized wave has its electric field oscillations defined along a line perpendicular to the direction of propagation, z. The field vector **E** and z define a *plane of polarization.* (b) The E-field oscillations are contained in the plane of polarization. (c) A linearly polarized light at any instant can be represented by the superposition of two fields E_x and E_y with the right magnitude and phase.

z-axis. The term **polarization** of an EM wave describes the behavior of the electric field vector in the EM wave as it propagates through a medium. If the oscillations of the electric field at all times are contained within a well-defined line then the EM wave is said to be **linearly polarized** as shown in Figure 7.1 (a). The field vibrations and the direction of propagation (z) define a plane of polarization (plane of vibration) so that linear polarization implies a wave that is **plane-polarized**. By contrast, if a beam of light has waves with the E-field in each in a random direction but perpendicular to z, then this light beam is *unpolarized.* A light beam can be linearly polarized by passing the beam through a *polarizer*, such as a polaroid sheet, a device that only passes electric field oscillations lying on a well defined plane at right angles to the direction of propagation.

Suppose that we arbitrarily place the x and y axes and describe the electric field in terms of its components E_x and E_y along x and y (we are justified to do this because E_x and E_y are perpendicular to z). To find the electric field in the wave at any space and time location, we add E_x and E_y *vectorially*. Both E_x and E_y can individually be described by a wave equation which must have the same angular frequency ω and wavenumber k. However, we must include a phase difference ϕ between the two

$$E_x = E_{xo}\cos(\omega t - kz) \tag{1}$$

and

$$E_y = E_{yo}\cos(\omega t - kz + \phi) \tag{2}$$

where ϕ is the phase difference between E_y and E_x; ϕ can arise if one of the components is delayed (retarded).

The linearly polarized wave in Figure 7.1 (a) has the **E** oscillations at $-45°$ to x-axis as shown in (b). We can generate this field by choosing $E_{xo} = E_{yo}$ and $\phi = \pm 180°(\pm\pi)$ in Eqs. (1) and (2). Put differently, E_x and E_y have the same magnitude but they are out of phase by 180°. If $\hat{\mathbf{x}}$ and $\hat{\mathbf{y}}$ are the unit vectors along x and y, using $\phi = \pi$ in Eq. (2), the field in the wave is

$$\mathbf{E} = \hat{\mathbf{x}}E_x + \hat{\mathbf{y}}E_y = \hat{\mathbf{x}}E_{xo}\cos(\omega t - kz) - \hat{\mathbf{y}}E_{yo}\cos(\omega t - kz)$$

or

$$\mathbf{E} = \mathbf{E}_o\cos(\omega t - kz) \tag{3}$$

where

$$\mathbf{E}_o = \hat{\mathbf{x}}E_{xo} - \hat{\mathbf{y}}E_{yo} \tag{4}$$

Equations (3) and (4) state that the vector $\mathbf{E}_o$ at $-45°$ to the x-axis and propagates along the z-direction.

There are many choices for the behavior of the electric field besides the simple linear polarization in Figure 7.1. For example, if the magnitude of the field vector **E** remains constant but its tip at a given location on z traces out a circle by rotating in a clockwise sense with time, as observed by the receiver of the wave, then the wave is said to be **right circularly polarized**[1] as in Figure 7.2. If the rotation of the tip of **E** is counterclockwise, the wave is said to be **left circularly polarized.** From Eqs. (1) and (2), it should be apparent that a right circularly polarized wave has $E_{xo} = E_{yo} = A$ (an amplitude), and $\phi = \pi/2$. This means that,

$$E_x = A\cos(\omega t - kz) \tag{5a}$$

and

$$E_y = -A\sin(\omega t - kz) \tag{5b}$$

It is relatively straightforward to show Eqs. (5a) and (5b) represent a circle that is

$$E_x^2 + E_y^2 = A^2 \tag{6}$$

as shown in Figure 7.2.

The snapshot of the circularly polarized light in Figure 7.2 shows that over a distance Δz, the field **E** rotates through an angle $\theta = k\Delta z$. Linear and circular polarization concepts are summarized in Figure 7.3 where for simplicity $E_{yo} = 1$ has been taken and the corresponding E_{xo} and ϕ are shown.

An **elliptically polarized** light, or elliptic light, has the tip of the **E**-vector trace out an ellipse as the wave propagates through a given location in space. As in circular polarization, light can be right and left elliptically polarized depending on clockwise or

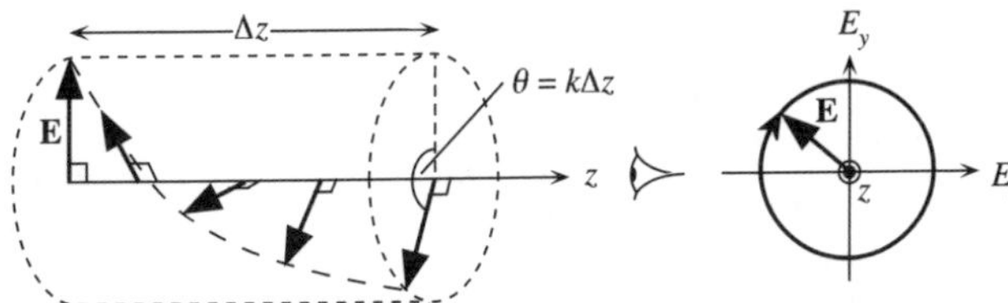

FIGURE 7.2 A *right circularly polarized light.* The field vector **E** is always at right angles to z, rotates clockwise around z with time, and traces out a full circle over one wavelength of distance propagated.

[1] There is a difference in this definition in optics and engineering. The definition here follows that in optics which is more prevalent in optoelectronics.

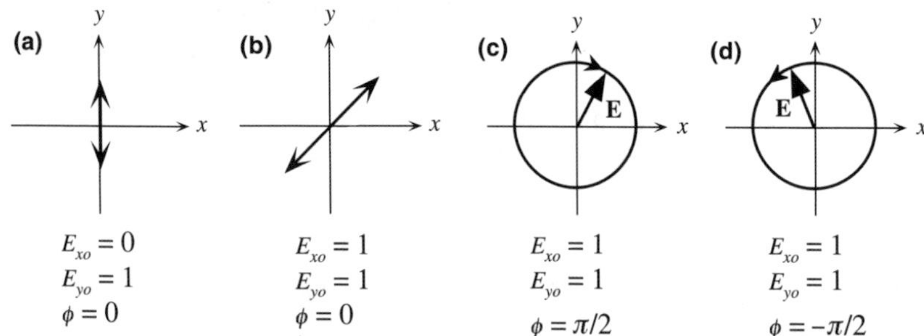

FIGURE 7.3 Examples of linearly, (a) and (b), and circularly polarized light (c) and (d); (c) is right circularly and (d) is left circularly polarized light (as seen when the wave directly approaches a viewer)

FIGURE 7.4 (a) Linearly polarized light with $E_{yo} = 2E_{xo}$ and $\phi = 0$. (b) When $\phi = \pi/4$ (45°), the light is right elliptically polarized with a tilted major axis. (c) When $\phi = \pi/2$ (90°), the light is right elliptically polarized. If E_{xo} and E_{yo} were equal, this would be right circularly polarized light.

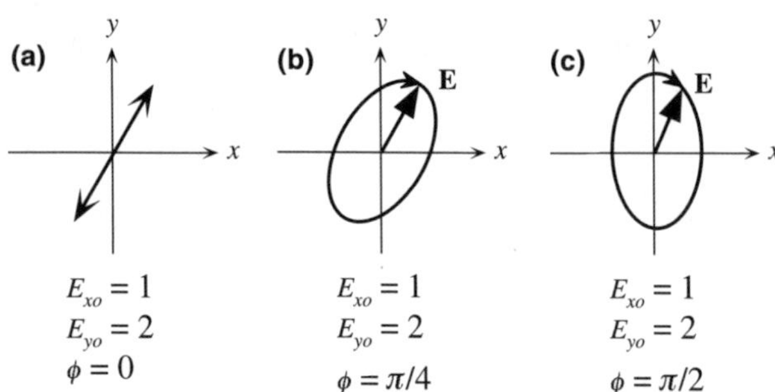

counterclockwise rotation of the **E**-vector. Figure 7.4 illustrates how an elliptically polarized light can result for any ϕ *not* zero or equal to any multiple of π, and when E_{xo} and E_{yo} are not equal in magnitude. Elliptic light can also be obtained when $E_{xo} = E_{yo}$ and the phase difference is $\pm\pi/4$ or $\pm 3\pi/4$ etc.

EXAMPLE 7.1.1 Elliptical and circular polarization

Show that if $E_x = A\cos(\omega t - kz)$ and $E_y = B\cos(\omega t - kz + \phi)$ that the amplitudes A and B are different and the phase difference ϕ is $\pi/2$, 90°, the wave is elliptically polarized.

Solution From the x and y components we have

$$\cos(\omega t - kz) = E_x/A$$

and

$$\cos(\omega t - kz + \pi/2) = -\sin(\omega t - kz) = E_y/B$$

Using, $\sin^2(\omega t - kz) + \cos^2(\omega t - kz) = 1$, we find,

$$\left(\frac{E_x}{A}\right)^2 + \left(\frac{E_y}{B}\right)^2 = 1$$

This is the equation that relates the instantaneous values E_x and E_y of the field along x and y. The equation is a *circle* when $A = B$, as in Figure 7.3 (c) and an *ellipse* when $A \neq B$ as in Figure 7.4 (c).

Further, at $z = 0$, and at $\omega t = 0, E = E_x = A$. Later when $\omega t = \pi/2, E = E_y = -B$. Thus the tip of electric field rotates in a clockwise sense and the wave is right circularly polarized.

B. Malus's Law

There are various optical devices that operate on the polarization state of a wave passing through it and thereby modify the polarization state. A linear polarizer will only allow electric field oscillations along some preferred direction, called the **transmission axis**, to pass through the device as illustrated in Figure 7.5. A *polaroid sheet* is a good example of a commercially available linear polarizer. **Dichroic crystals** such as tourmaline crystals are good polarizers because they are optically anisotropic and attenuate EM waves with fields that are *not* oscillating along the optical axis (hence the transmission axis) of the crystal. The emerging beam from the polarizer has its field oscillations along the transmission axis and hence it is *linearly polarized.*

Suppose that the linearly polarized light from the polarizer is now incident on a second identical polarizer. Then by rotating the transmission axis of this second polarizer we can analyze the polarization state of the incident beam; hence the second polarizer is called an **analyzer**. If the transmission axis of the second polarizer is at an angle θ to the **E**-field of the incident beam (*i.e.* to the first polarizer) then only the component $E \cos\theta$ of the field will be allowed to pass through the analyzer as in Figure 7.5. The irradiance (intensity) of light passing through the analyzer is proportional to the square of the electric field which means that the detected intensity varies as $(E\cos\theta)^2$. Since all the electric field will pass when $\theta = 0$ (**E** parallel to TA_2), this is the maximum irradiance condition. The irradiance I at any other angle θ is then given by **Malus's law**:

$$I(\theta) = I(0)\cos^2\theta \tag{6}$$

Malus's law

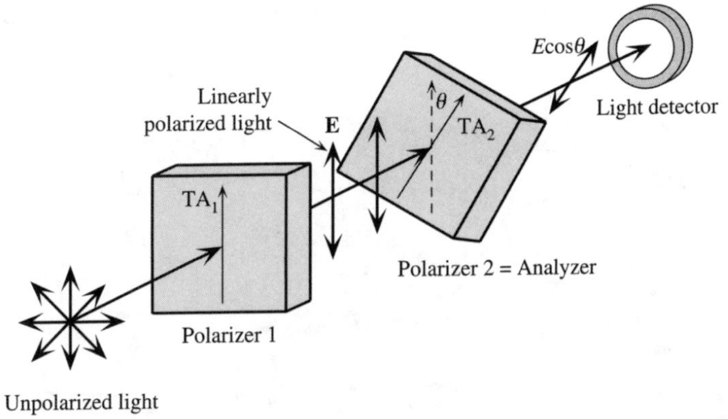

Linearly polarized light
E
Ecosθ
θ
TA_2
Light detector
TA_1
Polarizer 2 = Analyzer
Polarizer 1
Unpolarized light

FIGURE 7.5 Randomly polarized light is incident on a Polarizer 1 with a transmission axis TA_1. Light emerging from Polarizer 1 is linearly polarized with **E** along TA_1, and becomes incident on Polarizer 2 (called *analyzer*) with a transmission axis TA_2 at an angle θ to TA_1. A detector measures the intensity of the incident light. TA_1 and TA_2 are normal to the light direction.

Malus's law therefore relates the intensity of a linearly polarized light passing through a polarizer to the angle between the transmission axis and the electric field vector.

7.2 LIGHT PROPAGATION IN AN ANISOTROPIC MEDIUM: BIREFRINGENCE

A. Optical Anisotropy

An important characteristic of crystals is that many of their properties depend on the crystal direction, that is crystals are generally anisotropic. The dielectric constant ε_r depends on electronic polarization which involves the displacement of electrons with respect to positive atomic nuclei. Electronic polarization depends on the crystal direction inasmuch as it is easier to displace electrons along certain crystal directions. *This means that the refractive index n of a crystal depends on the direction of the electric field in the propagating light beam.* Consequently, the velocity of light in a crystal depends on the direction of propagation and on the state of its polarization, *i.e.* the direction of the electric field. Most noncrystalline materials such as glasses and liquids, and all cubic crystals are **optically isotropic**, that is the refractive index is the same in all directions. For all classes of crystals excluding cubic structures, the refractive index depends on the propagation direction and the state of polarization. The result of optical anisotropy is that, except along certain special directions, any unpolarized light ray entering such a crystal breaks into two different rays with different polarizations and phase velocities. When we view an image through a calcite crystal, an optically anisotropic crystal, we see two images, each constituted by light of different polarization passing through the crystal, whereas there is only one image through an optically isotropic crystal as depicted in Figure 7.6. Optically anisotropic crystals are called **birefringent** because an incident light beam may be doubly refracted.

Experiments and theories on "most anisotropic crystals", *i.e.* those with the highest degree of anisotropy, show that we can describe light propagation in terms of *three* refractive indices, called *principal refractive indices* n_1, n_2 and n_3, along three mutually

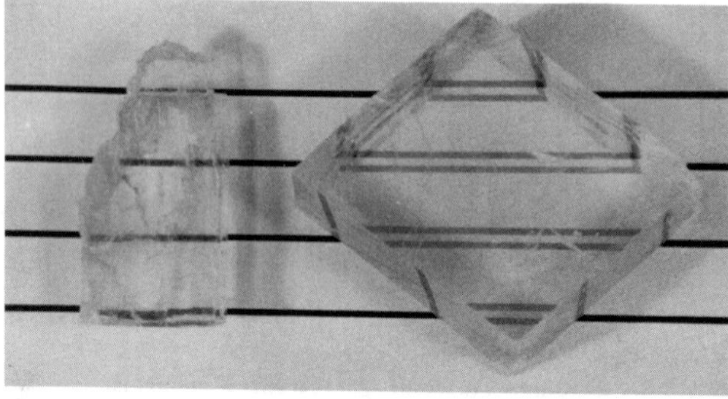

FIGURE 7.6 A line viewed through a cubic sodium chloride (halite) crystal (optically isotropic) and a calcite crystal (optically anisotropic).

TABLE 7.1 Principal refractive indices of some optically isotropic and anisotropic crystals (near 589 nm, yellow Na-D line)

Optically isotropic		$n = n_o$		
	Glass (crown)	1.510		
	Diamond	2.417		
	Fluorite (CaF_2)	1.434		
Uniaxial - Positive		n_o	n_e	
	Ice	1.309	1.3105	
	Quartz	1.5442	1.5533	
	Rutile (TiO_2)	2.616	2.903	
Uniaxial - Negative		n_o	n_e	
	Calcite ($CaCO_3$)	1.658	1.486	
	Tourmaline	1.669	1.638	
	Lithium niobate ($LiNBO_3$)	2.29	2.20	
Biaxial		n_1	n_2	n_3
	Mica (muscovite)	1.5601	1.5936	1.5977

orthogonal directions in the crystal, say x, y and z called *principal axes*. These indices correspond to the polarization state of the wave along these axes.

Crystals that have three distinct principal indices also have *two* optic axes and are called **biaxial crystals**. On the other hand, **uniaxial crystals** have two of their principal indices the same $(n_1 = n_2)$ and only have *one* optic axis. Table 7.1 summarizes crystal classifications according to optical anisotropy. Uniaxial crystals, such as quartz, that have $n_3 > n_1$ and are called **positive**, and those such as calcite that have $n_3 < n_1$ are called **negative** uniaxial crystals.

B. Uniaxial Crystals and Fresnel's Optical Indicatrix

For our discussions of optical anisotropy, we will consider uniaxial crystals such as calcite and quartz. All experiments and theories lead to the following basic principles.[2]

Any EM wave entering an anisotropic crystal splits into two orthogonal linearly polarized waves which travel with different phase velocities, that is they experience different refractive indices. These two orthoganally polarized waves in uniaxial crystals are called **ordinary** (o) and **extraordinary** (e) **waves**. The o-wave has the same phase velocity in all directions and behaves like an ordinary wave in which the field is perpendicular to the phase propagation direction. The e-wave has a phase velocity that depends on its direction of propagation and its state of polarization, and further the electric field in the e-wave is not necessarily perpendicular to the phase propagation direction. These two waves propagate with the same velocity only along a special direction called the **optic axis**. The o-wave is always perpendicularly polarized to the optic axis and obeys the usual Snell's law.

[2] These statements can be proved by solving Maxwell's equations in an anisotropic medium; see Further Reading.

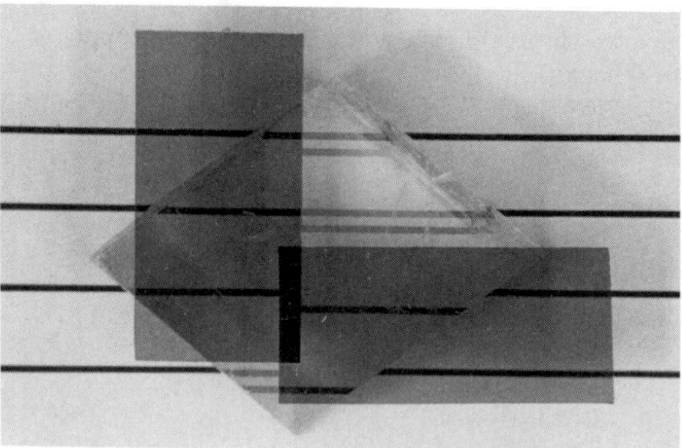

FIGURE 7.7 Two polaroid analyzers are placed with their transmission axes, along the long adges, at right angles to each other. The ordinary ray, undeflected, goes through the left polarizer whereas the extraordinary wave, deflected, goes through the right polarizer. The two waves therefore have orthogonal polarizations.

The two images observed in Figure 7.6 are due to *o*-waves and *e*-waves being refracted differently so that when they emerge from the crystal they have been separated. Each ray constitutes an image but the field directions are orthogonal. The fact that this is so is easily demonstrated by using two polaroid analyzers with their transmission axes at right angles as in Figure 7.7. If we were to view an object along the optic axis of the crystal, we would not see two images because the two rays would experience the same refractive index.

As mentioned above, we can represent the optical properties of a crystal in terms of three refractive indices along three orthogonal axes, the principal axes of the crystal, shown as x, y and z in Figure 7.8 (a). These are special axes along which the polarization vector and the electric field are parallel. (Put differently, the electric displacement[3]

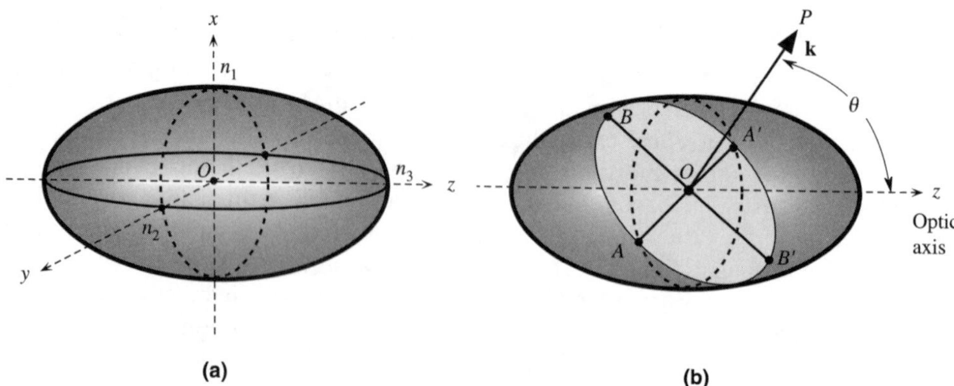

FIGURE 7.8 (a) Fresnel's ellipsoid (b) An EM wave propagating along OP at an angle θ to optic axis.

[3]Electric displacement **D** at any point is defined by $\mathbf{D} = \varepsilon_o\mathbf{E} + \mathbf{P}$ where **E** is the electric field and **P** is the polarization at that point.

D and the electric field **E** vectors are parallel). The refractive indices along these x, y and z axes are the principal indices n_1, n_2 and n_3 respectively for electric field oscillations along these directions (not to be confused with the wave propagation direction). For example, for a wave with a polarization parallel to the x-axis, the refractive index is n_1.

The refractive index associated with a particular EM wave in a crystal can be determined by using Fresnel's *refractive index ellipsoid*, called the **optical indicatrix**,[4] which is a refractive index surface placed in the center of the principal axes, as shown in Figure 7.8 (a), where the x, y and z axis intercepts are n_1, n_2, and n_3. If all three indices were the same, $n_1 = n_2 = n_3 = n_o$ we would have a spherical surface and all electric field polarization directions would experience the same refractive index, n_o. Such a spherical surface would represent an optically isotropic crystal. For positive uniaxial crystals such as quartz, $n_1 = n_2 < n_3$ which is the example in Figure 7.8 (a).

Suppose that we wish to find the refractive indices experienced by a wave traveling with an arbitrary wave vector **k**, which represents the direction of phase propagation. This phase propagation direction is shown as OP in Figure 7.1 (b) and is at an angle θ to the z-axis. We place a plane perpendicular to OP and passing through the center O of the indicatrix. This plane intersects the ellipsoid surface in a curve $ABA'B'$ which is an *ellipse*. The major (BOB') and minor (AOA') axes of this ellipse determine the field oscillation directions and the refractive indices associated with this wave. Put differently, the original wave is now represented by two orthogonally polarized EM waves.

The line AOA', the *minor axis*, corresponds to the polarization of the ordinary wave and its semiaxis AA' is the refractive index $n_o = n_2$ of this o-wave. The electric displacement and the electric field are in the same direction and parallel to AOA'. If we were to change the direction of OP we would always find the same minor axis, *i.e.* n_o is either n_1 or n_2 whatever the orientation of OP (try orientating OP to be along y and along x). This means that the o-wave always experiences the same refractive index in all directions. (The o-wave behaves just like an ordinary wave; hence the name.)

The line BOB' in Figure 7.8 (b), the *major axis*, corresponds to the electric displacement field (**D**) oscillations in the extraordinary wave and its semiaxis OB is the refractive index $n_e(\theta)$ of this e-wave. This refractive index is smaller than n_3 but greater than $n_2(=n_o)$. The e-wave therefore travels more slowly than the o-wave in this particular direction and in this crystal. If we change the direction of OP, we find that the length of the major axis changes with the OP direction. Thus, $n_e(\theta)$ depends on the wave direction, θ. As apparent, $n_e = n_o$ when OP is along the z-axis, that is, when the wave is traveling along z as in Figure 7.9 (a). This direction is the **optic axis** and all waves traveling along the optic axis have the same phase velocity whatever their polarization. When the e-wave is traveling along the y-axis, or along the x-axis, $n_e(\theta) = n_3 = n_e$ and the e-wave has its slowest phase velocity as shown in Figure 7.9 (b). Along any OP direction that is at an angle θ to the optic axis, the e-wave has a refractive index $n_e(\theta)$ given by

$$\frac{1}{n_e(\theta)^2} = \frac{\cos^2\theta}{n_o^2} + \frac{\sin^2\theta}{n_e^2} \quad (1)$$

Refractive index of the e-wave

Clearly, for $\theta = 0°$, $n_e(0°) = n_o$ and for $\theta = 90°$, $n_e(90°) = n_e$.

[4] There are various names in the literature with various subtle nuances: the Fresnel ellipsoid, optical indicatrix, index ellipsoid, reciprocal ellipsoid, Poinsot ellipsoid, ellipsoid of wave normals.

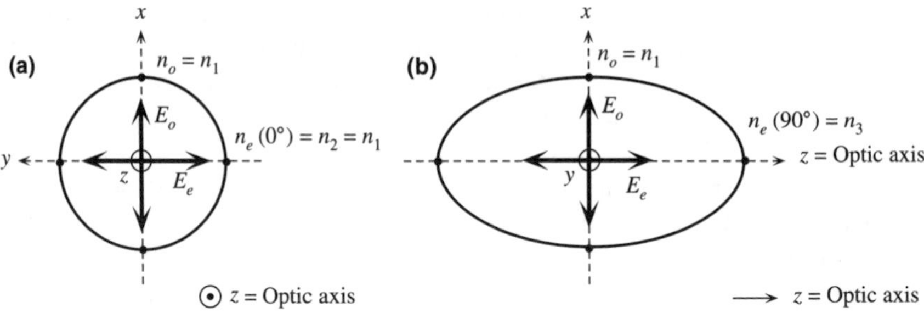

FIGURE 7.9 $E_o = E_{o\text{-wave}}$ and $E_e = E_{e\text{-wave}}$ (a) Wave propagation along the optic axis. (b) Wave propagation normal to optic axis.

The major axis BOB' in Figure 7.8 (b) determines the *e*-wave polarization by defining the direction of the displacement vector **D** and not **E**. Although **D** is perpendicular to **k**, this is not true for **E**. The electric field of the *e*-wave is orthogonal to that of the *o*-wave, and it is in the plane determined by **k** and the optic axis, as discussed below. **E** is orthogonal to **k** when the *e*-wave propagates along one of the principal axes.

From the indicatrix, or equivalently from Eq. (1), we can easily determine the refractive indices of the *o*- and *e*-waves in any direction and thereby calculate their wavevectors. We can then construct a **wavevector surface** for each of the *o* and *e*-waves which have the following property. The distance from the origin O to any arbitrary point P on a wavevector surface represents the value of **k** along the direction OP. Since the *o*-wave has the same refractive index in all directions, its wavevector surface will be a sphere of radius $n_o k_{\text{vacuum}}$, where k_{vacuum} is the wavevector in free space. This is shown as a circle in an "xz" cross section in Figure 7.10 (a). On the other hand, the wavevector of the *e*-wave depends on the propagation direction and is given by $n_e(\theta)k_{\text{vacuum}}$ so its surface is an ellipse

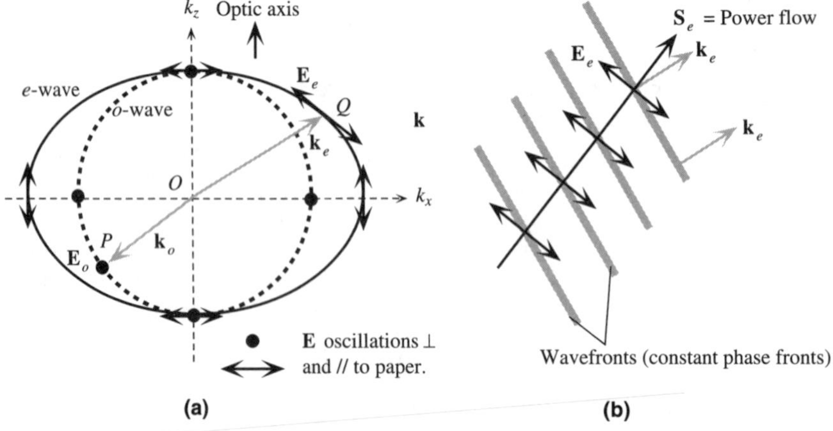

FIGURE 7.10 (a) Wavevector surface cuts in the xz plane for *o*- and *e*-waves. (b) An extraordinary wave in an anisotropic crystal with a $\mathbf{k}_e$ at an angle to the optic axis. The electric field is not normal to $\mathbf{k}_e$. The energy flow (group velocity) is along $\mathbf{S}_e$ wich is different than $\mathbf{k}_e$.

in the "xz" cross section in Figure 7.10 (a). Two example wavevectors $\mathbf{k}_o$ and $\mathbf{k}_e$ that represent an o-wave and an e-wave propagating along arbitrary directions OP and OQ respectively are depicted in Figure 7.10 (a) (Different directions chosen only for clarity).

The electric field $\mathbf{E}_o$ of the o-wave is always orthogonal to its wavevector direction $\mathbf{k}_o$ and also to the optic axis. This fact is shown as dots on the o-wave wavevector surface in Figure 7.10 (a) and is highlighted for an arbitrary $\mathbf{k}_o$ along OP. Since the electric and magnetic fields in the o-wave are normal to $\mathbf{k}_o$, the o-wave Poynting vector $\mathbf{S}_o$, *i.e.* the direction of energy flow, is along $\mathbf{k}_o$.

It may be thought that, as in normal EM wave propagation as in the o-wave, the electric field $\mathbf{E}_e$ in the e-wave should be normal to the wavevector $\mathbf{k}_e$. This is not generally true. The reason is that the polarization of the medium is not parallel to the inducing field in the e-wave and consequently the overall electric field $\mathbf{E}_e$ in the EM wave is *not* at right angles to the phase propagation direction $\mathbf{k}_e$ as indicated in Figure 7.10 (a). This means that the energy flow (group velocity) and the phase velocity directions are different (a phenomenon called the Poynting vector "walk-off" effect). The energy flow, *i.e.* the Poynting vector $\mathbf{S}_e$, direction is taken as the **ray direction** for the e-wave so that the wavefronts advance "sideways" as depicted in Figure 7.10 (b). The group velocity is in the same direction as energy flow ($\mathbf{S}_e$).

C. Birefringence of Calcite

Consider a calcite crystal ($CaCO_3$) which is a negative uniaxial crystal and also well known for its double refraction. When the surfaces of a calcite crystal have been cleaved, that is cut along certain crystal planes, the crystal attains a shape that is called a *cleaved form* and the crystal faces are rhombohedrons (parallelogram with 78.08° and 101.92°). A cleaved form of the crystal is called a *calcite rhomb*. A plane of the calcite rhomb that contains the optical axis and is normal to a pair of opposite crystal surfaces is called a *principal section*.

Consider what happens when an unpolarized or natural light enters a calcite crystal at normal incidence and thus also normal to a principal section to this surfaces, but at an angle to the optic axis as shown in Figure 7.11. The ray breaks into ordinary (o)

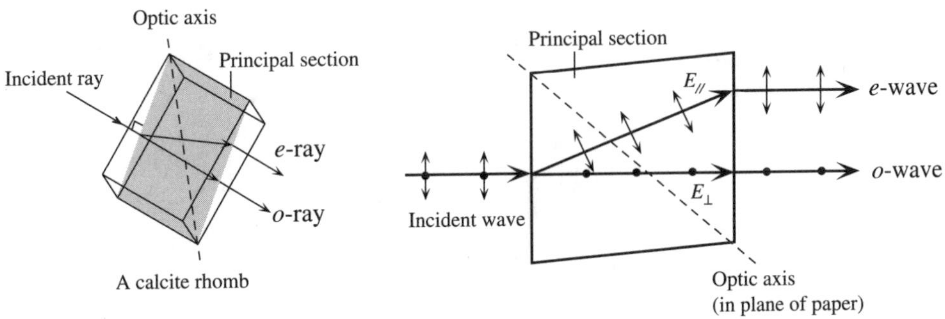

FIGURE 7.11 An EM wave that is off the optic axis of a calcite crystal splits into two waves called ordinary and extraordinary waves. These waves have orthogonal polarizations and travel with different velocities. The o-wave has a polarization that is always perpendicular to the optical axis.

and extraordinary (e) waves with mutually orthogonal polarizations. The waves propagate in the plane of the principal section as this plane also contains the incident light. The o-wave has its field oscillations perpendicular to the optic axis. It obeys Snell's law which means that it enters the crystal undeflected. Thus the direction of E-field oscillations must come out of the paper so that it is normal to the optic axis and also to the direction of propagation. The field $E_{\perp}$ in the o-ray is shown as dots, oscillating into and out of the paper.

The e-wave has a polarization orthogonal to the o-wave and in the principal section (which contains the optic axis and **k**). The e-wave polarization is in the plane of the paper, indicated as $E_{//}$, in Figure 7.11. It travels with a different velocity and diverges from the o-wave. Clearly, the e-wave does not obey the usual Snell's law inasmuch as the angle of refraction is not zero. We can determine the e-ray direction (power flow direction) by noting that the e-wave propagates sideways as in Figure 7.10 (b) at right angles to $E_{//}$.

If we were to cut a plate from a calcite crystal so that the optic axis (along z) would be parallel to two opposite faces of the plate as in Figure 7.12 (a), then a ray entering at normal incidence to one of these faces would not diverge into two separate waves. This is the case illustrated in Figure 7.9 (b) that is propagation along the y-direction except now $n_e < n_o$. The o- and e-waves would travel in the same direction but with different speeds. The waves emerge in the same direction as well which means that we would see no double refraction. This optical arrangement is used in the construction of various optical retarders and polarizers as discussed below. If we were to cut a plate so that the optic axis was perpendicular to the plate faces as in Figure 7.12 (b), then both the o and e-way would be traveling at the same speed (Figure 7.9 (a)) and along the same direction which means we would not again see any double refraction.

D. Dichroism

In addition to the variation in the refractive index, some anisotropic crystals also exhibit **dichroism**, a phenomenon in which the optical absorption in a substance depends on the direction of propagation and the state of polarization of the light beam. A dichroic crystal is an optically anisotropic crystal in which either the e-wave or the o-wave is heavily attenuated (absorbed). This means that a light wave of arbitrary polarization

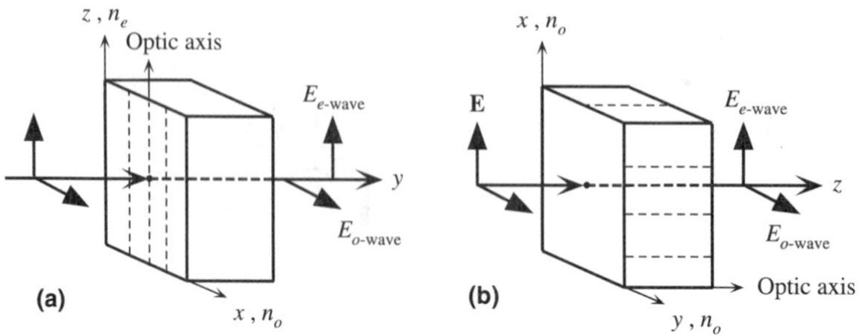

FIGURE 7.12 (a) A birefringent crystal plate with the optic axis parallel to the plate surfaces. (b) A birefringent crystal plate with the optic axis perpendicular to the plate surfaces.

entering a dichroic crystal emerges with a well defined polarization because the other orthogonal polarization would have been attenuated. Generally dichroism depends on the wavelength of light. For example, in a tourmaline (aluminum borosilicate) crystal, the *o*-wave is much more heavily absorbed with respect to the *e*-wave.

7.3 BIREFRINGENT OPTICAL DEVICES

A. Retarding Plates

Consider a positive uniaxial crystal such as quartz $(n_e > n_o)$ plate that has the optic axis (taken along z) parallel to the plate faces as in Figure 7.13. Suppose that a *linearly polarized* wave is incident at normal incidence on a plate face. If the field **E** is parallel to the optic axis (shown as $E_{//}$), then this wave will travel through the crystal as an *e*-wave with a velocity c/n_e slower than the *o*-wave since $n_e > n_o$. Thus, optic axis is the "slow axis" for waves polarized parallel to it. If **E** is right angles to the optic axis (shown as $E_\perp$) then this wave will travel with a velocity c/n_o, which will be the fastest velocity in the crystal. Thus, the axis perpendicular to the optic axis (say x) will be the "fast axis" for polarization along this direction. When a light ray enters a crystal at normal incidence to the optic axis and plate surface, as in Figure 7.12 (a), then the *o*- and *e*-waves travel along the same direction as shown in Figure 7.13. We can of course resolve a linear polarization at an angle α to z into $E_\perp$ and $E_{//}$. When the light comes out at the opposite face these two components would have been phase shifted by ϕ. Depending on the initial angle α of **E** and the length of the crystal, which determines the total phase shift ϕ through the plate, the emerging beam can have its initial linear polarization rotated, or changed into an elliptically or circularly polarized light as summarized in Figure 7.14.

If L is the thickness of the plate then the *o*-wave experiences a phase change $k_{o\text{-wave}}L$ through the plate where $k_{o\text{-wave}}$ is the wavevector of the *o*-wave: $k_{o\text{-wave}} = (2\pi/\lambda)n_o$, where λ is the free-space wavelength. Similarly, the *e*-wave experiences a phase change $(2\pi/\lambda)n_eL$ through the plate. Thus, the phase difference ϕ between the orthogonal components $E_\perp$ and $E_{//}$ of the emerging beam is,

$$\phi = \frac{2\pi}{\lambda}(n_e - n_o)L \qquad (1)$$

Relative phase through retarder plate

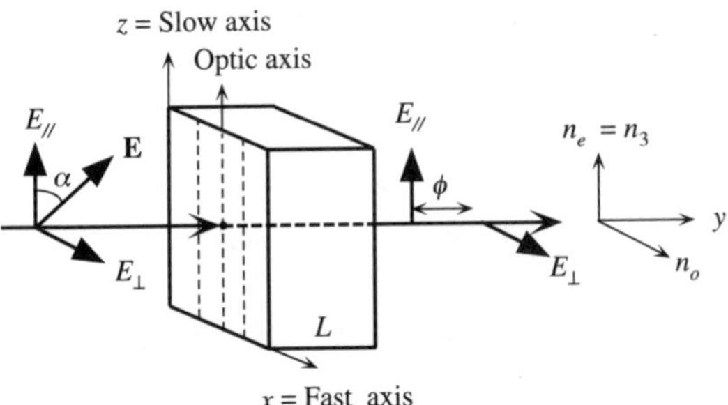

FIGURE 7.13 A retarder plate. The optic axis is parallel to the plate face. The *o*- and *e*-waves travel in the same direction but at different speeds.

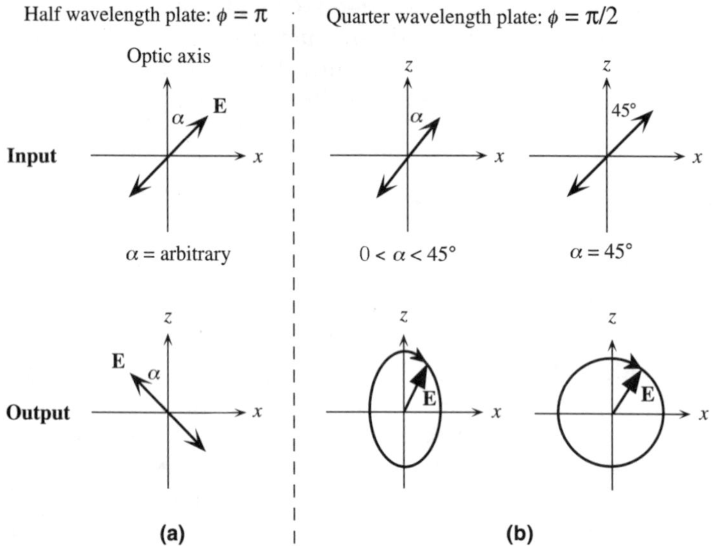

FIGURE 7.14 Input and output polarizations of light through (a) a half-wavelength plate and (b) through a quarter-wavelength plate.

The phase difference ϕ expressed in terms of full wavelengths is called the *retardation* of the plate. For example, a phase difference ϕ of 180° is a half-wavelength retardation.

The polarization of the through beam depends on the crystal type, $(n_e - n_o)$, and the plate thickness L. We know that depending on the phase difference ϕ between the orthogonal components of the field, the EM wave can be linearly, circularly or elliptically polarized as in Figure 7.3 and Figure 7.4.

A **half-wave plate retarder** has a thickness L such that the phase difference ϕ is π or 180°, corresponding to a half of wavelength $(\lambda/2)$ of retardation. The result is that, $E_{//}$ is delayed by 180° with respect to $E_{\perp}$. If we add the emerging $E_{\perp}$ and $E_{//}$ with this phase shift ϕ, **E** would be at an angle $-\alpha$ to the optic axis and still linearly polarized. **E** has been rotated counterclockwise through 2α.

A **quarter-wave plate retarder** has a thickness L such that the phase difference ϕ is $\pi/2$ or 90°, corresponding to a quarter of wavelength $(\lambda/4)$ of retardation. If we add the emerging $E_{\perp}$ and $E_{//}$ with this shift ϕ, the emerging light will be elliptically polarized if $0 < \alpha < 45°$ and circularly polarized if $\alpha = 45°$.

EXAMPLE 7.3.1 Quartz half-wave plate

What should be the thickness of a half-wave quartz plate for a wavelength $\lambda \approx 590$ nm given the ordinary and extraordinary refractive indices in Table 7.1.

Solution Half-wavelength retardation is a phase difference of π so that from equation Eq. (1)

$$\phi = \frac{2\pi}{\lambda}(n_e - n_o)L = \pi$$

giving,

$$L = \frac{\frac{1}{2}\lambda}{(n_e - n_o)} = \frac{\frac{1}{2}(590 \times 10^{-9}\text{ m})}{(1.5533 - 1.5442)} = 32.4\ \mu\text{m}$$

If we were to repeat the calculation for calcite, we would find about 1.7 micron thickness which is not very practical. Typically mica, quartz or polymeric substances are used as retarder plates because they have a $(n_e - n_o)$ difference that is not too large to result in an impractical thickness.

EXAMPLE 7.3.2 Circular polarization from linear polarization

Consider a linearly polarized light that is incident on a quarter-wavelength plate as in Figure 7.13 such that the polarization is 45° to the slow axis. Show that the output beam is circularly polarized.

Solution Both the x and z components of the electric field emerging from the retarder plate in Figure 7.13 are propagating along the y-axis with the same $\cos(\omega t - ky)$ harmonic term but with a phase difference ϕ. We are only interested in adding E_x and E_z vectorially at one location so that we can neglect the ky phase term. We can write the field components along slow (z) and fast (x) axes

$$\mathbf{E}_x = \hat{\mathbf{x}}E_\perp \cos(\omega t) \qquad \mathbf{E}_z = \hat{\mathbf{z}}E_{//}\cos(\omega t - \phi)$$

When the incident polarization is at 45°, then $E_{//} = E_\perp = E_o$. Putting $\phi = \pi/2$ for a quarter-wave plate, using $\cos(\omega t - \pi/2) = \sin(\omega t)$ and then $\sin^2(\omega t) + \cos^2(\omega t) = 1$, we find,

$$\cos^2(\omega t) + \sin^2(\omega t) = \left(\frac{E_x}{E_o}\right)^2 + \left(\frac{E_y}{E_o}\right)^2 = 1$$

which is the equation of a circle on the E_x and E_y axis (Figure 7.14 (b)) with a radius E_o.

B. Soleil-Babinet Compensator

An optical compensator is a device that allows one to control the retardation (*i.e.* the phase change) of a wave passing through it. In a wave plate retarder such as the half-wave plate, the relative phase change ϕ between the ordinary and extraordinary waves depends on the plate thickness and cannot be changed. In compensators, ϕ is adjustable. The Soleil-Babinet compensator described below is one such optical device that is widely used for controlling and analyzing the polarization state of light.

Consider the optical structure depicted in Figure 7.15 which has two quartz wedges touching over their large faces to form a "block" of adjustable height d. Sliding one

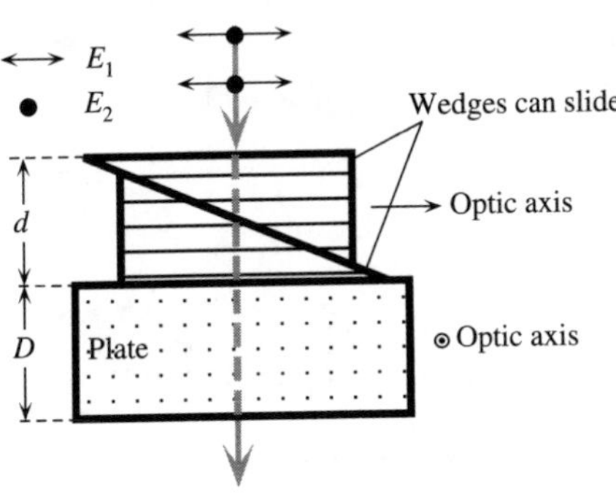

FIGURE 7.15 Soleil-Babinet Compensator.

wedge over the other wedge alters the "thickness" d of this block. The two-wedge block is placed on a parallel plate quartz slab with a fixed thickness D. The slab has its optic axis parallel to its surface face. The optical axes in the wedges are parallel but perpendicular to the optic axis of the slab as indicated in the figure.

Suppose that a linearly polarized light is incident on this compensator at normal incidence. We can represent this light by field oscillations parallel and perpendicular to the optic axis of the two-wedge block; these fields are E_1 and E_2 respectively. The polarization E_1 travels through the wedges (d) experiencing a refractive index n_e (E_1 is along the optic axis) and travels through the plate (D) experiencing an index n_o (E_1 perpendicular to the optic axis). Its phase change is

$$\phi_1 = \frac{2\pi}{\lambda}(n_e d + n_o D)$$

But the E_2 polarization wave first experiences n_o through the wedges (d) and then n_e through the plate (D) so that its phase change is,

$$\phi_2 = \frac{2\pi}{\lambda}(n_o d + n_e D)$$

The phase difference $\phi(=\phi_2 - \phi_1)$ between the two polarizations is,

Soleil-Babinet compensator

$$\phi = \frac{2\pi}{\lambda}(n_e - n_o)(D - d) \qquad \textbf{(2)}$$

It is apparent that as we can change d continuously by sliding the wedges (by using a micrometer screw), we can continuously alter the phase difference ϕ from 0 to 2π. We can therefore produce a quarter-wave or half-wave plates by simply adjusting this compensator. It should be emphasized that this control occurs over the surface region that corresponds to both the wedges and in practice this is a narrow region.

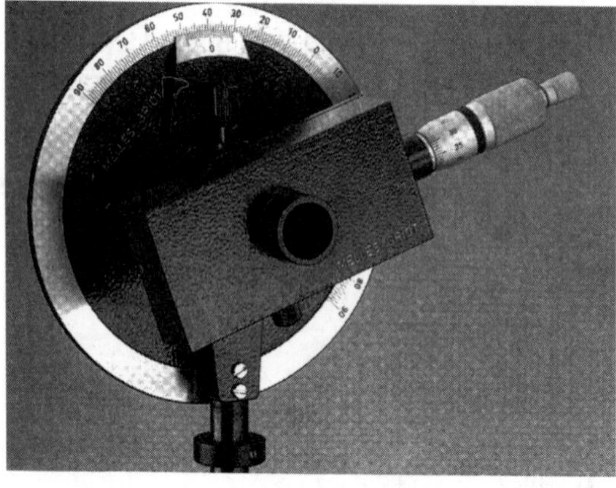

A Soleil-Babinet compensator *(Courtesy of Melles-Griot.)*

C. Birefringent Prisms

Prisms made from birefringent crystals are useful for producing a highly polarized light wave or polarization splitting of light. The Wollaston prism is a polarization splitter in which the split beam has orthogonal polarizations. Two calcite (or quartz) right angle prisms A and B are placed with their diagonal faces touching to form a rectangular block as shown in Figure 7.16. Looking at the cross section of this block, the optic axis in A is in the plane of the paper and that in B coming out of the paper; the two prisms have their optic axes mutually orthogonal. Further, as shown, the optic axes are parallel to the prism sides.

Consider a light wave of arbitrary polarization at normal incidence to prism A. The light beam entering prism A travels in this prism as two orthogonally polarized waves that have fields E_1 and E_2 as in Figure 7.16. E_1 (normal to the plane of paper) is orthogonal to the optic axis and corresponds to o-waves in A. E_2 (in the plane of the paper) is along the optic axis of A and corresponds to the e-waves. E_1 has a refractive index n_o and E_2 has n_e. However, in prism B, E_1 is the e-wave. This means that in going through the diagonal interface, E_1 experiences a *decrease* from n_o to n_e(for calcite). On the other hand, the e-wave in A becomes the o-wave in B and experiences an *increase* from n_e to n_o. Notice that E_2 is now orthogonal to the optic axis in B. These refractive index changes are opposite which means that the two waves are refracted in opposite directions at the interface as shown in Figure 7.16. The E_1-wave moves away from the normal to the diagonal face whereas the E_2-wave moves closer to this normal. The two orthogonal polarizations are therefore angularly separated out by this oppositely sensed refractions. The divergence angle depends on the prism wedge angle θ. Various Wollaston prisms with typical beam splitting angles of 15–45° are commercially available. It is left as an exercise to show that if we rotate the prism about the incident beam by 180°, we would see the two orthogonal polarizations E_1 and E_2 switched places, and if we use a quartz Wollaston prism, we would find E_1 and E_2 in the figure are again switched.

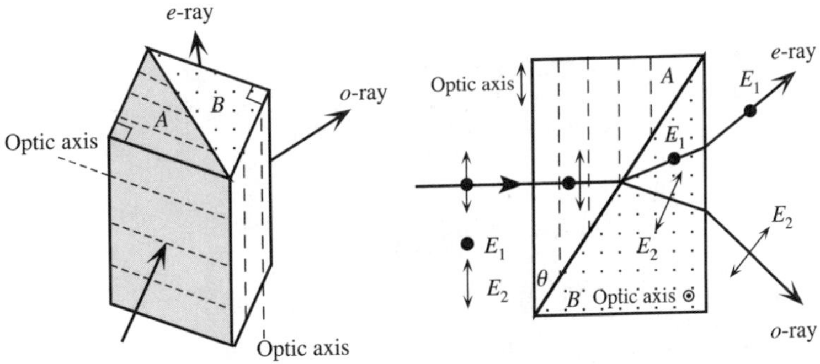

FIGURE 7.16 The Wollaston prism is a beam polarization splitter. E_1 is orthogonal to the plane of the paper and also to the optic axis of the first prism. E_2 is in the plane of the paper and orthogonal to E_1.

Commercial Wollaston prisms. The actual prim is held inside a cylindrical housing *(Courtesy of Melles Griot)*

7.4 OPTICAL ACTIVITY AND CIRCULAR BIREFRINGENCE

When a linearly polarized light wave is passed through a quartz crystal along its optic axis, it is observed that the emerging wave has its **E**-vector (plane of polarization) rotated, which is illustrated in Figure 7.17. This rotation increases continuously with the distance traveled through the crystal (about 21.7° per mm of quartz). The rotation of the plane of polarization by a substance is called **optical activity**. In very simply intuitive terms, optical activity occurs in materials in which the electron motions induced by the external electromagnetic field follows spiraling or helical paths (orbits).[5] Electrons flowing in helical paths resemble a current flowing in a coil and thus possess a magnetic moment. The optical field in light therefore induces oscillating magnetic moments which can be either parallel or antiparallel to the induced oscillating electric dipoles. Wavelets emitted from these oscillating induced magnetic and electric dipoles interfere to constitute a forward wave that has its optical field rotated either clockwise or counterclockwise.

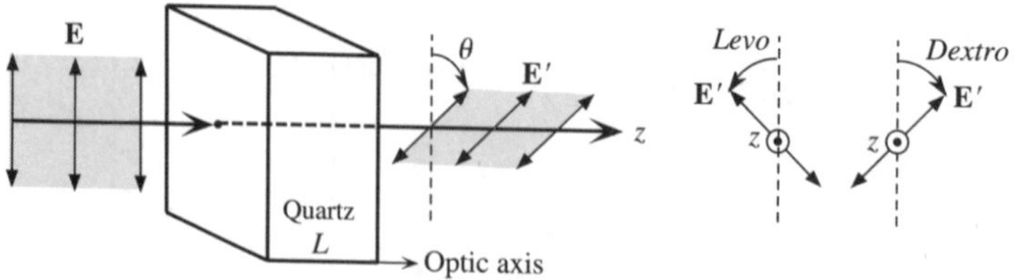

FIGURE 7.17 An optically active material such as quartz rotates the plane of polarization of the incident wave: The optical field **E** rotated to **E**′. If we reflect the wave back into the material, **E**′ rotates back to **E**.

[5] The explanation of optical activity involves examining both induced magnetic and electric dipole moments which will not be described here in detail. There are very readable qualitative explanations as mentioned under Further Reading.

If θ is the angle of rotation of **E**, then θ is proportional to the distance L propagated in the optically active medium as depicted in Figure 7.17. For an observer receiving the wave through quartz, the rotation of the plane of polarization may be *clockwise* (to the right) or *counterclockwise* (to the left) which are called *dextrorotatory* and *levorotatory* forms of optical activity. The structure of quartz is such that atomic arrangements spiral around the optic axis either in clockwise or counterclockwise sense. Quartz thus occurs in two distinct crystalline forms, right-handed and left-handed, which exhibit dextrorotatory and levorotatory types of optical activity respectively. Although we used quartz as an example, there are many substances that are optically active, including various biological substances and even some liquid solutions (*e.g.* corn syrup) that contain various organic molecules with a rotatory power.

The **specific rotatory power (θ/L)** is defined as the extent of rotation per unit length of distance traveled in the optically active substance. Specific rotatory power depends on the wavelength. For example, for quartz this is 49° per mm at 400 nm but 17° per mm at 650 nm.

Optical activity can be understood in terms of left and right circularly polarized waves traveling at different velocities in the crystal, *i.e.* experiencing different refractive indices. Due to the helical twisting of the molecular or atomic arrangements in the crystal, the velocity of a circularly polarized wave depends whether the optical field rotates clockwise or counterclockwise. A vertically polarized light with a field **E** at the input can be thought of as two right and left handed circularly polarized waves, $\mathbf{E}_L$ and $\mathbf{E}_R$, that are symmetrical with respect to the y-axis, *i.e.* at any instant $\alpha = \beta$, as shown in Figure 7.18 (α and β are ωt). If they travel at the same velocity through the crystal then they remain symmetrical with respect to the vertical ($\alpha = \beta$ remains the same) and the resultant is still a vertically polarized light. If however these travel at different velocities through a medium then at the output $\mathbf{E}'_L$ and $\mathbf{E}'_R$ are no longer symmetrical with respect to the vertical, $\alpha' \neq \beta'$, and their resultant is a vector $\mathbf{E}_R$ at an angle θ to y-axis.

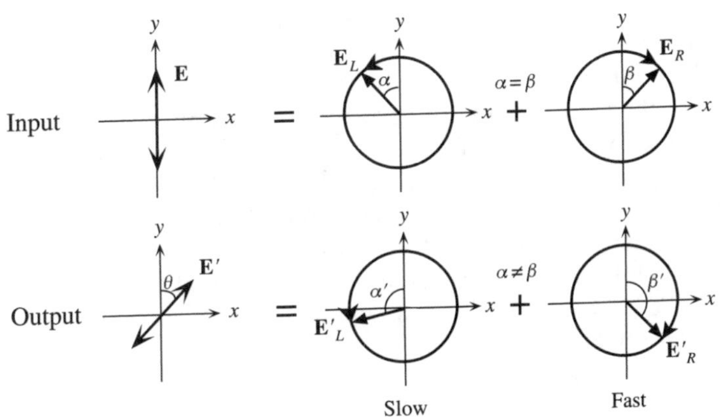

FIGURE 7.18 Vertically polarized wave at the input can be thought of as two right and left handed circularly polarized waves that are symmetrical, *i.e.* at any instant $\alpha = \beta$. If these travel at different velocities through a medium then at the output they are no longer symmetric with respect to y, $\alpha \neq \beta$, and the result is a vector $\mathbf{E}'$ at an angle θ to y.

Suppose that n_R and n_L are the refractive indices experienced by the right and left circularly polarized light respectively. After traversing the crystal length L, the phase difference between the two optical fields $\mathbf{E}'_L$ and $\mathbf{E}'_R$ at the output leads to a new optical field $\mathbf{E}'$ that is $\mathbf{E}$ rotated by θ, given by

Optical activity

$$\theta = \frac{\pi}{\lambda}(n_R - n_L)L \tag{1}$$

where λ is the free-space wavelength. For a left handed quartz crystal, and for 589 nm light propagation along the optic axis, $n_R = 1.54427$ and $n_L = 1.54420$ which means θ is about 21.4° per mm of crystal.

In a **circularly birefringent** medium, the right and left handed circularly polarized waves propagate with different velocities and experience different refractive indices n_R and n_L. Since optically active materials naturally rotate the optical field, it is not unreasonable to expect that a circularly polarized light with its optical field rotating in the same sense as the optical activity will find it easier to travel through the medium. Thus, an optically active medium possesses different refractive indices for right and left circularly polarized light and exhibits circular birefringence. It should be mentioned that if the direction of the light wave is reversed in Figure 7.17, the ray simply retraces itself and $\mathbf{E}'$ becomes $\mathbf{E}$.

7.5 ELECTRO-OPTIC EFFECTS

A. Definitions

Electro-optic effects refer to changes in the refractive index of a material induced by the application of an external electric field, which therefore "modulates" the optical properties; the applied field is not the electric field of any light wave, but a separate external field. We can apply such an external field by placing electrodes on opposite faces of a crystal and connecting these electrodes to a battery. The presence of such a field distorts the electron motions in the atoms or molecules of the substance, or distorts the crystal structure resulting in changes in the optical properties. For example, an applied external field can cause an optically isotropic crystal such as GaAs to become birefringent. In this case, the field induces principal axes and an optic axis. Typically changes in the refractive index are small. The frequency of the applied field has to be such that the field appears static over the time scale it takes for the medium to change its properties, that is respond, as well as for any light to cross the substance. The electro-optic effects are classified according to first and second order effects.

If we were to take the refractive index n to be a function of the applied electric field E, that is $n = n(E)$, we can of course expand this as a Taylor series in E. The new refractive index n' would be

Field induced refractive index

$$n' = n + a_1E + a_2E^2 + \dots \tag{1}$$

where the coefficients a_1 and a_2 are called the *linear* electro-optic effect and *second* order electro-optic effect coefficients. Although we would expect even higher terms in the expansion in Eq. (1), these are generally very small and their effects negligible within highest practical fields. The change in n due to the first E term is called the **Pockels effect**. The change in n due to the second E^2 term is called the **Kerr effect**,[6] and the coefficient a_2 is generally written as λK where K is called the **Kerr coefficient**. Thus, the two effects are,

$$\Delta n = a_1 E \tag{2}$$

Pockels effect

and

$$\Delta n = a_2 E^2 = (\lambda K) E^2 \tag{3}$$

Kerr effect

All materials exhibit the Kerr effect. It may be thought that we will always find some (non-zero) value for a_1 for all materials but this is not true and only certain crystalline materials exhibit the Pockels effect. If we apply a field **E** in one direction and then reverse the field and apply $-$**E** then according to Eq. (2), Δn should change sign. If the refractive index increases for **E** it must decrease for $-$**E**. Reversing the field should *not* lead to an identical effect (the same Δn). The structure has to respond differently to **E** and $-$**E**. There must therefore be some *asymmetry* in the structure to distinguish between **E** and $-$**E**. In a noncrystalline material, Δn for **E** would be the same as Δn for $-$**E** as all directions are equivalent in terms of dielectric properties. Thus $a_1 = 0$ for all noncrystalline materials (such as glasses and liquids). Similarly, if the crystal structure has a center of symmetry then reversing the field direction has an identical effect and a_1 is again zero. Only crystals that are **noncentrosymmetric**[7] exhibit the Pockels effect. For example a NaCl crystal (centrosymmetric) exhibits no Pockels effect but a GaAs crystal (noncentrosymmetric) does.

B. Pockels Effect

Friedrich Carl Alwin Pockels (1865–1913) son of Captain Theodore Pockels and Alwine Becker, was born in Vincenza (Italy). He obtained his doctorate from Göttingen University in 1888. From 1900 until 1913, he was a professor of theoretical physics in the Faculty of Sciences and Mathematics at the University of Heidelberg where he carried out extensive studies on electro-optic properties of crystals - the Pockels effect is the basis of many practical electro-optic modulators *(Courtesy of the Department of Physics and Astronomy, University of Heidelberg, Germany.)*

---

[6] *John Kerr* (1824–1907) was a Scottish physicist who was a faculty member at Free Church Training College for Teachers, Glasgow (1857–1901) where he set-up an optics laboratory and demonstrated the Kerr effect (1875).

[7] A crystal has a center of symmetry about a point O, if any atom with a position vector **r** from O also appears when we invert **r**, that is take $-$**r**.

The Pockels effect expressed in Eq. (2) is an over-simplification because in reality we have to consider the effect of an applied field along a particular crystal direction on the refractive index for light with a given propagation direction and polarization. For example, suppose that x, y and z are the principal axes of a crystal with refractive indices n_1, n_2 and n_3 along these directions. For an optically isotropic crystal, these would be the same whereas for a uniaxial crystal $n_1 = n_2 \neq n_3$ as depicted in the xy cross section in Figure 7.19 (a). Suppose that we suitably apply a voltage across a crystal and thereby apply an external dc field E_a along the z-axis. In Pockels effect, the field will modify the optical indicatrix. The exact effect depends on the crystal structure. For example, a crystal like GaAs, optically isotropic with a spherical indicatrix, becomes birefringent, and a crystal like KDP (KH_2PO_4-potassium dihydrogen phosphate) that is uniaxial becomes biaxial. In the case of KDP, the field E_a along z rotates the principal axes by 45° about z, and changes the principal indices as indicated in Figure 7.19 (b). The new principal indices are now n_1' and n_2' which means that the cross section is now an ellipse. Propagation along the z-axis under an applied field in Figure 7.19 (b) now occurs with different refractive indices n_1' and n_2'. As apparent in Figure 7.19 (b), the applied field induces new principal axes $x'$and y' for this crystal. In the case of $LiNbO_3$ (lithium niobate), an optoelectronically important uniaxial crystal, a field E_a along the y-direction does not significantly rotate the principal axes but rather changes the principal refractive indices n_1 and n_2 (both equal to n_o) to n_1' and n_2' as illustrated in Figure 7.19 (c).

As an example consider a wave propagating along the z-direction (optic axis) in a $LiNbO_3$ crystal. This wave will experience the same refractive index $(n_1 = n_2 = n_o)$ whatever the polarization as in Figure 7.19 (a). However, in the presence of an applied field E_a parallel to the principal y axis as in Figure 7.19 (c), the light propagates as two orthogonally polarized waves (parallel to x and y) experiencing different refractive indices n_1' and n_2'. The applied field thus *induces a birefringence* for light traveling along the z-axis. The field induced rotation of the principal axes in this case, though present, is small and can be neglected. Before the field E_a is applied, the refractive indices n_1

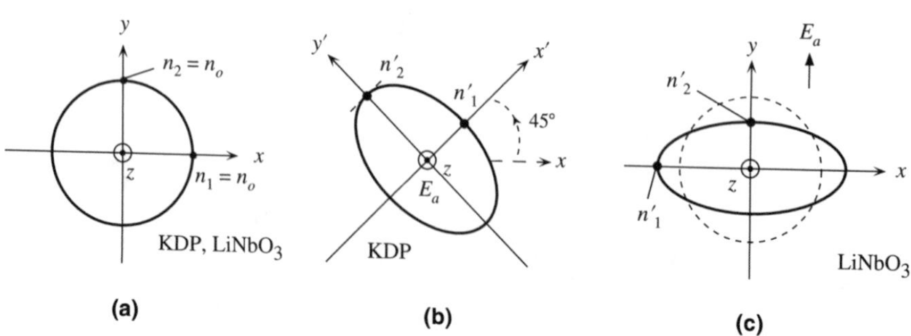

FIGURE 7.19 (a) Cross section of the optical indicatrix with no applied field, $n_1 = n_2 = n_o$ (b) The applied external field modifies the optical indicatrix. In a KDP crystal, it rotates the principal axes by 45° to x' and y' and n_1 and n_2 change to n_1' and n_2'. (c) Applied field along y in $LiNbO_3$ modifies the indicatrix and changes n_1 and n_2 to n_1' and n_2'.

and n_2 are both equal to n_o. The Pockels effect then gives the new refractive indices n_1' and n_2' in the presence of E_a as

$$n_1' \approx n_1 + \tfrac{1}{2}n_1^3 r_{22} E_a \quad \text{and} \quad n_2' \approx n_2 - \tfrac{1}{2}n_2^3 r_{22} E_a \tag{4}$$

Pockels effect

where r_{22} is a constant, called a **Pockels coefficient**, that depends on the crystal structure and the material. The reason for the seemingly unusual subscript notation is that there are more than one constant and these are elements of a tensor that represents the optical response of the crystal to an applied field along a particular direction with respect to the principal axes (the exact theory is more mathematical than intuitive). We therefore have to use the correct Pockels coefficients for the refractive index changes for a given crystal and a given field direction.[8] If the field were along z, the Pockels coefficient in Eq. (4) would be r_{13}.

It is clear that the control of the refractive index by an external applied field (and hence a voltage) is a distinct advantage that enables the phase change through a Pockels crystal to be controlled or modulated; such a **phase modulator** is called a *Pockels cell*. In the *longitudinal Pockels cell phase modulator* the applied field is in the direction of light propagation whereas in the *transverse phase modulator*, the applied field is transverse to the direction of light propagation. For light propagation along z, the longitudinal and transverse effects are illustrated in Figure 7.19 (b) and (c) respectively.

Consider the transverse phase modulator in Figure 7.20. In this example, the applied electric field, $E_a = V/d$, is applied parallel to the y-direction, normal to the direction of light propagation along z. Suppose that the incident beam is linearly polarized (shown as **E**) say at 45° to the y axes. We can represent the incident light in terms of polarizations ($\mathbf{E}_x$ and $\mathbf{E}_y$) along the x and y axes. These components $\mathbf{E}_x$ and $\mathbf{E}_y$ experience refractive indices n_1' and n_2' respectively. Thus when $\mathbf{E}_x$ traverses the lenght distance L, its phase changes by ϕ_1,

$$\phi_1 = \frac{2\pi n_1'}{\lambda}L = \frac{2\pi L}{\lambda}\left(n_o + \tfrac{1}{2}n_o^3 r_{22}\frac{V}{d}\right)$$

When the component $\mathbf{E}_y$ traverses the distance L, its phase changes by ϕ_2, given by a similar expression except that r_{22} changes sign. Thus the phase change $\Delta\phi$ between the two field components is

$$\Delta\phi = \phi_1 - \phi_2 = \frac{2\pi}{\lambda}n_o^3 r_{22}\frac{L}{d}V \tag{5}$$

Tranverse Pockels effect

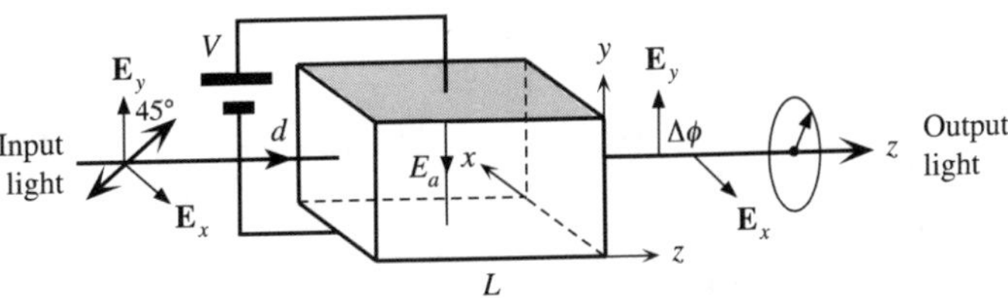

FIGURE 7.20 Tranverse Pockels cell phase modulator. A linearly polarized input light into an electro-optic crystal emerges as a circularly polarized light.

[8] The reader should not be too concerned with the subscripts but simply interpret them as identifying the right Pockels coefficient value for the particular electro-optic problem at hand.

The applied voltage thus inserts an adjustable phase difference $\Delta\phi$ between the two field components. The polarization state of output wave can therefore be controlled by the applied voltage and the Pockels cell is a **polarization modulator**. We can change the medium from a quarter-wave to a half-wave plate by simply adjusting V. The voltage $V = V_{\lambda/2}$, the **half-wave voltage**, corresponds to $\Delta\phi = \pi$ and generates a half-wave plate. The advantage of the transverse Pockels effect is that we can independently reduce d, and thereby increases the field, and increase the crystal length L, to build-up more phase change; $\Delta\phi$ is proportional to L/d. This is not the case in the longitudinal Pockels effect. If L and d were the same, typically $V_{\lambda/2}$ would be a few kilovolts but tailoring d/L to be much smaller than unity would bring $V_{\lambda/2}$ down to desirable practical values.

From the polarization modulator in Figure 7.20, we can build an **intensity modulator** by inserting a polarizer P and an analyzer A before and after the phase modulator as in Figure 7.21 such that they are cross-polarized, *i.e.* P and A have their transmission axes at 90° to each other. The transmission axis of P is at 45° to the y-axis (hence A also has its transmission axis at 45° to y) so that the light entering the crystal has equal $\mathbf{E}_x$ and $\mathbf{E}_y$ components. In the absence of an applied voltage, the two components travel with the same refractive index and polarization output from the crystal is the same as its input. There is no light detected at the detector as A and P are at right angles ($\theta = 90°$ in Malus's law).

An applied voltage inserts a phase difference $\Delta\phi$ between the two electric field components. The light leaving the crystal now has an elliptical polarization and hence a field component along the transmission axis of A. A portion of this light will therefore pass through A to the detector. The transmitted intensity now depends on the applied voltage V. The field components at the analyzer will be out of phase by an amount $\Delta\phi$. We have to find the total field $\mathbf{E}$ and the component of this field along the transmission axis of A. Suppose that E_o is the amplitude of the wave incident on the crystal face. The amplitudes along x and y-axis will be $E_o/\sqrt{2}$ each (notice that $\mathbf{E}_x$ is along the $-x$ direction). The total field $\mathbf{E}$ at the analyzer is,

$$\mathbf{E} = -\hat{\mathbf{x}}\frac{E_o}{\sqrt{2}}\cos(\omega t) + \hat{\mathbf{y}}\frac{E_o}{\sqrt{2}}\cos(\omega t + \Delta\phi)$$

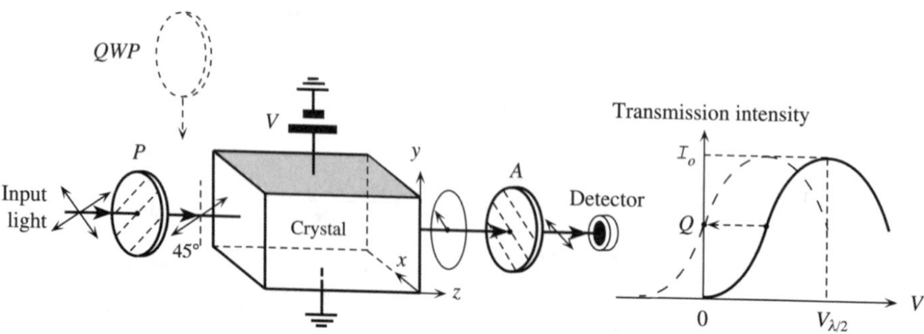

FIGURE 7.21 Left: A tranverse Pockels cell intensity modulator. The polarizer P and analyzer A have their transmission axis at right angles and P polarizes at an angle 45° to y-axis. Right: Transmission intensity vs. applied voltage characteristics. If a quarter-wave plate (QWP) is inserted after P, the characteristic is shifted to the dashed curve.

A factor cos(45°) of each component passes through A. We can resolve $\mathbf{E}_x$ and $\mathbf{E}_y$ along A's transmission axis and then add these components and use a trigonometric identity to obtain the field emerging from A. The final result is:

$$E = E_o \sin\left(\tfrac{1}{2}\Delta\phi\right)\sin\left(\omega t + \tfrac{1}{2}\Delta\phi\right)$$

The intensity I of the detected beam is then

$$I = I_o \sin^2\left(\tfrac{1}{2}\Delta\phi\right) \tag{6a}$$

or

$$I = I_o \sin^2\left(\frac{\pi}{2}\cdot\frac{V}{V_{\lambda/2}}\right) \tag{6b}$$

Pockels intensity modulator

where I_o is the light intensity under full transmission.

An applied voltage of $V_{\lambda/2}$ is needed to allow full transmission. The transmission characteristics of this modulator is shown in Figure 7.21. In digital electronics, we would switch a light pulse on and off so that the non-linear dependence of transmission intensity on V in Eq. (6b) would not be a problem. However, if we wish to obtain a linear modulation between the intensity I and V we need to bias this structure about the apparent "linear region" of the curve at half-height. This is done by inserting a quarter-wave plate after the polarizer P which provides a circularly polarized light as input. That means $\Delta\phi$ is already shifted by $\pi/4$ before any applied voltage. The applied voltage then, depending on the sign, increases or decreases $\Delta\phi$. The new transmission characteristic is shown as a dashed curve in Figure 7.21. We have, effectively, optically biased the modulator at point Q on this new characteristic (will it matter if instead we were to insert the quarter-wave plate before the analyzer?).

EXAMPLE 7.5.1 Pockels Cell Modulator

What should be the aspect ratio d/L for the transverse $LiNiO_3$ phase modulator in Figure 7.20 that will operate at a free-space wavelength of 1.3 μm and will provide a phase shift $\Delta\phi$ of π (half wavelength) between the two field components propagating through the crystal for an applied voltage of 24 V? (Use Table 7.2.)

Solution From Eq. (5), putting $\Delta\phi = \pi$ for the phase difference between the field components $\mathbf{E}_x$ and $\mathbf{E}_y$ in Figure 7.20 and letting $V = V_{\lambda/2}$ gives,

$$\Delta\phi = \frac{2\pi}{\lambda} n_o^3 r_{22} \frac{L}{d} V_{\lambda/2} = \pi$$

or

$$\frac{d}{L} = \frac{1}{\Delta\phi}\cdot\frac{2\pi}{\lambda} n_o^3 r_{22} V_{\lambda/2} = \frac{1}{\pi}\cdot\frac{2\pi}{(1.3\times 10^{-6})}(2.2)^3(3.4\times 10^{-12})(24)$$

giving

$$d/L = 1.3\times 10^{-3}$$

This particular transverse phase modulator has the field applied along the y-direction and light traveling along the z-direction as in Figure 7.20. If we were to use the transverse arrangement in which the field is applied along the z-axis, and the light travels along the y-axis, the relevant Pockels coefficients would be greater and the corresponding aspect ratio d/L would be $\sim 10^{-2}$. We cannot arbitrarily set d/L to any ratio we like for the simple reason that when d becomes too small, the light will suffer diffraction effects that will prevent it from passing through the device. d/L ratios 10^{-3}–10^{-2} in practice can be implemented by fabricating an integrated optical device.

C. Kerr Effect

Suppose that we apply a strong electric field to an otherwise optically isotropic material such as glass (or liquid). The change in the refractive index will be due to the Kerr effect, a second order effect. We can arbitrarily set the z-axis of a Cartesian coordinate system along the applied field as depicted in Figure 7.22 (a). The applied field distorts the electron motions (orbits) in the constituent atoms and molecules, including those valence electrons in covalent bonds, in such a way that it becomes "more" difficult for the electric field in the light wave to displace electrons parallel to the applied field direction. Thus a light wave with a polarization parallel to the z-axis will experience a smaller refractive index, reduced from its original value n_o to n_e. A light waves with polarizations orthogonal to the z-axis will experience the same refractive index n_o. The applied field thus *induces birefringence* with an optic axis parallel to the applied field direction as shown in Figure 7.22 (a). The material becomes birefringent for waves traveling off the z-axis.

The polarization modulator and intensity modulator concepts based on the Pockels cell can be extended to the Kerr effect as illustrated in Figure 7.22 (b) which shows a **Kerr cell phase modulator**. In this case, the applied field again induces birefringence. The phase modulator in Figure 7.22 (b) uses the Kerr effect whereupon the applied field along z induces a refractive index n_e parallel to the z-axis whereas that along the x-axis will still be n_o. The light components $\mathbf{E}_x$ and $\mathbf{E}_z$ then travel along the material with different velocities and emerge with a phase difference $\Delta\phi$ resulting in an elliptically polarized light. However, the Kerr effect is small as it is a second order effect, and therefore only accessible for modulation use at high fields. The advantage, however, is that, all materials, including glasses and liquids, exhibit the Kerr effect and the response time in solids is very short, much less than nanoseconds leading to high modulation frequencies (greater than GHz).

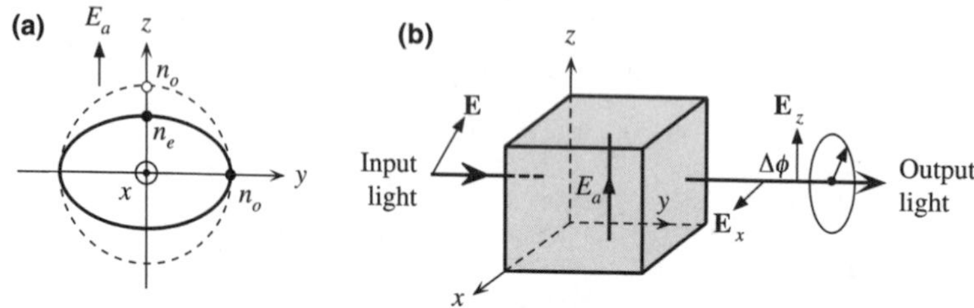

FIGURE 7.22 (a) An applied electric field, via the Kerr effect, induces birefringences in an otherwise optically istropic material. (b) A Kerr cell phase modulator.

TABLE 7.2 Pockels (r) and Kerr (K) coefficients in various materials.

Material	Crystal	Indices	Pockels Coefficients $\times 10^{-12}$ m/V	K m/V^2	Comment
$LiNbO_3$	Uniaxial	$n_o = 2.272$ $n_e = 2.187$	$r_{13} = 8.6; r_{33} = 30.8$ $r_{22} = 3.4; r_{51} = 28$		$\lambda = 500$ nm
KDP	Uniaxial	$n_o = 1.512$ $n_e = 1.470$	$r_{41} = 8.8; r_{63} = 10.5$		$\lambda \approx 546$ nm
GaAs	Isotropic	$n_o = 3.6$	$r_{41} = 1.5$		$\lambda \approx 546$ nm
Glass	Isotropic	$n_o \approx 1.5$	0	3×10^{-15}	
Nitrobenzene	Isotropic	$n_o \approx 1.5$	0	3×10^{-12}	

If E_a is the applied field, then the change in the refractive index for polarization parallel to the applied field as in Figure 7.22 (a) can be shown to be given by

$$\Delta n = \lambda K E_a^2 \tag{7}$$

Kerr effect

where K is the **Kerr coefficient**. We can now use Eq. (3) to find the induced phase difference $\Delta\phi$ and hence relate it to the applied voltage.

Table 7.2 summarizes the Pockels and Kerr coefficients for various materials. The Kerr effect also occurs in anisotropic crystals but the effect there cannot be simply characterized by a single coefficient K.

EXAMPLE 7.5.2 Kerr Effect Modulator

Suppose that we have a glass rectangular block of thickness (d) 100 μm and length (L) 20 mm and we wish to use the Kerr effect to implement a phase modulator in a fashion depicted in Figure 7.22. The input light has been polarized parallel to the applied field E_a direction, along the z-axis. What is the applied voltage that induces a phase change of π (half-wavelength)?

Solution The phase change $\Delta\phi$ for the optical field E_z is

$$\Delta\phi = \frac{2\pi\Delta n}{\lambda}L = \frac{2\pi(\lambda K E_a^2)}{\lambda}L = \frac{2\pi L K V^2}{d^2}$$

For $\Delta\phi = \pi, V = V_{\lambda/2}$,

$$V_{\lambda/2} = \frac{d}{\sqrt{2LK}} = \frac{(100 \times 10^{-6})}{\sqrt{2(20 \times 10^{-3})(3 \times 10^{-15})}} = 9.1 \text{ kV!}$$

Although the Kerr effect is fast, it comes at a costly price. Note that K depends on the wavelength and so does $V_{\lambda/2}$.

7.6 INTEGRATED OPTICAL MODULATORS

A. Phase and Polarization Modulation

Integrated optics refers to the integration of various optical devices and components on a single common substrate, for example, lithium niobate, just as in integrated electronics all the necessary devices for a given function are integrated in the same semiconductor crystal substrate (chip). There is a distinct advantage to implementing various

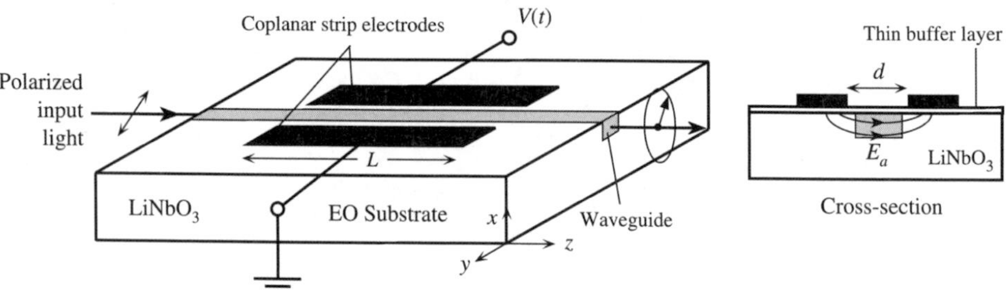

FIGURE 7.23 Integrated tranverse Pockels cell phase modulator in which a waveguide is diffused into an electro-optic (EO) substrate. Coplanar strip electrodes apply a transverse field E_a through the waveguide. The substrate is an x-cut $LiNbO_3$ and typically there is a thin dielectric buffer layer (*e.g.* ~200 nm thick SiO_2) between the surface electrodes and the subtrate to separate the electrodes away from the waveguide.

optically communicated devices, *e.g.* laser diodes, waveguides, splitters, modulators, photodetectors and so on, on the same substrate as it leads to miniaturization and also to an overall enhancement in performance and usability (typically).

One of the simplest examples is the **polarization modulator** shown in Figure 7.23 where an embedded waveguide has been fabricated by implanting a $LiNbO_3$ substrate with Ti atoms which increase the refractive index. Two coplanar strip electrodes run along the waveguide and enable the application of a transverse field E_a to light propagation direction z. The external modulating voltage $V(t)$ is applied between the coplanar drive electrodes and, by virtue of the Pockels effect, induces a change Δn in the refractive index and hence a voltage dependent phase shift through the device. We can represent light propagation along the guide in terms of two *orthogonal modes*, E_x along x and E_y along y. These two modes experience symmetrically opposite phase changes.[9] The phase shift $\Delta\phi$ between the E_x and E_y polarized waves would normally be given by Eq. (5) in §7.5. However, in this case the applied (or induced) field is not uniform between the electrodes and, further, not all applied field lines lie inside the waveguide. The electro-optic effect takes place over the spatial overlap region between the applied field and the optical fields. This *spatial overlap efficiency* is lumped into a coefficient Γ and the phase shift $\Delta\phi$ is written as

$$\Delta\phi = \Gamma\frac{2\pi}{\lambda}n_o^3 r_{22}\frac{L}{d}V$$

where typically $\Gamma \approx 0.5 - 0.7$ for various integrated polarization modulators of this type. Since the phase shift depends on the product of V and L, a comparative device parameter would be the $V \times L$ product for a phase shift of π (half-wavelength), *i.e.* $V_{\lambda/2}L$. At $\lambda = 1.5\ \mu$m for an x-cut $LiNbO_3$ modulator as in Figure 7.23, with $d \approx 10\ \mu$m, $V_{\lambda/2}L \approx 35$ V cm. For example, a modulator with $L = 2$ cm has a half-wave voltage

[9] These are called transverse electric (TE) and transverse magnetic (TM) modes.

$V_{\lambda/2} = 17.5$ V. By comparison, for a z-cut $LiNbO_3$ plate, that is for light propagation along the y-direction and E_a along z, the relevant Pockels coefficients (r_{13} and r_{33}) are much greater than r_{22} which leads to $V_{\lambda/2}L \approx 5$ V cm.

B. Mach-Zehnder Modulator

Induced phase shift by applied voltage can be converted to an amplitude variation by using an **interferometer**, a device that interferes two waves of the same frequency but different phase. Consider the structure shown in Figure 7.24 which has implanted single-mode waveguides in a $LiNbO_3$ (or other electro-optic) substrate in the geometry shown. The waveguide at the input branches out at C to two arms A and B and these arms are later combined at D to constitute the output. The splitting at C and combining at D involve simple Y-junction waveguides. In the ideal case, the power is equally split at C so that the field is scaled by a factor $\sqrt{2}$ going into each arm. The structure acts as an interferometer because the two waves traveling through the arms A and B interfere at the output port D and the output amplitude depends on the phase difference (optical path difference) between the A and B-branches. Two back-to-back identical phase modulators enable the phase changes in A and B to be modulated. Notice that the applied field in branch A is in the opposite direction to that in branch B. The refractive index changes are therefore opposite which means the phase changes in arms A and B are also opposite. For example, if the applied voltage induces a phase change of $\pi/2$ in arm A, this will be $-\pi/2$ in arm B so that A and B would be out of phase by π. These two waves will then interfere destructively and cancel each other at D. The output intensity would then be zero. Since the applied voltage controls the phase difference between the two interfering waves A and B at the output, this voltages also controls the output light intensity, though the relationship is not linear.

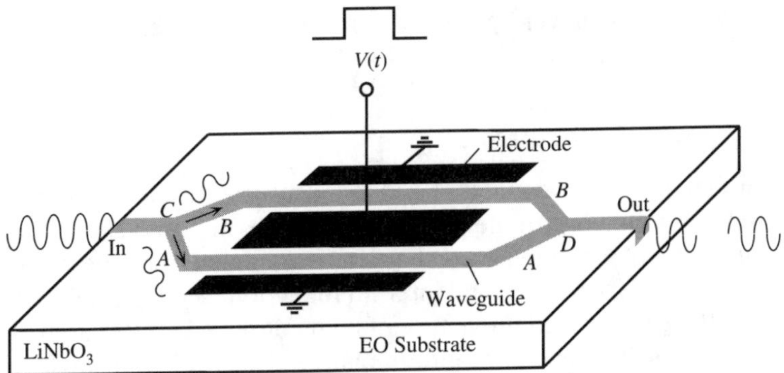

FIGURE 7.24 An integrated Mach-Zehnder optical intensity modulator. The input light is split into two coherent waves A and B, which are phase shifted by the applied voltage, and then the two are combined again at the output.

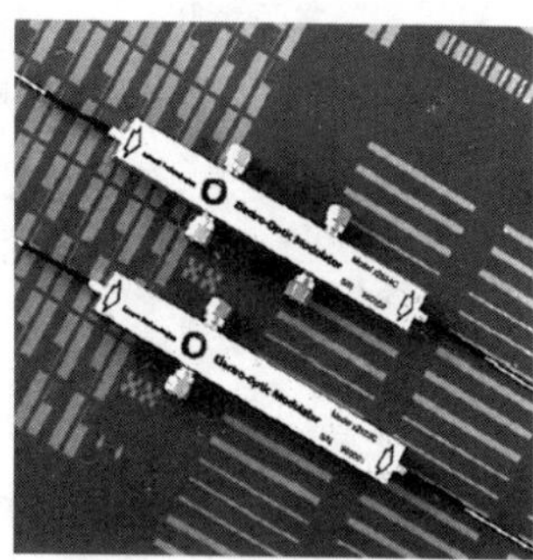

Ti diffused lithium niobate electro-optic (Pockels effect) modulators for use in high-speed optical fiber communications up to 16 GHz. Operates at 1550 nm. Maximum modulation voltage is ±20V. *(Courtesy of Lucent Technologies.)*

It is apparent that the relative phase difference between the two waves A and B is therefore doubled with respect to a phase change ϕ in a single arm. We can predict the output intensity by adding waves A and B at D. If A is the amplitude of wave A and B (assumed equal power spitting at C), the optical field at the output is

$$E_{\text{output}} \propto A\cos(\omega t + \phi) + A\cos(\omega t - \phi) = 2A\cos\phi\cos(\omega t)$$

The output power is proportional to E^2_{output} which is maximum when $\phi = 0$. Thus,

$$\frac{P_{\text{out}}(\phi)}{P_{\text{out}}(0)} = \cos^2\phi \qquad \textbf{(1)}$$

Although the derivation is oversimplified,[10] it nonetheless represents approximately the right relationship between the power transfer and the induced phase change per modulating arm. The power transfer is zero when $\phi = \pi/2$ as expected. In practice, the Y-junction losses and uneven splitting results in less than ideal performance; A and B do not totally cancel out when $\phi = \pi/2$.

C. Coupled Waveguide Modulators

When two parallel optical waveguides A and B are sufficiently close to each other, then the electric fields associated with the propagation modes in A and B can overlap, as depicted in Figure 7.25 (a). This implies that light can be coupled from one guide to another in a reminiscent way to frustrated total internal reflection (see Ch. 1). We can use qualitative arguments to understand the nature of light coupling between these two guides. Suppose that we launch a light wave into the guide A operating in single mode. Since the separation d of the two guides is small, some of the electric field in the evanescent wave of this mode extends into guide B, and therefore some electromagnetic energy

[10] See R. Syms and J. Cozens, *Optical Guided Waves and Devices* (McGraw-Hill UK, 1992) Ch. 9.

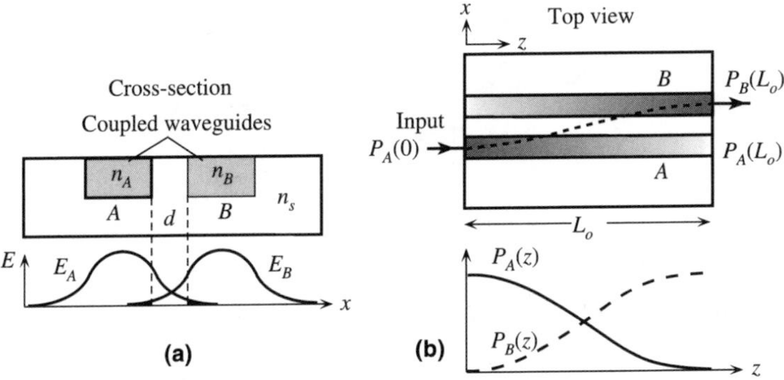

FIGURE 7.25 (a) Cross section of two closely spaced waveguides A and B (separated by d) embedded in a substrate. The evanescent field from A extends into B and vice versa. Note: n_A and $n_B > n_s$ (= substrate refractive index). (b) Top view of the two guides A and B that are coupled along the z-direction. Light is fed into A at $z = 0$, and it is gradually transferred to B along z. At $z = L_o$, all the light has been transferred to B. Beyond this point, light begins to be transferred back to A in the same way.

will be transferred from guide A to B. This energy transfer will depend on the efficiency of coupling between the two guides and the nature of the modes in A and B, which in turn depend on the geometries and refractive indices of guides and the substrate (acting as a cladding material).

As light in guide A travels along z, it leaks into B and, if the mode in B has the right phase, the transferred light waves build up along z as a propagating mode in B as indicated in Figure 7.25 (b). By the same token, the light now traveling in B along z can be transferred back into A if the mode in A has the right phase. The efficient transfer of energy back and forth between the two guides A and B requires that the two modes be in phase to allow the transferred amplitude to build-up along z. If the two modes are out of phase, the waves transferred into a guide do not reinforce each other and the coupling efficiency is poor. Suppose that β_A and β_B are the propagation constants of the fundamental modes in A and B, then there is a phase mismatch per unit length along z that is $\Delta\beta = \beta_A - \beta_B$. The efficiency of energy transfer between the two guides depends on this phase mismatch. If the phase mismatch $\Delta\beta = 0$ then full transfer of power from A to B will require a coupling distance L_o, called the **transfer distance**, as shown in Figure 7.25 (b). This transfer distance depends on the efficiency of coupling C between the two guides A and B, which in turn depends on the refractive indices and geometries of the two guides. C depends on the extent of overlap of the mode fields E_A and E_B in Figure 7.25 (a). The transmission length L_o is inversely proportional to C (in fact, theory shows that $L_o = \pi/C$).

In the presence of no mismatch, $\Delta\beta = 0$, full transmission would occur over the distance L_o. However, if there is a mismatch $\Delta\beta$ then the transferred power ratio over the distance L_o becomes a function of $\Delta\beta$. Thus, if $P_A(z)$ and $P_B(z)$ represent the light power in the guides A and B at z, then

$$\frac{P_B(L_o)}{P_A(0)} = f(\Delta\beta) \tag{2}$$

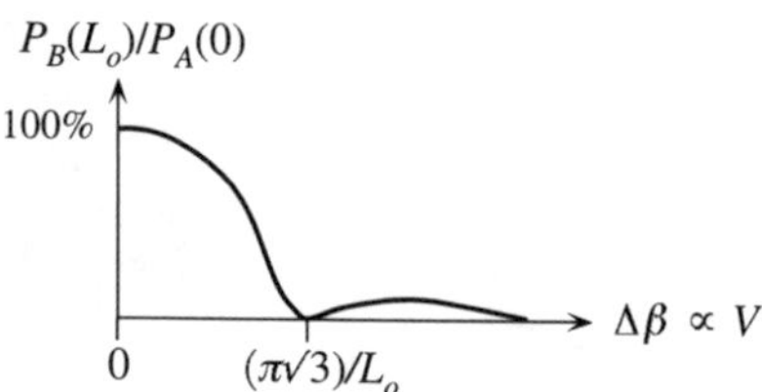

FIGURE 7.26 Transmission power ratio from guide A to guide B over the transmission length L_o as a function of mismatch $\Delta\beta$.

This function is shown in Figure 7.26 which has its maximum when $\Delta\beta = 0$ (no mismatch) and then decays to zero at $\Delta\beta = \pi\sqrt{3}/L_o$. If we could induce a **phase mismatch** of $\Delta\beta = \pi\sqrt{3}/L_o$ by applying an electric field, which modulates the refractive indices of the guides, we could then prevent the transmission of light power from A to B. Light is then *not* transferred to B.

Figure 7.27 shows an integrated directional coupler where two implanted symmetrical guides A and B are coupled over a transmission length L_o and also have electrodes placed on them. In the absence of an applied field, $\Delta\beta = 0$ (no mismatch) and there is a full transmission from guide A to B. If we apply a voltage between the electrodes, the two guides experiences an applied field E_a in opposite directions and hence experience opposite changes in their refractive indices. Suppose that $n(= n_A = n_B)$ is the refractive index of each guide, and Δn is the induced index change in each guide by the Pockels effect. The induced index difference Δn_{AB} between the guides is $2\Delta n$. Taking, as a first approximation, $E_a \approx V/d$, the mismatch $\Delta\beta$ is,

$$\Delta\beta = \Delta n_{AB}\left(\frac{2\pi}{\lambda}\right) \approx 2\left(\frac{1}{2}n^3 r\frac{V}{d}\right)\left(\frac{2\pi}{\lambda}\right)$$

where r is the appropriate Pockels coefficient. Setting $\Delta\beta = \pi\sqrt{3}/L_o$ for the prevention of transfer, the corresponding switching voltage V_o is,

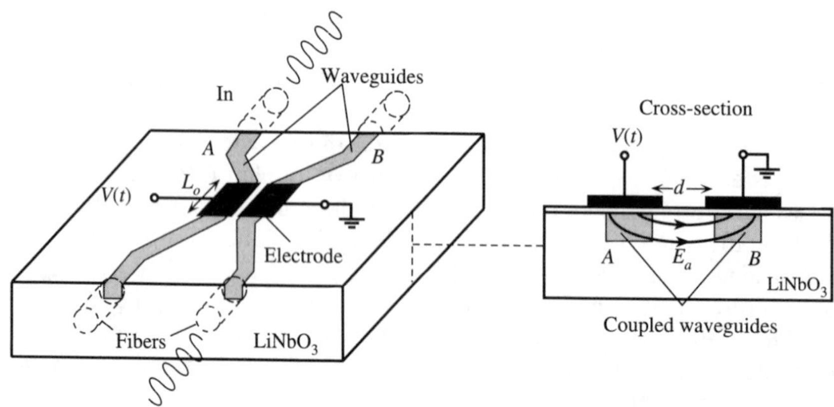

FIGURE 7.27 An integrated directional coupler. Applied field E_α alters the refractive indices of the two guides and changes the strength of coupling.

$$V_o = \frac{\sqrt{3}\lambda d}{2n^3 r L_o} \tag{3}$$

Switching voltage

Since L_o depends inversely on the coupling efficiency C, for a given wavelength, V_o depends on the refractive indices and the geometry of the guides.

EXAMPLE 7.6.1 Modulated Directional Coupler

Suppose that two optical guides embedded in a substrate such as $LiNbO_3$ are coupled as in Figure 7.27, and the transmission length $L_o = 10$ mm. If the coupling separation d is ~10 μm, $n \approx 2.2$, the operating wavelength is 1.3 μm, taking a rough Pockels coefficient $r \approx 10 \times 10^{-12}$ m/V, what is the switching voltage?

Solution Using Eq. (3), we can calculate the switching voltage,

$$V_o \approx \frac{\sqrt{3}\lambda d}{2n^3 r L_o} = \frac{\sqrt{3}(1.3 \times 10^{-6})(10 \times 10^{-6})}{2(2.2)^3(10 \times 10^{-12})(10 \times 10^{-3})} = 10.6 \text{ V}$$

7.7 ACOUSTO-OPTIC MODULATOR

It is found that an induced strain (S) in a crystal changes its refractive index n. This is called the **photoelastic effect**. The strain changes the density of the crystal and distorts the bonds (and hence the electron orbits) which lead to a change in the refractive index n. If we were to examine the change in $1/n^2$ instead of n, then we would find that this is proportional to the induced strain S and the proportionality constant is the **photoelastic coefficient** p, *i.e.*

$$\Delta\left(\frac{1}{n^2}\right) = pS \tag{1}$$

Photoelastic effect

The relationship is not as naive as stated in Eq. (1) since we must consider the effect of a strain S along one direction in the crystal on the induced change in n for a particular light propagation direction and some specific polarization. Equation (1) in reality is a tensor relationship.

We can generate traveling acoustic or ultrasonic waves on the surface of a piezoelectric crystal (such as $LiNbO_3$) by attaching interdigital electrodes onto its surface, as shown in Figure 7.28, and applying a modulating voltage at radio frequencies (RF). The **piezoelectric effect** is the phenomenon of generation of strain in a crystal by the application of an external electric field. The modulating voltage $V(t)$ at electrodes will therefore generate a *surface acoustic wave* (SAW) via the piezoelectric effect. These acoustic waves propagate by rarefactions and compressions of the crystal surface region which lead to a periodic variation in the density and hence a periodic variation in the refractive index in synchronization with the acoustic wave amplitude. Put differently, the periodic variation in the strain S leads to a periodic variation in n owing to the photoelastic effect.

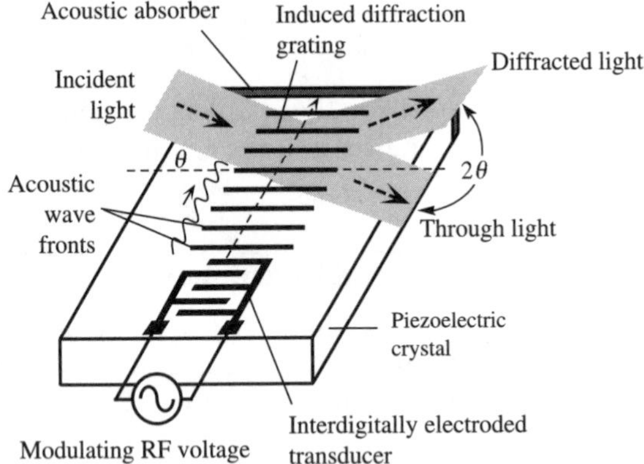

FIGURE 7.28 Traveling acoustic waves create a harmonic variation in the refractive index and thereby create a diffraction grating that diffracts the incident beam through an angle 2θ.

We can simplistically view the crystal surface region as alternations in the refractive index from minimum n_{min} to maximum n_{max} as depicted in Figure 7.29. We can treat the incident optical beam as a stream of parallel coherent waves, two of which, A and B, are shown in Figure 7.29. These waves will be reflected at O and O' by the index changes to become A' and B'. If A' and B' are in phase then they would reinforce each other and thereby constitute a *diffracted beam*. Suppose that the acoustic wavelength is Λ which represents the separation of index boundaries. The optical path difference between A' and B' is $PO'Q$ which is $2\Lambda \sin\theta$. This must be a whole optical wavelength in the medium. *i.e.* it must be λ/n, where λ is the free-space wavelength. Thus, the condition that gives the angle θ for a diffracted beam to exist is,

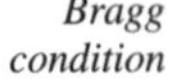
Bragg condition

$$2\Lambda \sin\theta = \lambda/n \quad \textbf{(1)}$$

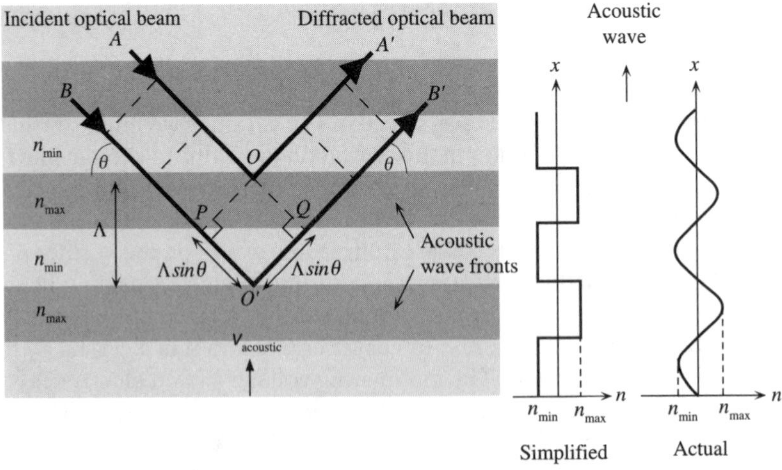

FIGURE 7.29 Consider two coherent optical waves A and B being "reflected" (strictly, scattered) from two adjacent acoustic wavefronts to become A' and B'. These reflected waves can only constitute the diffracted beam if they are in phase. The angle θ is exaggerated (typically this is a few degrees).

Provided that the incident beam makes an angle θ satisfying Eq. (1), called the *Bragg angle*, it will be diffracted. We therefore have a means of deflecting an incoming optical beam by simply choosing the acoustic wavelength Λ. It is apparent from Eq. (1) that modulating the acoustic wavelength leads to modulating the diffraction angle θ.

Suppose that ω is the angular frequency of the incident optical wave. The optical wave reflections occur from a moving diffraction pattern which moves with a velocity $v_{acoustic}$ as shown in Figure 7.29. As a result of the *Doppler effect*, the diffracted beam has either a slightly higher or slightly lower frequency depending on the direction of the traveling acoustic wave. If Ω is the frequency of the acoustic wave then the diffracted beam has a Doppler shifted frequency given by

$$\omega' = \omega \pm \Omega \qquad \textbf{(2)}$$

Doppler shift

When the acoustic wave is traveling towards the incoming optical beam as in Figure 7.29, then the diffracted optical beam frequency is up-shifted, *i.e.* $\omega' = \omega + \Omega$. If the acoustic wave is traveling away from the incident optical beam then the diffracted frequency is down-shifted, $\omega' = \omega - \Omega$. It is apparent that we can modulate the frequency (wavelength) of the diffracted light beam by modulating the frequency of the of the acoustic waves. (The diffraction angle is then also changed.)

Although the Bragg condition specifies the incidence angle θ for beam deflection, it does not say anything about the diffracted optical intensity. Not all the wave amplitudes become reflected at O and O' in Figure 7.29 which means that there is also a through (undeflected) beam. The amplitudes of the reflections at the refractive index boundaries in Figure 7.29 Figure 7.29 depend on the induced photoelastic change Δn which depends on the amplitude of the acoustic waves. The intensity of the diffracted beam is proportional to Δn^2 and hence to the intensity of the acoustic wave. Changing the intensity of the acoustic wave (by changing the RF voltage) modulates the intensity of the diffracted light beam.

EXAMPLE 7.7.1 Modulated Directional Coupler

Suppose that we generate 250 MHz acoustic waves on a $LiNbO_3$ substrate. The acoustic velocity in $LiNbO_3$ is 6.57 km/s and the refractive index is about 2.2. Consider modulating a red-laser beam from a He-Ne laser, $\lambda = 632.8$ nm. Calculate the acoustic wavelength and hence the Bragg deflection angle. What is the Doppler shift in the wavelength?

Solution If f is the frequency of the acoustic waves, the acoustic wavelength Λ is

$$\Lambda = \frac{v_{acoustic}}{f} = \frac{(6.57 \times 10^3 \text{ m s}^{-1})}{(250 \times 10^6 \text{ s}^{-1})} = 2.63 \times 10^{-5} \text{ m} = 26.3 \ \mu\text{m}.$$

The Bragg angle is

$$\sin\theta = \frac{\lambda/n}{2\Lambda} = \frac{(632.8 \times 10^{-9} \text{ m})/(2.2)}{2(2.63 \times 10^{-5} \text{ m})} = 0.00547$$

so that $\theta = 0.31°$ or a deflection angle 2θ of 0.62°.

The Doppler shift in frequency from Eq. (2) is simply the acoustic frequency 250 MHz

---

7.8 MAGNETO-OPTIC EFFECTS

When an optically inactive material such as glass is placed in a strong magnetic field and then a plane polarized light is sent *along* the direction of the magnetic field, it is found that the emerging light's plane of polarization has been rotated. This is called the **Faraday effect** as originally observed by Michael Faraday (1845). The magnetic field can be applied, for example, by inserting the material into the core of a magnetic coil, a solenoid. The induced specific rotatory power (θ/L) has been found to be proportional to the magnitude of applied magnetic field, B. The amount of rotation θ is given by

Faraday effect

$$\theta = \vartheta BL \tag{1}$$

where B is the magnetic field (flux density), L is the length of the medium, and ϑ is the so-called *Verdet* constant. It depends on the material and the wavelength. The Faraday effect is typically small. A magnetic field of $\sim$0.1 T causes a rotation of about 1° through a glass rod of length 20 mm.

It seems to appear that an "optical activity" has been induced by the application of a strong magnetic field to an otherwise optically inactive material. There is however an important distinction between the natural optical activity in Figure 7.17 and the Faraday effect. The sense of rotation θ in the Faraday effect, for a given material (Verdet constant), depends only on the direction of the magnetic field **B**. If ϑ is positive, for light propagating parallel to **B**, the optical field **E** rotates in the same sense as an advancing right-handed screw pointing in the direction of **B**, as in Figure 7.30. The direction of light propagation, as in Figure 7.30, does not change the absolute sense of rotation of θ. If we reflect the wave to pass through the medium again, the rotation increases to 2θ.

An **optical isolator** allows light to pass in one direction and not in the opposite direction. For example, the light source can be isolated from various reflections by placing a polarizer and a Faraday rotator that rotates the field by 45° as in Figure 7.30. The reflected light will have a $2\theta = 90°$ rotation and will not pass through the polarizer back to the source. The magnetic field is typically applied by enclosing the Faraday medium in a rare-earth magnet ring.

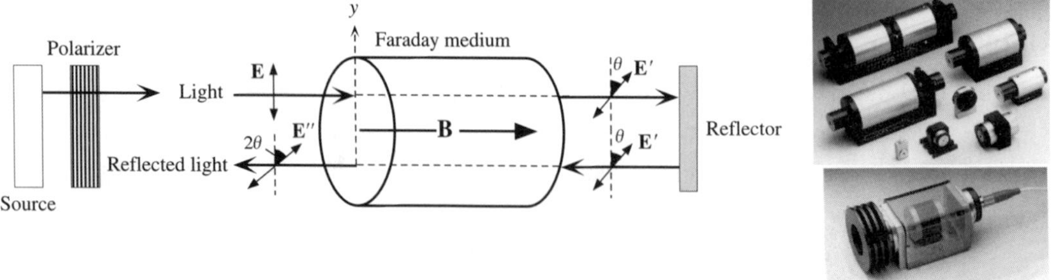

FIGURE 7.30 The sense of rotation of the optical field **E** depends only on the direction of the magnetic field for a given medium (given Verdet constant). If light is reflected back into the Faraday medium, the field rotates a further θ in the same sense to come out as **E**″ with a 2θ rotation with respect to **E**. Right: Faraday effect optical isolators: Air-path isolators (top) and laser diode-to-fiber isolator (below). The cylindrical case contains a rare-earth magnet. (Courtesy of OFR)

7.9 NON-LINEAR OPTICS AND SECOND HARMONIC GENERATION

The application of an electric field E to a dielectric material causes the constituent atoms and molecules to become polarized. The medium responds to the field E by developing a *polarization* **P** which represents the net induced dipole moment per unit volume. In a linear dielectric medium the induced polarization **P** is proportional to the electric field E at that point and the two are related by $P = \varepsilon_o \chi E$ where χ is the *electric susceptibility*. However, the linearity fails at high fields and the **P** vs. E behavior deviates from the linear relationship as shown in Figure 7.31 (a). **P** becomes a function of E which means that we can expand it in terms of increasing powers of E. It is customary to represent the induced polarization as

Induced polarization

$$P = \varepsilon_o \chi_1 E + \varepsilon_o \chi_2 E^2 + \varepsilon_o \chi_3 E^3 \qquad (1)$$

where χ_1, χ_2 and χ_3 are the linear, second-order and third-order susceptibilities. The coefficients decrease rapidly for higher terms and are not shown in Eq. (1). The importances of the second and third terms, *i.e.* nonlinear effects, depend on the field strength E. Nonlinear effects begin to become observable when fields are very large, *e.g.* $\sim 10^7$ V m^{-1}. Such high fields require light intensities ($\sim$1000 kW cm^{-2}) that invariably require lasers. All materials, whether crystalline or noncrystalline, possess a finite χ_3 coefficient. However, only certain classes of crystals have a finite χ_2, and the reason is the same as for the observation of the Pockels effect.[11] Only those crystals, such as quartz, that have no center of symmetry have non-zero χ_2 coefficient; these crystals are also piezoelectric. One of the most important consequences of the nonlinear effect is the **second harmonic generation (SHG)**[12], when an intense light beam of angular frequency ω passing through an appropriate crystal (*e.g.* quartz) generates a light beam of double the frequency, 2ω. SHG is based on a finite χ_2 coefficient in which the effect of χ_3 is negligible.

Consider a beam of monochromatic light, with a well-defined angular frequency ω, passing through a medium. The optical field E at any point in the medium will polarize the medium at the same point in synchronization with the optical field oscillations. An oscillating dipole moment is well known as an electromagnetic emission source (just like an antenna). These secondary electromagnetic emissions from the dipoles in the medium interfere and constitute the actual wave traveling through the medium (Huygen's construction from secondary waves). Suppose that the optical field is oscillating sinusoidally between $\pm E_o$ as shown in Figure 7.31 (b). In the linear regime (E_o is "small"), **P** oscillations will also be sinusoidal with a frequency ω.

If the field strength is sufficiently large, the induced polarization will not be linear as indicated in Figure 7.31 (a), and **P** will not oscillate in a simple sinusoidal fashion as indicated Figure 7.31 (b). The polarization now oscillates between P_+ and P_- and is not symmetrical. The oscillations of the dipole moment **P** now emit waves not only at the fre-

---

[11] See §7.5 A on the Pockels Effect.

[12] Second harmonic generation was first experimentally demonstrated by Peter A. Franken and coworkers at the University of Michigan (1961) where they focused a 3 kW pulse of red light (694.3 nm) from a ruby laser onto a quartz crystal and observed some ultraviolet light (347.15 nm) coming out. The conversion efficiency was 1 in 10^8 (*Optics, Second Edition*, Eugene Hecht, Addison Wesley, 1987, p. 612).

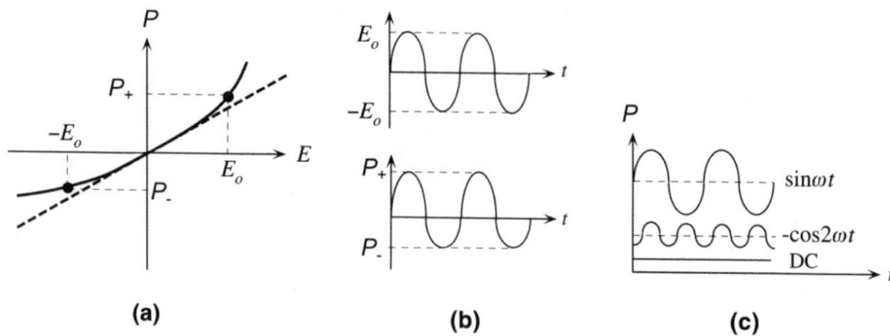

FIGURE 7.31 (a) Induced polarization vs. optical field for a nonlinear medium. (b) Sinusoidal optical field oscillations between $\pm E_o$ result in polarization oscillations between P_+ and P_-. (c) The polarization oscillation can be representad by sinusoidal oscillations at angular frequencies ω (fundamental), 2ω (second harmonic) and a small DC component.

quency[13] ω but also at 2ω. In addition there is a dc component (light is "rectified", *i.e.* gives rise to a small permanent polarization). The *fundamental* ω and the *second harmonic,* 2ω, components, along with the dc, are shown in Figure 7.31 (c).

If we write the optical field as $E = E_o \sin(\omega t)$ and substitute into Eq. (1), after some trigonometric manipulation,[14] and neglecting χ_3 terms, we would find the induced **P** as

$$P = \varepsilon_o \chi_1 E_o \sin(\omega t) - \tfrac{1}{2}\varepsilon_o \chi_2 E_o \cos(2\,\omega t) + \tfrac{1}{2}\,\varepsilon_o \chi_2 E_o \qquad \textbf{(2)}$$

The first term is the fundamental, second is the second harmonic and third is the dc term as summarized in Figure 7.31 (c). The second harmonic (2ω) oscillation of local dipole moments generates secondary second harmonic (2ω) waves in the crystal. It may be thought that these secondary waves will interfere constructively and result in a second harmonic beam just as the fundamental (ω) secondary waves interfere and give rise to the propagating light beam. However, the crystal will normally possess different refractive indices $n(\omega)$ and $n(2\omega)$ for frequencies ω and 2ω which means that the ω- and 2ω-waves propagate with different phase velocities v_1 and v_2 respectively. As the ω-wave propagates in the crystal it generates secondary 2ω waves along its path, depicted as $S_1, S_2, S_3, \ldots$, in Figure 7.32. When wave S_2 is generated, S_1 must arrive there in phase

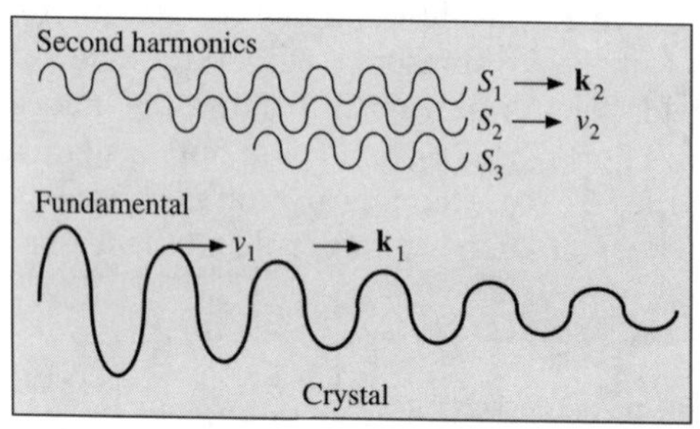

FIGURE 7.32 As the fundamental wave propagates, it periodically generates second harmonic waves $(S_1, S_2, S_3, \ldots)$ and if these are in phase then the amplitude of the second harmonic light builds up.

[13] In this section, the adjective *angular* is dropped from the frequency though implied.

[14] That is, $\sin^2(\omega t) = \frac{1}{2}[1 - \cos(2\,\omega t)]$

which means S_1 must travel with the same velocity as the fundamental wave; and so on as in Figure 7.32. It is apparent that only if these 2ω-waves are in phase, that is they propagate with the same velocity as the ω-wave, that they can interfere constructively and constitute a second harmonic beam. Otherwise, the S_1, S_2 and S_3.. will eventually fall out of phase and destroy each other and there will be either no or very little second harmonic beam. The condition that the second harmonic waves must travel with the same phase velocity as the fundamental wave to constitute a second harmonic beam is called **phase matching** and requires $n(\omega) = n(2\omega)$. For most crystals this is not possible as n is dispersive; depends on the wavelength.

SHG efficiency depends on the extent of phase matching, $n(\omega) = n(2\omega)$. One method is to use a birefringent crystal as these have two refractive indices: ordinary index n_o and extraordinary index n_e. Suppose that along a certain crystal direction at an angle θ to the optic axis, $n_e(2\omega)$ at the second harmonic is the same as $n_o(\omega)$ at the fundamental frequency: $n_e(2\omega) = n_o(\omega)$. This is called **index matching** and the angle θ is the **phase matching angle**. Thus, the fundamental wave would propagate as an ordinary wave and the second harmonic as an extraordinary wave and both would be in phase. This would maximize the conversion efficiency, though this would still be limited by the magnitude of second term with respect to the first in Eq. (1). To separate the second harmonic beam from the fundamental beam, something like a diffraction grating, a prism or an optical filter will have to be used at the output as depicted in Figure 7.33. The phase matching angle θ depends on the wavelength (or ω) and is sensitive to temperature.

It is instructive to consider the SHG process in terms of photon interactions as shown in Figure 7.34. Two fundamental mode photons interact with the dipoles moments to produce a single second harmonic photon. The photon momentum is $\hbar\mathbf{k}$ and

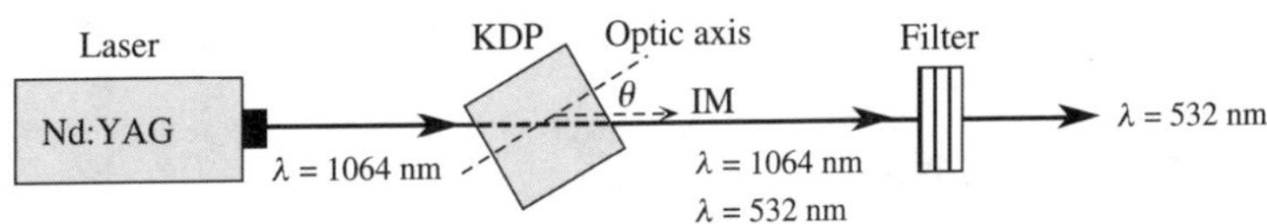

FIGURE 7.33 A simplified schematic illustration of optical frequency doubling using a KDP (potassium dihydrogen phosphate) crystal. IM is the index matched direction at an angle θ (about 35°) to the optic axis along which $n_e(2\omega) = n_o(\omega)$. The focusing of the laser beam onto the KDP crystal and the collimation of the light emerging from the crystal are not shown.

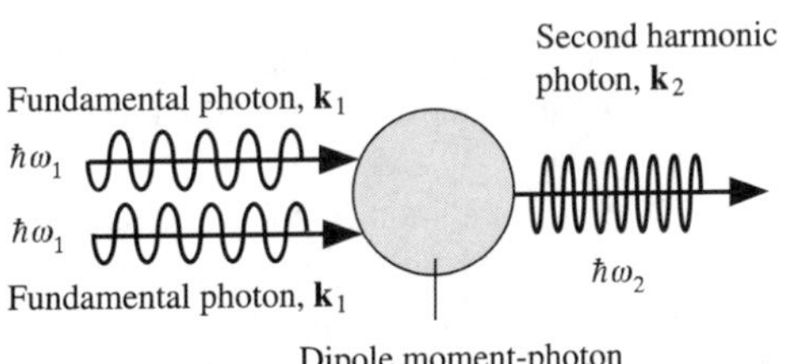

FIGURE 7.34 Photonic interpretation of second harmonic generation (SHG)

energy is $\hbar\omega$. Suppose that subscripts 1 and 2 refer to fundamental and second harmonic photons. In general terms, we can write the following two equations. Conservation of momentum requires that

$$\hbar\mathbf{k}_1 + \hbar\mathbf{k}_1 = \hbar\mathbf{k}_2 \tag{3}$$

The conservation of energy requires that

$$\hbar\omega_1 + \hbar\omega_1 = \hbar\omega_2 \tag{4}$$

We tacitly assumed that the interaction does not result in phonon (lattice vibration) generation or absorption. We can satisfy Eq. (4) by taking $\omega_2 = 2\omega_1$, so that the frequency of the second harmonic is indeed twice the fundamental. To satisfy Eq. (3) we need $\mathbf{k}_2 = 2\mathbf{k}_1$. The phase velocity v_2 of the second harmonic waves is

$$v_2 = \frac{\omega_2}{k_2} = \frac{2\omega_1}{2k_1} = \frac{\omega_1}{k_1} = v_1$$

Thus, the fundamental and the second harmonic photons are required to have the same phase velocity which is tantamount to the phase matching criterion developed above in terms of pure waves. If $\mathbf{k}_2$ is not exactly $2\mathbf{k}_1$, *i.e.* $\Delta k = k_2 - 2k_1$ is not zero, *i.e.* there is a mismatch, then SHG is only effective over a limited length l_c which can be shown to be given by $l_c = \pi/\Delta k$. This length l_c is essentially the *coherence length* of the second harmonic (depending on the index difference, this may be quite short, *e.g.* $l_c \approx 1 - 100$ μm). If the crystal size is longer than this, the second harmonics will interfere randomly with each other and the SHG efficiency will be very poor, if not zero. Phase matching is therefore an essential requirement for SHG. The conversion efficiency depends on the intensity of exciting laser beam, the materials χ_2 coefficient and the extent of phase matching and can be substantial (as high as 70–80%) if well-engineered by, for example, placing the converting crystal into the cavity of the laser itself.

QUESTIONS AND PROBLEMS

7.1 Polarization Suppose that we write the E_x and E_y components of a light wave generally as:

$$E_x = E_{xo}\cos(\omega t - kz) \quad \text{and} \quad E_y = E_{yo}\cos(\omega t - kz + \phi)$$

Show that at any instant E_x and E_y satisfy the ellipse equation on the E_y vs. E_x coordinate system:

$$\left(\frac{E_x}{E_{xo}}\right)^2 + \left(\frac{E_y}{E_{yo}}\right)^2 - 2\left(\frac{E_x}{E_{xo}}\right)\left(\frac{E_y}{E_{yo}}\right)\cos\phi = \sin^2\phi$$

Sketch schematically what this ellipse looks like assuming $E_{xo} = 2\,E_{yo}$. When would this ellipse form an (a) ellipse with its major axis on the x-axis, (b) a linearly polarized light at 45°, (c) right and left circularly polarized light?

7.2. Linear and circular polarization Show that a linearly polarized light wave can be represented by two circularly polarized light waves with opposite rotations. Consider the simplest case of a wave linearly polarized along the y-axis. What is your conclusion?

7.3 Wire-grid polarizer Figure 7.35 shows a wire grid-polarizer which consists of closely spaced parallel thin conducting wires. The light beam passing through the wire-grid is observed to be linearly polarized at right angles to the wires. Can you explain the operation of this polarizer?

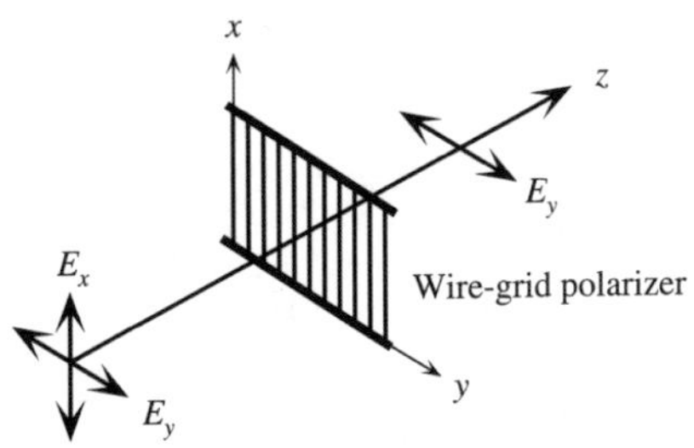

FIGURE 7.35 The wire grid-acts as a polarizer.

7.4 Anisotropic scattering and polarization It is well known that an oscillating electric dipole emits electromagnetic radiation. Figure 7.36 (a) shows a snap shot of the electric field pattern around an oscillating electric dipole moment $p(t)$ parallel to the y-axis. There is no field radiation along the dipole axis y. The irradiance I of the radiation along a direction at an angle θ to the perpendicular to the dipole axis, is proportional to $\cos^2\theta$. Sketch the relative radiation intensity pattern around the oscillating dipole. Suppose that the electric field in an incoming electromagnetic wave induces dipole oscillations in a molecule of the medium. Explain how scattering of an incident unpolarized electromagnetic wave by this molecule leads to waves with different polarizations along the x and y axes in Figure 7.36 (b).

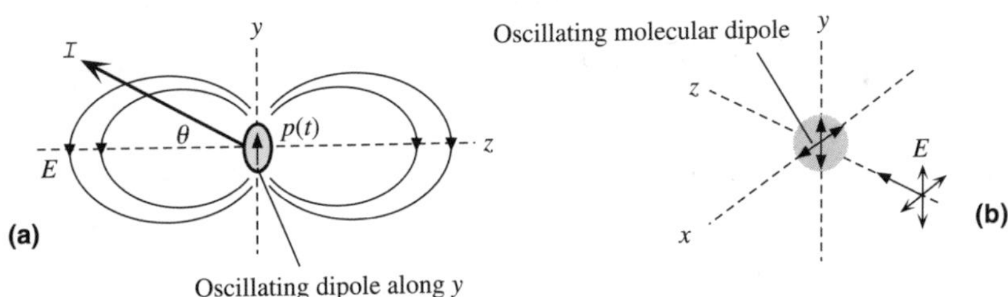

FIGURE 7.36 (a) A snap shot of the field pattern around an oscillating dipole moment in the y-direction. Maximum electromagnetic radiation is perpendicular to the dipole axis and there is no radiation along the dipole axis. (b) Scattering of electromagnetic waves from induced molecular dipole oscillations is anisotropic.

7.5 Jones Matrices When we represent the state of polarization of a light wave using a matrix, called a Jones matrix[15] (or vector) then various operations on the polarization state correspond to multiplying this matrix with another matrix that represents the optical operation. Consider a light wave traveling along z with field components E_x and E_y along x and y. These components are orthogonal and, in general, would be of different magnitude and have a phase difference ϕ between them. If we use the exponential notation then

$$E_x = E_{xo}\exp[j(\omega t - kz + \phi_x)] \quad \text{and} \quad E_y = E_{yo}\exp[j(\omega t - kz + \phi_y)]$$

[15] R. Clark Jones introduced these matrices circa 1941.

Jones matrix is a column matrix whose elements are E_x and E_y without the common $\exp j(\omega t - kz)$ factor

$$\mathbf{E} = \begin{bmatrix} E_x \\ E_y \end{bmatrix} = \begin{bmatrix} E_{x0}\exp(j\phi_x) \\ E_{y0}\exp(j\phi_y) \end{bmatrix} \quad \mathbf{(1)}$$

Usually Eq. (1) is normalized by dividing by the total amplitude $E_o = (E_{xo}^2 + E_{yo}^2)^{1/2}$. We can also factor out $\exp(j\phi_x)$ to further simplify to obtain the Jones matrix:

$$\mathbf{J} = \frac{1}{E_o}\begin{bmatrix} E_{xo} \\ E_{yo}\exp(j\phi) \end{bmatrix} \quad \mathbf{(2)}$$

where $\phi = \phi_y - \phi_x$.

(a) Table 7.3 shows Jones vectors for various polarizations. Identify the state of polarization for each matrix.

(b) Passing a wave of given Jones vector $\mathbf{J}_{\text{in}}$ through an optical device is represented by multiplying $\mathbf{J}_{\text{in}}$ by the **transmission matrix T** of the device. If $\mathbf{J}_{\text{out}}$ is the Jones vector for the output light through the device, then $\mathbf{J}_{\text{out}} = \mathbf{T}\,\mathbf{J}_{\text{in}}$. Given

$$T = \begin{bmatrix} 1 & 0 \\ 0 & j \end{bmatrix} \quad \mathbf{(3)}$$

determine the polarization state of the output wave given the Jones vectors in Table 7.3, and the optical operation represented by **T**. Hint: Use $\begin{bmatrix} 1 \\ 1 \end{bmatrix}$ as input for determining **T**.

TABLE 7.3 Jones vectors

Jones vector $\mathbf{J}_{in}$	$\begin{bmatrix} 1 \\ 0 \end{bmatrix}$	$\frac{1}{\sqrt{2}}\begin{bmatrix} 1 \\ 1 \end{bmatrix}$	$\begin{bmatrix} \cos\theta \\ \sin\theta \end{bmatrix}$	$\frac{1}{\sqrt{2}}\begin{bmatrix} 1 \\ j \end{bmatrix}$	$\frac{1}{\sqrt{2}}\begin{bmatrix} 1 \\ -j \end{bmatrix}$
Polarization	?	?	?	?	?
Transmission matrix T	$\begin{bmatrix} 1 & 0 \\ 0 & 0 \end{bmatrix}$	$\begin{bmatrix} e^{j\phi} & 0 \\ 0 & e^{j\phi} \end{bmatrix}$	$\begin{bmatrix} 1 & 0 \\ 0 & j \end{bmatrix}$	$\begin{bmatrix} 1 & 0 \\ 0 & -1 \end{bmatrix}$	$\begin{bmatrix} 1 & 0 \\ 0 & e^{-j\Gamma} \end{bmatrix}$
Optical operation	?	?	?	?	?

7.6 Modulation by Malus's law Suppose that a linearly polarized light is passed through a polarizer placed with its transmission axis at angle $\pi/4$ (45°) to the incoming optical field. Suppose now that we "rotationally modulate" the transmission axis of the polarizer by small amounts of ϕ about $\pi/4$ (45°). Show that the change in the transmission intensity is

$$\Delta I = -\phi + \frac{2}{3}\phi^3 - \ldots$$

where ϕ is in radians. What is the extent of change (in degrees) in ϕ so that the second term is only 1% of the first term? What is your conclusion?

7.7 Birefringence Consider a negative uniaxial crystal such as calcite ($n_e < n_o$) plate that has the optic axis (taken along z) parallel to the plate faces. Suppose that a linearly polarized wave is incident at normal incidence on a plate face. If the optical field is at an angle 45° to the optic axis, sketch the rays through the calcite plate.

7.8 Wave plates Calculate and compare the thickness of quarter-wave plates made from calcite, quartz and $LiNbO_3$ crystals all operating at a wavelength of $\lambda \approx 590$ nm. What is your conclusion? Assuming little relative change in the indices, what are the thicknesses at double the wavelength?

7.9 Soleil Compensator Consider a Soleil compensator as in Figure 7.15 that uses a quartz crystal. Given a light wave with a wavelength $\lambda \approx 600$ nm, a lower plate thickness of 5 mm, calculate the range of d values in Figure 7.15 that provide a retardation from 0 to π (half-wavelength).

7.10 Quartz Wollaston prism Draw a quartz Wollaston prism and clearly show and identify the directions of orthogonally polarized waves traveling through the prisms. How would you test the polarization states of the emerging rays? Consider two identical Walloston prisms, one from calcite and the other from quartz. Which will have a greater beam splitting ability? (Explain).

7.11 Glan-Foucault prism Figure 7.37 shows the cross section of a Glan-Foucault prism which is made of two right angle calcite prisms with a prism angle of 38.5°. Both have their optic axes parallel to each other and to the block faces as in the figure. Explain the operation of the prisms and show that the o-wave does indeed experience total internal reflection.

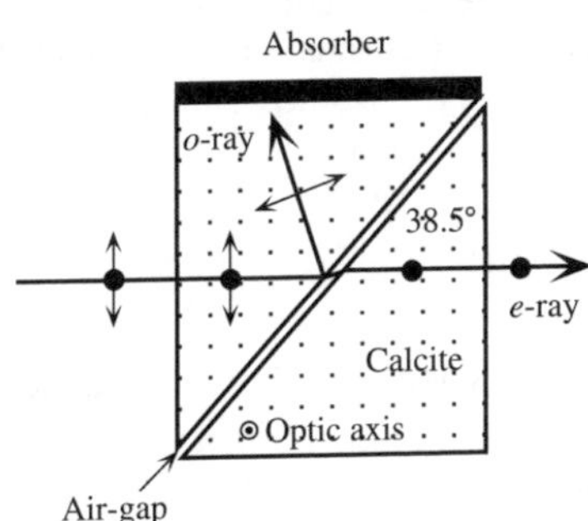

FIGURE 7.37 The Glan-Foucault prism provides linearly polarized light.

7.12 Faraday Effect Application of a magnetic field along the direction of propagation of a linearly polarized light wave through a medium results in the rotation of the plane of polarization. The amount of rotation θ is given by

$$\theta = \vartheta BL$$

where B is the magnetic field (flux density), L is the length of the medium, and ϑ is the so-called *Verdet* constant. It depends on the material and the wavelength. In contrast to optical activity, sense of rotation of the plane of polarization is independent of the direction of light propagation. Given that glass and ZnS have Verdet constants of about 3 and 22 minutes of arc Tesla^{-1} meter^{-1} at 589 nm respectively, calculate the necessary magnetic field for a rotation of 1° over a length 10 mm. What is the rotation per unit magnetic field for a medium of length 1 m? (Note 60 minutes of arc = 1°).

7.13 Optical activity

(a) Consider an optically active medium. The experimenter A (Alan) sends a vertically polarized light into the this medium as in Figure 7.17. The light that emerges from the back of the crystal is received by an experimenter B (Barbara). B observes that the optical field $\boldsymbol{E}$ has been rotated to $\boldsymbol{E}'$ counterclockwise. She reflects the wave back into the medium so that A can receive it. Describe the observations of A and B. What is your conclusion?

(b) Figure 7.38 shows a simplified version of the Fresnel prism that converts an incoming unpolarized light into two divergent beams that have opposite circular polarizations. Explain the principle of operation.

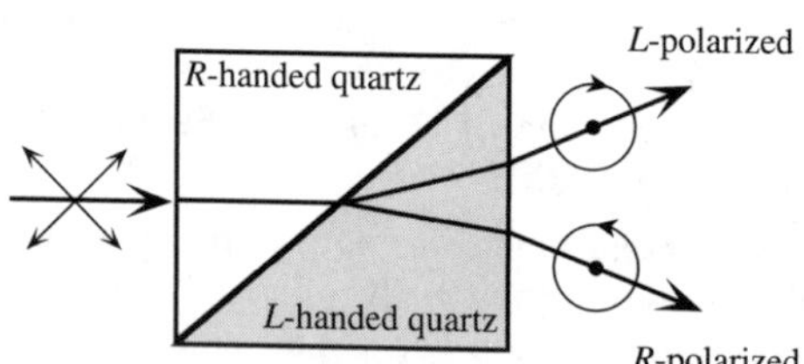

FIGURE 7.38 The Fresnel prism for separating unpolarized light into two divergent beams with opposite circular polarizations (R = right, L = left; divergence is exaggerated).

7.14 $LiNbO_3$ phase modulator What should be the aspect ratio d/L for the transverse $LiNiO_3$ phase modulator in Figure 7.20 that will operate at a free-space wavelength of 1.3 μm and will provide a phase shift $\Delta\phi$ of π (half wavelength) between the two field components propagating through the crystal for an applied voltage of 12 V?

We cannot arbitrarily set d/L to any ratio we like for the simple reason that when d becomes too small, the light will suffer diffraction effects that will prevent it from passing through the device. Consideration of diffraction effects leads to (see, for example, A.K. Ghatak and K. Thyagarajan, *Optical Electronics*, Cambridge University Press, Cambridge, 1989, pp. 477–479)

$$d \approx 2\left(\frac{\lambda L}{n_o \pi}\right)^{1/2}$$

Taking the crystal length $L \approx 20$ mm, calculate d and hence the new aspect ratio.

7.15 Transverse Pockels cell with $LiNbO_3$ Suppose that instead of the configuration in Figure 7.20, the field is applied along the z-axis of the crystal, the light propagates along the y-axis. The x-axis is the polarization of the ordinary wave and z-axis that of the extraordinary wave. Light propagates through as o- and e-waves. Given that $E_a = V/d$, where d the crystal length along z, the indices are

$$n'_o \approx n_o - \tfrac{1}{2}n_o^3 r_{13} E_a \quad \text{and} \quad n'_e \approx n_e - \tfrac{1}{2}n_e^3 r_{33} E_a$$

Show that the phase difference between the o- and e-waves emerging from the crystal is,

$$\Delta\phi = \phi_e - \phi_o = \frac{2\pi L}{\lambda}(n_e - n_o) - \frac{2\pi L}{\lambda}\frac{1}{2}\left(n_e^3 r_{33} - n_o^3 r_{13}\right)\frac{V}{d}$$

where L is the crystal length along the y-axis.

Explain the first and second terms. How would you use two such Pockels cells to cancel the first terms in the total phase shift for the two cells.

If the light beam entering the crystal is linearly polarized in the z-direction, show that

$$\Delta\phi = \frac{2\pi n_e L}{\lambda} + \frac{2\pi L}{\lambda}\frac{(n_e^3 r_{33})}{2}\frac{V}{d}$$

Consider a nearly monochromatic light beam of the free-space wavelength $\lambda = 500$ nm and polarization along z-axis. Calculate the voltage V_π needed to *change* the output phase $\Delta\phi$ by π given a $LiNbO_3$ crystal with $d/L = 0.01$ (see Table 7.2).

7.16 Longitudinal Pockels cell

(a) Sketch schematically the structure of a longitudinal Pockels cell in which the applied field is along the direction of light propagation, both parallel to z-axis (optic axis). Suggest schemes that would allow light to enter the crystal along the applied field direction.

(b) Suppose that a $LiNbO_3$ crystal is used. $LiNbO_3$ is uniaxial and $n_1 = n_2 = n_o$ (polarizations parallel to x and y) and $n_3 = n_e$ (polarization parallel to z). Neglecting the rotation of the axes (same principal axes in the presence of an applied field), if E_a is the field along z, then the new refractive index is

$$n'_o \approx n_o - \tfrac{1}{2}n_o^3 r_{13} E_a$$

Calculate the half-wave voltage required to induce a retardation of π between the emerging and incident waves if the free space wavelength is 1 μm. What are their polarizations? [Note: For $LiNbO_3$ at 633 nm, $n_o \approx 2.28$, $r_{13} \approx 9 \times 10^{-12}$ m/V.]

(c) Suppose that a KDP crystal is used. KDP is uniaxial and $n_1 = n_2 = n_o$ (polarizations parallel to x and y) and $n_3 = n_e$ (polarization parallel to z). The principal axes x and y are rotated by 45° to become x' and y' as in Figure 7.19 (b) and

$$n'_1 \approx n_o - \tfrac{1}{2}n_o^3 r_{63} E_a \quad n'_2 \approx n_o + \tfrac{1}{2}n_o^3 r_{63} E_a \quad \text{and} \quad n'_3 = n_3 = n_e$$

Calculate the half-wave voltage required to induce a retardation of π between the emerging components of the electric field, for free space wavelength of 633 nm if for KDP at 633 nm, $n_o \approx 1.51$, $r_{63} \approx 10.5 \times 10^{-12}$ m/V.

***7.17 Response time and bandwidth of an EO phase modulator** Consider the electro-optic modulator in Figure 7.39 (a). Suppose that we suddenly apply a step-voltage from a supply with an output resistance $R_s = 50\ \Omega$ (or this may be the impedance of a coaxial cable bringing in the pulse). The EO (electro-optic) crystal between the electrodes is a dielectric with a relative permittivity ε_r. The time required to charge (and discharge) the capacitance between the electrodes is determined by the time constant of the electric circuit, $\tau = R_s C_{EO}$, where C_{EO} is the capacitance of the EO crystal. On the other hand, light has to propagate through the length L of the EO crystal. This transit time τ_{light} of light is determined by the length L and the refractive index n. Thus,

$$\tau = \frac{R_s L W \varepsilon_o \varepsilon_r}{D} \quad \text{and} \quad \tau_{\text{light}} = \frac{L}{c/n}$$

Calculate these characteristic times for a GaAs crystal assuming an operation at $\lambda = 1.3$ μm, length $L = 20$ mm, W = 2 mm and thickness $D = 2$ mm given that $\varepsilon_r \approx 12$ and $n \approx 3.6$.

If we were to apply an ac signal V_s from a source of output resistance R_s, then the maximum modulation frequency f_m would be $(2\pi R_s C_{\text{EO}})^{-1}$ and above this frequency most of the voltage would drop across R_s. We can improve the high frequency performance of the modulator by connecting an inductance L with a certain equivalent parallel resistance $R_p\ (R_p \gg R_s)$ as in Figure 7.39 (b). The source now sees an impedance Z formed by the parallel combination of C_{EO}, L, and R_p. The voltage across the modulator is maximum, V_m, at the resonant frequency f_o. The bandwidth Δf is the frequencies over which the voltage across the crystal is greater than $V_m/\sqrt{2}$. Show that f_o and Δf are:

$$f_o = \frac{1}{2\pi\sqrt{LC_{\text{EO}}}} \quad \text{and} \quad \Delta f = \frac{1}{2\pi C_{\text{EO}} R_p}$$

At resonance, the power delivered to the crystal modulator is V_m^2/R_p. Suppose that our application requires a maximum phase change of Φ (between field components) through the EO crystal. From Eq. (5) in § 7.5, V_m induces a phase change $\Phi = (2\pi/\lambda)(n_o^3 r_{12})(L/D)V_m$, where r_{12} is the appropriate Pockels coefficient. Show that power delivered per unit frequency bandwidth is

$$\frac{P}{\Delta f} = \frac{\lambda^2 \varepsilon_o \varepsilon_r}{4\pi n_o^6 r_{12}^2} \frac{WD}{L} \Phi^2$$

Recall that we are applying an ac voltage so that V_m is the *amplitude* of this signal and not peak-to-peak value (which is $2V_m$). If we wish to modulate from zero to full intensity, then peak-to-peak voltage $2V_m$ should span a phase change π or V_m should span $\Phi = \pi/2$. Thus, on and off corresponds to $\Phi = \pi/2$. Consider a GaAs EO modulator configured as in Figure 7.39 (a) which has $n \approx 3.6$, $\varepsilon_r \approx 12$, $r_{12} \approx 1.5 \times 10^{-12}$ m/V and a volume of material with length $L = 10$ mm, width $W = 15\ \mu$m, thickness $D = 5\ \mu$m, calculate the power per unit frequency bandwidth assuming $\Phi = \pi/2$ is required to turn the intensity on and off.

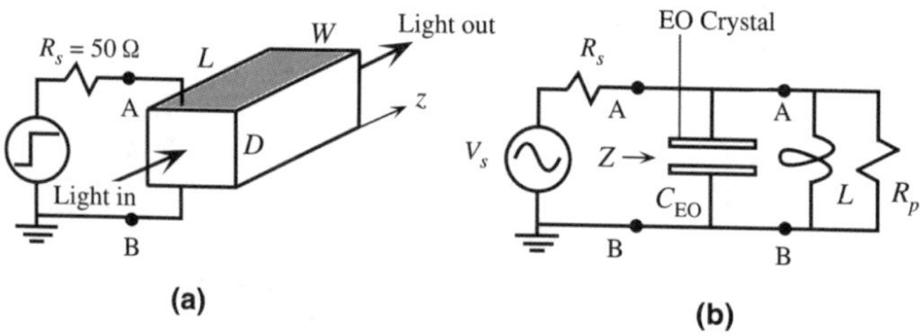

FIGURE 7.39 (a) A step voltage is suddenly applied to an EO modulator. (b) An inductance L with an equivalent parallel resistance R_p is placed across the EO crystal modulator to match the capacitance C_{EO}.

7.18 Bragg acousto-optic modulator diffraction Consider the acousto-optic modulator in Figure 7.28. We can represent the incident and diffracted optical waves in terms of their wavevectors **k** and **k′**. The incident and the diffracted photons will have energies $\hbar\omega$ and $\hbar\omega'$ and momenta $\hbar k$ and $\hbar k'$. An acoustic wave consists of **lattice vibrations** (vibrations of the crystal atoms) and these vibrations are quantized just like electromagnetic waves are quantized in terms of photons. A quantum of lattice vibration is called a **phonon**. A traveling lattice wave is essentially a strain wave and can be represented by $S = S_o \cos(\Omega t - Kx)$ where S is the instantaneous strain at x, Ω is the angular acoustic frequency, and K is the wavevector, $K = 2\pi/\Lambda$ and S_o is the amplitude of the strain wave. A phonon has an energy $\hbar\Omega$ and a momentum $\hbar K$. When an incoming photon is diffracted, it does so by interacting with a phonon; it can absorb or generate a phonon. We can treat the interactions as we do between any two particles; they must obey the conservation of momentum and energy rules:

$$\hbar\mathbf{k}' = \hbar\mathbf{k} \pm \hbar\mathbf{K}$$

$$\hbar\omega' = \hbar\omega \pm \hbar\Omega$$

The positive sign case is illustrated in Figure 7.40 which involves absorbing a phonon. Since the acoustic frequencies are orders of magnitude smaller than optical frequencies ($\Omega << \omega$), we can assume that in magnitude $k' \approx k$. Hence using the above rules, which correspond to Figure 7.40, derive the Bragg diffraction condition.

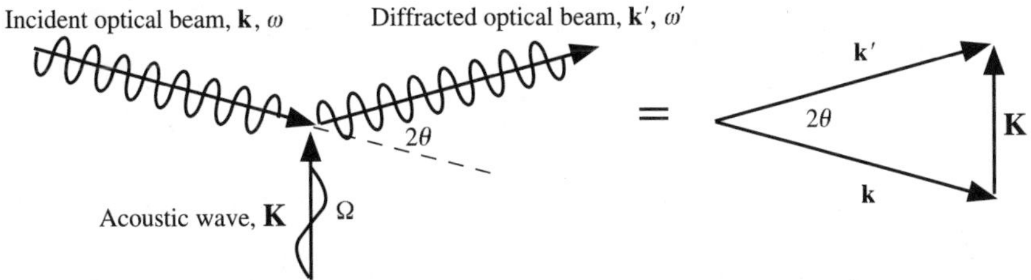

FIGURE 7.40 Wavevectors for the incident and diffracted optical waves and the acoustic wave.

7.19 Magneto-optic modulator and isolator

(a) Sketch schematically how you would construct a light intensity modulator using the Faraday effect. What would be its advantages and disadvantages?

(b) Sketch schematically how you would construct an optical isolator, using a Faraday rotator and two polarizers, that allows light to travel in one direction but not in the opposite direction.

7.20 SHG The mismatch between k_2 for the second harmonic wavevector and k_1 for the fundamental wave is defined by $\Delta k = k_2 - 2k_1$. Perfect match means $k_2 = 2k_1$ and $\Delta k = 0$. When $\Delta k \neq 0$, then the coherence length l_c is given by $l_c = \pi/(\Delta k)$. Show that

$$l_c = \frac{\lambda}{4(n_2 - n_1)}$$

where λ is the free-space wavelength of the fundamental wave. Suppose that a light with wavelength 1000 nm is passed through KDP crystal along its optic axis. Given that $n_o = 1.509$ at $\lambda = 1000$ nm and $n_o = 1.530$ nm at 2λ, what is the coherence length l_c? Find the percentage difference between n_2 and n_1 for a coherence length of 2 mm?

7.21 Optical Kerr effect Consider a material in which the polarization does not have the second order term:

$$P = \varepsilon_o\chi_1 E + \varepsilon_o\chi_3 E^3 \quad \text{or} \quad P/(\varepsilon_o E) = \chi_1 + \chi_3 E^2$$

The first term with the electric susceptibility χ_1 corresponds to the relative permittivity ε_r and hence to the refractive index n_o of the medium in the absence of the third order term, *i.e.* under low fields. The E^2 term represents the irradiance I of the beam. Thus, the refractive index depends on the intensity of the light beam, a phenomenon called the **optical Kerr effect**:

$$n = n_o + n_2 I \quad \text{and} \quad n_2 = \frac{3\eta\chi_3}{4n_o^2}$$

and $\eta = (\mu_o/\varepsilon_o)^{1/2} = 120\pi = 377\ \Omega$, is the impedance of free space.

(a) Typically, for many glasses, $\chi_3 \approx 10^{-21}$ m^2/W; for many doped glasses, $\chi_3 \approx 10^{-18}$ m^2/W; for many organic substances, $\chi_3 \approx 10^{-17}$ m^2/W; for semiconductors, $\chi_3 \approx 10^{-14}$ m^2/W. Calculate n_2 and the intensity of light needed to change n by 10^{-3} for each case.

(b) The phase ϕ at a point z is given by

$$\phi = \omega_o t - \frac{2\pi n}{\lambda} z = \omega_o t - \frac{2\pi[n_o + n_2 I]}{\lambda} z$$

It is clear that the phase depends on the light intensity I and the change in the phase along Δz due to light intensity alone is

$$\Delta\phi = \frac{2\pi n_2 I}{\lambda}\Delta z$$

As the light intensity modulates the phase, this is called **self-phase modulation**. Obviously light is controlling light.
When the light intensity is small $n_2 I \ll n_o$, *obviously the instantaneous frequency*

$$\omega = \partial\phi/\partial t = \omega_o.$$

Suppose we have an intense beam and the intensity I is time dependent $I = I(t)$. Consider a pulse of light traveling along the z-direction and the light intensity vs. t shape is a "Gaussian" (this is approximately so when a light pulse propagates in an optical fiber, for example). Find the instantaneous frequency ω. Is this still ω_o? How does the frequency change with "time", or across the light pulse? The change in the frequency over the pulse is called **chirping**. Self-phase modulation therefore changes the frequency spectrum of the light pulse during propagation. What is the significance of this result?

(c) Consider a Gaussian beam in which intensity across the beam cross section falls with radial distance in a Gaussian fashion. Suppose that the beam is made to pass through a plate of nonlinear medium. Explain how the beam can become **self-focused**? Can you envisage a situation where diffraction effects trying to impose divergence are just balanced by self-focusing effects?

Michael Faraday (English physicist, 1791–1867) *(Courtesy of AIP Emilio Segrè Visual Archives, Zeleny Collection.)*

Notation and Abbreviations

A	area; cross sectional area
a	half dimension of the core region of an optical waveguide; fiber core radius; lattice parameter of a crystal, side of a cubic crystal
a (subscript)	acceptor, *e.g.* N_a = acceptor concentration
A_{21}	Einstein coefficient for spontaneous emission
ac	alternating current
AM	air mass, describes the integrated solar intensity and depends on location
APD	avalanche photodiode
AR	antireflection coating
B	maximum bit-rate; frequency bandwidth; direct recombination capture coefficient
$\mathbf{B}$, B	magnetic field vector (T), magnetic field, magnetic flux density
b	normalized propagation constant (Ch. 2)
B_{12} B_{21}	Einstein absorption and stimulated emission coefficients for the rate transitions between energy levels 1 and 2
BJT	bipolar junction transistor
BL	bit-rate $\times$ distance product (Gb s^{-1}) $\times$ km
C	capacitance
c	speed of light in free-space $(3 \times 10^8 \text{ m s}^{-1})$
C^3	cleaved coupled cavity
CB	conduction band
C_{dep}	depletion region capacitance of a *pn* junction; junction capacitance of *pin*
CVD	chemical vapor deposition
CW	continuous wave
D	diameter of a circular aperture; diffusion coefficient (m^2 s^{-1}); detectivity (Ch. 6); thickness
$\mathbf{D}$, D	displacement vector (C m^{-2})
D	direct (bandgap)
d	differential in mathematics; distance; diameter
d (subscript)	donor, *e.g.* N_d = donor concentration
DBR	distributed Bragg reflector
dc	direct current
D_{ch}	chromatic dispersion coefficient

DFB	distributed feedback
DH	double-heterostructure
D_m	material dispersion coefficient
DOS	density of states, number of electronic states per unit energy per unit volume in a given energy band
D_p	polarization dispersion coefficient
D_w	waveguide dispersion coefficient
E	electric field in Ch. 1, 2 and 7; energy in Ch. 3, 4 and 5
E	electric field in Ch. 3, 4 and 5
e	electronic charge $(1.60218 \times 10^{-19}\ \mathrm{C})$
$\boldsymbol{e}$	Napierian base, 2.71828...
e (subscript)	electron, *e.g.* μ_e = electron drift mobility
E_a	applied external field (non optical field) in an electro-optic crystal
E_a, E_d	acceptor and donor energy levels
E_c, E_v	conduction band edge energy, valence band edge energy
EDFA	erbium doped fiber amplifier
E_F	Fermi energy
E_{Fi}	Fermi energy in the intrinsic semiconductor
E_{Fn}	Fermi energy in the *n*-side of a *pn* junction
E_{Fp}	Fermi energy in the *p*-side of a *pn* junction
E_g	bandgap energy
EHP	electron hole pair
ELED	edge light emitting diode
EM	electromagnetic
EMF, emf	electromagnetic force (V)
E_n	energy corresponding to quantum number n
EO	electro-optic
E_{ph}	photon energy, $h\upsilon$ (usually stated in eV)
eV	electron volt
$\exp(x)$	exponential function of x
F	Finesse of a Fabry-Perot cavity (Ch. 1); excess noice factor (Ch. 5)
$f(E)$	Fermi-Dirac function
FF	fill factor of a solar cell
f_{op}	optical bandwidth
FTIR	frustrated total internal reflection
FWHP	full width at half power
G	gain; rate of generation ($\mathrm{m^{-3}\ s^{-1}}$); photoconductive gain (Ch. 6)
g	optical gain coefficient due to stimulated emissions exceeding absorption between the same two energy levels ($\mathrm{m^{-1}}$)

$g(E)$	density of states, number of electronic states (possible electron wavefunctions) per unit volume per unit energy
$g(\upsilon)$	optical gain lineshape; optical gain vs. frequency curve shape
G_{op}	net optical gain
g_{ph}	photogeneration rate per unit volume
G_{ph}	rate of photogeneration (m^{-3} s^{-1})
GRIN	graded index
g_{th}	threshold optical gain coefficient for achieving a lasing emission from a laser device (for laser oscillations)
$G_{thermal}$	thermal generation rate of electron hole pairs (m^{-3} s^{-1})
H	height of the optical cavity in a laser diode
$\mathbf{H}, H$	magnetic field intensity (strength) or magnetizing field (A m^{-1})
h	Planck's constant $(6.6261 \times 10^{-34}$ J s$)$
h (subscript)	hole, *e.g.* μ_h = hole drift mobility
$\hbar$	Planck's constant divided by 2π $(1.0546 \times 10^{-34}$ J s$)$, $\hbar = h/2$.
HF	high frequency
I	electric current (A)
I	irradiance or intensity of light (W m^{-2})
I	indirect (bandgap)
i (subscript)	initial; intrinsic, *e.g.* n_i = intrinsic concentration
IC	integrated circuit
I_d	dark current
I_{diff}	forward diffusion (Shockley) current of a *pn* junction
I_{ph}	photocurrent; multiplied APD photocurrent
I_{pho}	unmultiplied APD photocurrent
I_{recom}	forward recombination current of a *pn* junction
I_{ro}	reverse recombination current of a *pn* junction
I_{sc}	short circuit current in light (solar cells)
I_{so}	reverse Shockley saturation current of a *pn* junction
I_{th}	threshold current
i_n	noise current, rms fluctuations *in the photodetector current*
i_{ph}	photocurrent
J	current density
j	imaginary constant: $\sqrt{-1}$
J_{gen}	reverse current density in a *pn* junction due to thermal generation of EHPs in the SCL
J_{rev}	total reverse current density in a *pn* junction
J_{th}	threshold current density value
K	Kerr coefficient
$\mathbf{K}, K$	phonon wavevector

$\mathbf{k}, k$	wavevector, wavenumber, free-space wavevector : $k = 2\pi/\lambda$
k_B	Boltzmann constant $(k_B = R/N_A = 1.3807 \times 10^{23}\ \text{J K}^{-1})$
KDP	potassium dihydrogen phosphate
KE	kinetic energy
L	distance, length; optical cavity length; inductance
ℓ	length
ℓ_n, ℓ_p	lengths of the neutral n and p regions outside depletion region in a pn junction
l	orbital angular momentum quantum number of an electron, $0, 1, 2, \ldots n - 1$; n is the principal quantum number
LASER	light amplification by stimulated emission of radiation
LD	laser diode
L_e, L_h	electron and hole diffusion lengths: $L_e = \sqrt{(D_e\tau_e)}$ and $L_h = \sqrt{(D_h\tau_h)}$
LED	light emitting diode
LF	low frequency
$\ln(x)$	natural logarithm of x
L_o	transfer distance for full light transfer from one to another coupled optical guide
LP	linearly polarized
M	number of modes (Ch. 2); multiplication in avalanche effect (Ch. 7); mass
m	integer; mode number; mass
m	numerical constant characterizing the photon energy width of the LED output spectrum
MBE	molecular beam epitaxy
m_e	mass of the electron in free space $(9.10939 \times 10^{-31}\ \text{kg})$
m_e^*, m_h^*	electron and hole effective masses in a crystal
MFD	mode field distance/diameter, $2w_o$
MMF	miltimode fiber
MQW	multiple quantum well
N	number of molecules per unit volume; n-type semiconductor with a wider bandgap
n	electron concentration in the conduction band in Ch. 3, 4, 5, 6; principal quantum number $(0, 1, 2 \ldots)$
n	pn junction ideality factor in Ch. 6
n	refractive index in Ch. 1, 2 and 7
n	refractive index in Ch. 3, 4, 5 and 6
N_A	Avagodro's number $(6.022 \times 10^{23}\ \text{mol}^{-1})$
NA	numerical aperture
N_a, N_d	acceptor and donor dopant concentrations in a semiconductor
N_c, N_v	effective density of states at the conduction and valence band edges respectively

n_e, n_o	refractive index for the extraordinary and ordinary waves in a birefringent crystal
NEP	noise equivalent power, 1/Detectivity
N_g	group index, $N_g = c/v_g$
n_i	intrinsic concentration
n_L, n_R	refractive index experienced by left circularly polarized and right circularly polarized waves in a crystal
n_n, p_p	instantaneous majority carrier concentrations (*e.g.* electrons in an *n*-type semiconductor)
n_{no}, p_{po}	equilibrium majority carrier concentrations (*e.g.* electrons in an *n*-type semiconductor)
n_p, p_n	instantaneous minority carrier concentrations (*e.g.* electrons in a *p*-type semiconductor)
n_{po}, p_{no}	equilibrium minority carrier concentrations (*e.g.* electrons in a *p*-type semiconductor)
$n_p(0)$	electron concentration just outside the depletion region in the *p*-side of a *pn* junction
N_{ph}	photon concentration in the optical cavity
NRZ	nonreturn to zero (pulses)
OVD	outside vapor deposition
$\mathbf{P}$	polarization vector, polarization per unit volume in a dielectric
P	power, energy flow per unit time; *p*-type doping in a larger bandgap material.
p	hole concentration in the valence band; photoelastic coefficient (Ch. 7)
PD	photodiode, photodetector
PE	potential energy
pin	*p*-type/intrinsic/*n*-type semiconductor photodiode
$p_n(0)$	hole concentration just outside the depletion region in the *n*-side of a *pn* junction
Q	charge
q	an integer; diffraction order; charge
QE	quantum efficiency
R	reflectance (fractional reflected light intensity at a dielectric-dielectric interface); spectral responsivity or radiant sensitivity of a photodetector
R	resistance
R_L	load resistance (external to device)
R_p, R_s	parallel and series resistances of a solar cell
r	Pockels coefficient
r	radial distance
$\mathbf{r}$	position vector
r	EM wave reflection coefficient
RZ	return to zero (pulses)

rms	root mean square
S	strain (Ch. 7)
S, S	Poynting vector that quantifies the rate of electromagnetic energy flow
s, p, f	atomic subshells
SAGM	separate absorption, grading and multiplication
SAM	separate absorption and multiplication
SCL	space charge layer, depletion region, of a *pn* junction
SHG	second harmonic generation
SLED	surface emitting LED
SMF	single mode fiber
SNR, S/N	signal to noise ratio
SQW	single quantum well
T	temperature in Kelvin
T	transmittance, fractional intensity transmitted through a dielectric-dielectric interface
t	time; thickness
t	transmission coefficient
t_h	transit time of holes
TA	transmission axis of a polarizer
t_{drift}	drift time of a carrier
TE	transverse electric field modes
TEM	transverse electric and magnetic (TEM) modes
T_f	fictive temperature, softening temperature of glass (K)
TIR	total internal reflection
TM	transverse magnetic field modes
t_r	rise time of a signal (usually from 10% to 90%)
UV	ultraviolet
V	V-number, V-parameter, normalized thickness
V	applied voltage , potential energy in the Schrödinger equation (Ch. 3)
v	velocity
V_{br}	avalanche breakdown voltage
$V_{\lambda/2}$	applied voltage in a Pockels cell that introduces a relative phase change $\Delta\phi$ of π (Ch. 7)
V_o	built-in voltage, potential
V_{oc}	open circuit voltage of a solar cell
V_r	reverse bias voltage
VB	valance band
VCSEL	vertical cavity surface emitting laser
v_d	drift velocity

v_e, v_h	electron and hole drift velocities
v_g	group velocity
v_{th}	thermal velocity of an electron in the conduction band and a hole in the valence band
w	width
W	width; thickness; width of depletion region in a *pn* junction
W_n, W_p	widths of depletion region on the *n*-side and on the *p*-side of the *pn* junction
w_o	half-waiste of a Gaussian beam (Ch. 1); mode field radius (MFD ÷ 2) (Ch. 2)
W_o	width of depletion region of a *pn* junction with no applied voltage
x	distance; excess avalanche noise index for APDs (Ch. 5)
$\hat{\mathbf{x}}$	unit vector along the *x*-direction
$\hat{\mathbf{y}}$	unit vector along the *y*-direction
$\hat{\mathbf{z}}$	unit vector along the *z*-direction
$\perp$ (subscript)	perpendicular
// (subscript)	parallel
α	optical attenuation; attenuation coefficient for the electric field in medium 2; absorption coefficient of a semiconductor; dielectric polarizability per molecule
α_{dB}	attenuation in dB
α_R	Rayleigh scattering attenuation
α_t	total attenuation coefficient in the optical cavity of a semiconductor laser
α_{max}	maximum acceptance angle for an optical fiber
β	propagation constant; base-to-collector bipolar transistor current gain
β_m	propagation constant along a planar dielectric waveguide for mode m
β_T	isothermal compressibility (m^2/N)
Γ, Γ_{ph}	flux ($m^{-2}\ s^{-1}$), photon flux (photons $m^{-2}\ s^{-1}$)
γ	loss coefficient in a laser optical cavity
Δ	difference, change; normalized index difference, $\Delta = (n_1 - n_2)/n_1$
$\Delta\tau$	propagation time difference along an optical guide
$\Delta\tau_{1/2}$	time width of the output light pulse between half intensity points
$\Delta\beta$	phase mismatch per unit length between two propagating waves
$\Delta\phi$	phase change
$\Delta\lambda$	spectral width of the output spectrum (intensity vs. wavelength) of a light source. This can be between half-intensity points $(\Delta\lambda_{1/2})$ or between rms points.
$\Delta\lambda_m$	wavelength separation of allowed modes in an optical cavity
Δn	$n - n_o$, excess electron concentration (above equilibrium)
ΔN_{th}	$(N_2 - N_1)_{th}$: threshold population for threshold gain (lasing emission)
$\Delta\theta$	divergence angle due to diffraction effects

δ	small change; penetration depth of an EM wave from one medium to another
$\delta\lambda_m$	spectral wavelength width of an allowed mode in an optical cavity (Ch. 1); wavelength separation between two consecutive optical cavity modes (Ch. 4)
$\varepsilon = \varepsilon_o\varepsilon_r$	permittivity of a medium where ε_o = permittivity of free space or absolute permittivity, ε_r = relative permittivity or dielectric constant
ε_n	energy in a quantum well with respect to the bottom of the conduction band, n is a quantum number 1, 2,...
η	efficiency; quantum efficiency
η_{external}	external efficiency
η_{slope}	slope efficiency of a laser diode
θ	angle ; angular spherical coordinate
θ_c	critical angle for total internal reflection
θ_i	angle of incidence
θ_p	Brewster's polarization angle
κ	propagation constant along the transverse direction of a planar dielectric waveguide
Λ	periodicity of a corrugated Bragg grating (Ch. 4); acoustic wavelength (Ch. 7)
λ	free space wavelength
λ_B	Bragg wavelength of a grating type structure that satisfies the Bragg condition
λ_c	critical or cut-off wavelength
λ_g	threshold wavelength for photoexcitation; bandgap wavelength
$\mu = \mu_o\mu_r$	magnetic permeability (H m^{-1}); μ_o = absolute permeability, μ_r = relative permeability
μ_d	drift mobility (m^2 V^{-1} s^{-1})
μ_h, μ_e	hole drift mobility, electron drift mobility (m^2 V^{-1} s^{-1})
μm	micron, micrometer
π	pi, 3.14159...
ρ	resistivity; charge density (C m^{-3})
ρ_{net}	net space change density
$\rho(h\upsilon)$	photon energy density per unit frequency (Ch. 4)
$\rho_{eq}(h\upsilon)$	Planck's black body equilibrium radiation distribution as a function of frequency (Ch. 4)
σ	conductivity; rms deviation of dispersion of a light pulse going through a fiber
τ	group delay time in a fiber; recombination time of a charge carrier; lifetime (τ_e for electrons, τ_h for holes); time constant
τ_g	group delay time per unit length, $\tau_g = 1/v_g$; mean time to thermally generate an electron-hole pair in a semiconductor
τ_{ph}	photon decay time in an optical cavity, time it takes for the photon to be lost from the cavity
τ_{sp}	average time for an electron-hole pair recombination by spontaneous emission

ϑ	phonon frequency; lattice vibration frequency; Verdet constant
υ	frequency
Φ	phase change; work function
φ	an angular spherical coordinate
ϕ	phase ($\phi = \omega t - kz + \phi_o$); phase change
ϕ_o	phase constant
χ	electron affinity; electric susceptibility
χ_1, χ_2, χ_3	linear, second-order and third-order electric susceptibilities
$\psi(x)$	spatial dependence of the electron wavefunction under steady state conditions satisfying the Schrödinger equation
$\psi_k(x)$	Bloch wavefunction, electron wavefunction in a crystal with a wavevector k.
Ω	phonon angular freqeuncy
ω	angular frequency $2\ \pi\ \upsilon$ (of light)

Index

D

E

J

K

L

M

N

O

P

Q

R

S

W

"Errors using inadequate data are much less than those using no data at all"—Charles Babbage